Effective Professional and Technical Writing

SECOND EDITION

MICHAEL L. KEENE
University of Tennessee, Knoxville

D. C. Heath and Company
Lexington, Massachusetts   Toronto

This book is dedicated to all my family.

Address editorial correspondence to:

D. C. Heath
125 Spring Street
Lexington, MA 02173

Acquisitions Editor: Paul A. Smith
Developmental Editor: Linda Bieze
Production Editor: Ron Hampton
Designer: Cornelia Boynton
Production Coordinator: Charles Dutton
Permissions Editor: Margaret Roll

International Standard Book Number: 0–669–20595–8

Library of Congress Catalog Number: 92-81215

10   9   8   7   6   5   4   3   2   1

PREFACE

All over the country, students are turning to upper-division (or advanced) writing courses to gain the communication skills they will need as professionals in engineering, business, science, government, and all the other fields college graduates enter. Although such writing courses often carry titles like Technical Writing, Professional Writing, or Business and Technical Writing, the courses usually involve many more kinds of students and fields of writing than these titles suggest. Thus, *Effective Professional and Technical Writing*, Second Edition, addresses the kinds of writing tasks and problems professionals in *all* fields may encounter.

Students come to such courses with high expectations: that the writing assignments will be somehow real in ways the assignments in their earlier courses may not have been, and that the writing problems they have been bothered by for years can be solved in months. Most important, students in such writing courses seek a kind of knowledge that can be integrated into their professional lives. Pursuing such knowledge, these students are willing to work hard and often on their writing, and they frequently will seize the initiative to learn more than the texts and their teachers present.

Effective Professional and Technical Writing, Second Edition, has grown out of the continuing success of one such class that I teach for students majoring in professional, technical, and business fields. Thus, I define this book in terms of the students for whom it is designed, the philosophy it is based on, the kinds of assignments the students are asked to write, and the structure of the book.

The Students

This book is intended for students in widely varied fields of study—such as engineering, science, business, agriculture, computer science, or health sciences—who typically have not had a writing course since their freshman year. By now, most of these students have begun to internalize the concepts, vocabularies, and values of their major fields. The only audience they have written for is an academic one, and the only writing situation they have experienced is in the classroom.

The Philosophy

Effective Professional and Technical Writing, Second Edition, has a definite philosophy.

Student Versus Professional Writing. For students to become effective communicators, the most important thing they need to learn is to recognize and adjust to the differences between student and professional writing. Perhaps the major failing of graduates entering professional life is their assumption that all writing situations mirror those encountered in the classroom. They thus fail to recognize the need to tailor their writing to different audiences, purposes, roles for the writer, and communication situations. To change this assumption, students need explicit practice in audience analysis and adaptation, which many college courses spend relatively little time on.

Because they may always have written for experts—the scholar-teachers who evaluate what students know based on what they write—these students may not have developed the ability to communicate their fields' important subject matter to people outside those fields. For example, such students may unconsciously assume that all readers have the same level of knowledge or interest in the subject that they do. But in professional life, the writer who assumes this is usually the writer who fails to communicate.

Problem Solving. Recent graduates who have done only classroom writing may also be unable to see writing as *problem solving* for *decision makers*, a function that is rare in student writing but common in professional settings. Thus, the structure of papers by recent graduates often mirrors the processes students went through while researching the paper, arriving finally at some conclusion, while the "real world" reader actually needs "the bottom line" presented first.

Finally, anyone writing a report with scientific, technical, or business subject matter may feel tempted to give in to the subject matter, pouring out page after page with no regard for how the report's unique situation needs to shape and guide the selection, presentation, and arrangement of that material. Building on their classroom experiences, entry-level professionals are powerfully tempted to make their writing overly *content-driven*, while their readers need more explanation, analysis, and awareness on the writer's part of his or her responsibility for going beyond the quantitative base and qualifying details so as to provide clear, professional judgments and projections. As Mary Coney and James Souther, who have popularized this analysis of the problems professional writers face, put it:

> Like the sorcerer's apprentice whose broom machine goes out of control, the writer can easily get caught up in the data production and lose authorial direction. If in the past we faced the dangers of making decisions with too little information, we may well now face the dangers of making decisions with too much. Both situations promote poor decision-making.[1]

The Communication Situation. This book's emphasis on the communication situation helps students develop a strong concern for audience and purpose in their writing. Audience and purpose are the key intangible human factors needed to balance the natural attractiveness of technical subject matter's apparent concreteness. Students learn that as their writing becomes more content-driven, they must become more concerned with providing explanations and clear examples, distinguishing between important material and trivia, and making key relationships explicit.

Thus, although this book is concerned with all the traditional features of good writing—grammar, organization, style, and content—whenever possible, it connects those features to the needs and values of each specific communication situation—each specific purpose, message, audience, and writer's role. In that sense, this approach may be said to be broadly *rhetorical*. Although this kind of writing characteristically occurs in situations where there are facts, data, and other tangible subjects, situations which would place this writing outside of the realm of much traditional rhetoric, still this kind of writing occurs always and only within human contexts and for human ends, thus making all professional writing situations also rhetorical situations. Or, as Edmond Weiss puts it:

> Problems are not facts; conclusions are not facts; recommendations are not facts. Even when they are based on hard fact, the problems, conclusions, and recommendations must be put into a convincing case or a moving presentation. To be effective in their jobs, engineers and scientists must do much more than inform; they must explain, prove, evaluate, justify, defend, attack, choose, advocate, and refute. Facts alone can do none of these. Clear, well-argued reports can.[2]

The Assignments

Effective Professional and Technical Writing, Second Edition, offers instructors a wide variety of exercises and writing assignments to choose from. Depending on the teacher's choice, the typical papers in one semester may include a formal letter of introduction, a job application letter and résumé, an extended definition-and-description report, a process explanation, a proposal, and a major report, but the chapters can also lead to a number of other assignments—situation-based cases (in every chapter), progress reports, problem solving, short speeches, and poster presentations, among others. While the content for most of the student writing is drawn from each student's major field of study, the specific target audience may be an executive with layperson knowledge, another professional in the student's field, the public, scientifically educated nonexperts, or any of a number of possibilities created by the assignment or the teacher. In the book's later chapters (Parts Four and Five) especially, the purposes and processes unique to writing beyond college receive primary attention; the concern shifts away from helping students translate their college-level writing into professional writing and toward presenting the characteristics of effective professional writing directly.

The Structure

This book is divided into five parts and an appendix:

Part One: An Introduction to the Writing Professionals Do explains key differences between writing in classrooms and writing on the job. It also explains how to write in a clear, economical style, and the ways in which working writers deal with the challenges of presenting their material effectively.

Part Two: Letters and Memos is a short course on these important writing tasks that affect every professional's workday. Special emphasis is given to writing successful job-application letters and résumés.

Part Three: Elements of Professional Writing presents the building blocks of professional writing, the elements out of which any report is constructed.

Part Four: Processes in Professional Writing describes the constructive processes by which those building blocks become effective professional communication.

Part Five: Applications presents typical kinds of professional reports and other communication tasks.

Appendix: A Short Course in Writing presents a brief review of the writing process, including a detailed system for revising, ranging from the document's largest, most abstract features to the structure of individual sentences; a generative review of sentence grammar; and a brief look at solutions to common grammatical and stylistic writing problems.

In general, the sequence of chapters within each of the five parts (as well as the sequence of parts) proceeds from simple to complex and from the basic (or purely rhetorical) to the applied. Thus Parts One, Two, and Three present fundamental principles and techniques of writing—from how to write letters of request to how to explain processes. Parts Four and Five deal with writing on a more applied level—from how to design a professional report to how to write a proposal. Similarly, Parts One through Three discuss writing from the student's viewpoint, while Parts Four and Five focus on writing from the professional's point of view. The accompanying *Instructor's*

Guide offers several syllabi for both semester- and quarter-length courses.

Changes in the Second Edition

Readers familiar with the first edition will notice

- The prose style has been tightened, and examples, especially more technical ones, have been added throughout the book.
- Many chapters now feature situational case writing assignments (indicated by ➤) in addition to a number of other writing and speaking tasks.
- Chapter 3 has been broadened into a general discussion of professional styles, including sections about avoiding sexism in writing, maintaining an objective stance, and using readability measures as an aid to effective writing.
- The chapter on visuals (now Chapter 10) has been redesigned and expanded, in addition to being moved so that the order of chapters in Part Three follows more closely the order of elements of an actual report.
- Section 4 in Chapter 11 on computer technology has been updated and expanded.
- Chapter 20 on oral presentations now features a section on making poster presentations, which are becoming more important in many fields.
- The Appendix has been expanded to take the systematic process of revising from the sentence level all the way up to the level of the whole document.
- A list of proofreading symbols and a document checklist have been added to the inside covers.

My method in all of these changes has been to consider feedback from the students, teachers, and professionals who have used this book, in addition to my own understanding of professional writing and how best to teach it; my goal has been to make *Effective Professional and Technical Writing*, Second Edition, clearer, easier to use, and more appropriately suited to its audience and their situations as writers. If you have suggestions for how I might do a better job of achieving those goals, I invite you to write to me. My address is Dr. Michael L. Keene, Department of English, The University of Tennessee, Knoxville, Tennessee 37996-0430.

Acknowledgments

I want to acknowledge and express my appreciation for the support of all the people who helped me become a professional teacher of writing who specializes in rhetorical approaches to business and technical communication, and especially for those who have assisted me in writing and revising this book   students, colleagues, reviewers, friends, and family. No one successfully navigates such a long course alone. Special thanks, then, to Kate Adams, Chuck Anderson, David Armbruster, Greg Cowan, Maxine Hairston, Kitty Locker, Gordon Mills, Ralph Voss, and Merrill Whitburn.

Thanks, also, to the reviewers who made suggestions for the revision of this book: Thomas T. Barker, Texas Tech University; Warren Cushine, Hudson Valley Community College; Kathryn Harris, Arizona State University, Tempe; Carol Johnston, Clemson University; David M. Locke, University of Florida; E. Anne Martin, Western Michigan University; Lawrence Milbourn, El Paso Community College; Charles Nelson, Youngstown State University; and Robert Shenk, University of New Orleans.

My appreciation, also, to the staff of D. C. Heath, especially acquisitions editor Paul A. Smith, developmental editor Linda Bieze, production editor Ron Hampton, and designer Cornelia Boynton. Finally, my very special thanks to my wife, Claire, and my children Amy and Ben.

Michael L. Keene

NOTES

1. Mary B. Coney and James W. Souther, "Analytical Writing Revisited: An Old Cure for a Worsening Problem," *Technical Communication*, 31, 1, 1984.
2. Edmond H. Weiss, *The Writing System for Engineers and Scientists*, Prentice-Hall, Englewood Cliffs, NJ, 1982.

CONTENTS

PART FOUR Processes in Professional Writing   268

18. Proposals   465

19. Written Reports   500

Index

Introduction

By your junior or senior year in college, you may well have begun to realize that you need to be able to write better. In the last two years of college, more and more classes require short reports, long reports, and term papers of all kinds. Some of you, such as those in engineering or management, may have already realized that throughout your professional life you will be reading and writing reports. For others of you, simply facing graduation and writing job-application (or law-school or medical-school application) letters brings the realization that all your life you may be held back by your inability to handle even a one-page letter (much less a 20-page report) quickly and effectively.

Just how important is writing to a professional? One professional engineer, a member of a large and successful consulting engineering firm, wrote:

> There are two aspects to the role writing plays in this company. Reports are the company's only product. Regardless of what level of talent we employ, or how sophisticated our analyses are, the primary conduit for us to reach our customer is through the interim (if any) and final report. If those documents don't satisfy the customer's need and place us in a good light, this company cannot be successful. The second aspect of writing's role follows from the importance of the first. Regardless of an individual's technical competence, if his writing ability is inadequate, his long-term value to the company, and therefore his salary and progression in the company, are limited.

Another engineer put it this way:

> Maybe I could have gotten this job without that writing class. But I never could have kept it.

This book is designed to help all of you—from all majors—learn how to handle the writing tasks you will face in business, industry, government, science, or whatever other field you enter. This book can help you learn to solve your writing problems faster, more effectively, and with less stress.

Everyone can profit by learning more and better communication skills. A middle manager in a high-technology manufacturing firm puts it more succinctly:

> Anyone around here who can't write doesn't stay around here long.

PART ONE

An Introduction to the Writing Professionals Do

Writing at work can be both more challenging and more rewarding than writing in school. But there may not be very much in your background (or in the academic backgrounds of most students) that can prepare you for the demands professional life will make on your writing skills. This book's purpose is to give you the writing skills you will need when you leave college. Part One, the first three chapters, provides the groundwork in theory and principles for the rest of the book.

An article in *Consulting Engineer* vividly describes the writing problem engineering students face when they begin their professional lives. The same argument could be made concerning recent college graduates in nearly any profession:

> Many engineering managers assume that engineers hired right out of undergraduate or graduate school have adequate writing skills. After all, haven't all engineering graduates written the obligatory lab reports, freshman English essays, and long technical reports? No doubt some have. But very few have written technical writing that counts in ways that it does on the job. Few of these former students completed writing projects that were used to solve actual problems, were written for someone who paid for the project, or were written for readers often too busy to read the entire report.(1)

Another article adds depth to this picture. Two writing specialists at a major university surveyed 200 college-educated working people to find out how much of their time on the job they spent writing. The occupations of the people selected for the survey were carefully chosen to reflect accurately what government statistics say is the composition of the college-educated work force, and the other features of the study were equally carefully controlled. Here is one paragraph

1. The Communication Situation

2. How Writers at Work Write and Revise

3. Style and Readability

from the conclusion:

When respondents were asked what percentage of work time they spend writing, 193 of the 197 who answered this question said that they write on the job. Furthermore, 145 of the 197 write at least 10% of their total work time or for four hours in a 40-hour week; 98 of the 197 write 20% of total work time or eight hours in a 40-hour week. People in professional and technical occupations—the types of occupations in which over half of college-trained people are employed—on the average write nearly 30% of total work time.(2)

This book is designed to help you learn to meet the writing demands that your postcollege life will make on you. Chapter 1 explains in detail how college and professional writing are different in important ways. Chapters 2 and 3 begin the process of teaching you how to cope with those important differences, starting you on the way to becoming a more effective writer.

NOTES

1. William S. Pfeiffer, "A Short Course in Report Guide Preparation," *Consulting Engineer* (October 1982): 77–83.
2. Lester Faigley and Thomas P. Miller, "What We Learn from Writing on the Job," *College English* (October 1982): 557–569.

1. *The Communication Situation*

Everyone can profit from learning better writing skills. For a professional, good writing skills are *essential*. Whatever your profession—doctor, engineer, accountant, or wildlife biologist—you will rely more on your writing skills once you have graduated than you did while you were in college. Effective writing—and effective communication in general—is the lifeblood of business and industry (Fig. 1.1).

The main product (in some cases, the *only* product) of many management organizations, research and testing laboratories, engineering firms, and financial institutions is written documents. In many companies the ability to write good, readable reports distinguishes those who move up the corporate ladder from those who don't. A just-graduated engineer may be surprised to find that 50 or 60 percent of each workday is taken up by reading and writing reports or hearing and delivering oral presentations. It's even more surprising to learn that as one moves up the corporate ladder, the percentage of time spent in such communication activities increases.

Figure 1.1 Lines of Communication

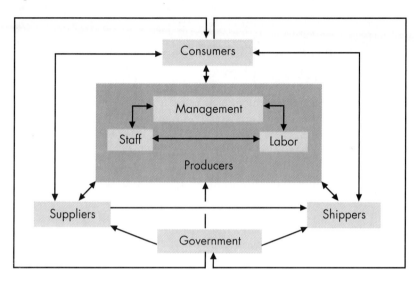

1. *Effective Writing*

Consider a typical professional's morning:

In the carpool to work, you read the day's trade paper to learn the new trends. You get to work and check your mail, reading the letters that look clear, simple to read, and easy to handle, putting the rest aside for later. After checking your E-mail, you review your notes for the 10 A.M. meeting with the production staff. The meeting reminds you that your boss is one of the most effective communicators you've ever met; he organizes his thoughts well and presents them clearly and effectively in both speech and writing.

During the meeting you are put in charge of a special project, and you realize this means you will have to write another thirty-page report. Preparing your last one took evenings and weekends for a month and nearly wrecked your marriage. You wish you could handle report writing as well as Lee, across the table, does. You're convinced you've got the know-how and the ideas you need to succeed, but you just can't communicate them effectively. *Nothing you do ever looks as good on paper as it seemed in your thoughts.*

By the time the meeting ends, it's 11:30 and you've spent the whole morning worrying over communication—especially writing. As the afternoon begins, you dive into the stack of paperwork on your desk. Letters and memos to read and to write, reports to look over, and decisions to make based on all that written information. And the FAX keeps bringing in more.

Whether you are a doctor, engineer, accountant, or biologist, if you wish to succeed in business, industry, government, or research, you will need to be able to communicate effectively. The goal of this book is to help you learn to do so.

1.1   What Is Effective Writing?

Effective professional writing gets its point across. It moves the reader in the direction the author desires. It *communicates*. If your goal is to convey information, then you will want to do so clearly and quickly. If your goal is to persuade, you will want the reader to follow your line of reasoning willingly.

Effective writing repays your reader for the time it took to read your report (or manual, brochure, letter, etc.). *In*effective writing *costs* time and money. If your report is not effective, more than time has been lost. In professional life, wasted time is wasted money. If you waste your reader's time and money, you've wasted your own and your company's time and money as well. No professional can afford that kind of waste.

1.2   Why Is Effective Writing Important?

In college you may pay a relatively low price for ineffective writing. If your reports take twice as long as they should to write or if they never turn out nearly as well as you want them to, you may not feel you've lost too much. Your major may require little writing in college, and the effectiveness of your writing may rarely determine the course of your college career. If you do happen to receive a low grade on a writing assignment every so often, you may feel you have plenty of company among your classmates. You may also feel that as a student you are not expected to produce really polished writing but rather to focus on learning the subject matter.

On the job all of that changes. From the moment you send in a job application letter and résumé, from the moment you stand up to introduce yourself in the first staff meeting, people around you are evaluating you and the quality of your work. They do that evaluation largely on the basis of your communication skills. You get the job or you don't; you impress your boss that you are the best one to represent your company or you don't; you sell the customer or you don't; you get the contract or you don't. Therefore you keep your job—maybe get a promotion—or you don't.

In your professional life, the image of you projected by your writing and speaking is often more visible than you are. Thus, you want your writing to be effective because it will reflect on both you and your employer. But you can do much more than this; you want your words to reflect the true quality of your thoughts. If you only think about avoiding major blunders in your writing, you're like a soccer team stuck on defense or a volleyball team with no spikers. There may be no more frustrating professional experience than to have a good idea but not be able to communicate it effectively in the form your job requires. If you follow the suggestions offered here for developing effective writing skills, that kind of frustration may be one big professional problem you will never have.

2. College Writing Versus Professional Writing

In the writing professionals do, many things are different from the way they are in the writing students do:

The kinds of people you write for.

Their positions relative to yours.

Their reasons for reading.

The kind of reading they do.

The images of themselves they want to see in your reports.

Most of all, the people you write for and the situations you write in are *real* in a way college writing seldom is, and the way you adapt to your professional situation and your professional audience must reflect that reality.

What changes when you move from college writing to professional writing? It's easier to start by naming what stays the same. Generally you will still be writing within the same subject area (engineers write about engineering, accountants about accounting, and so on), although even that may not always be the case. But while the message, or content, remains the same, just about everything else about your writing changes. The purposes your writing serves, your audiences, and your own role as writer—all of these key elements of the communication situation change in important ways. It's popular among communication specialists to focus on successful handling of four intangibles as keys to effective communication: *audience, purpose, message,* and *writer's role.* When you move from student writing to professional writing, at least three of these may change drastically (Fig. 1.2).

3. Audience

During your college years you are always writing for experts—people who know more about your subject than you do. After college the roles are often reversed. Often *you* are the expert, writing to someone who knows much less about the subject than you do. Or you may

Figure 1.2   The Changing Writer's Universe: From Student to Professional

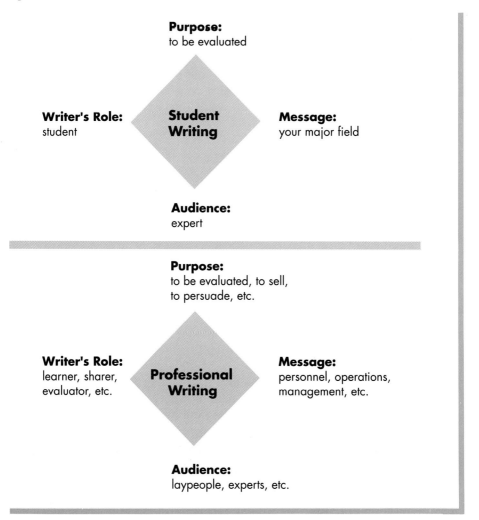

be writing to your co-workers in the company—to equals. Or to total strangers. Or to technicians who work for you. These different audiences make demands on your writing skills that little in your experience as a student writer may have prepared you for. Your audience outside college is different from your audience inside college in at least these three important ways: its *kind*, its *position*, and its *role*. Professionals who are successful writers adjust to those factors almost automatically. If you continue to grow as a writer, the strategies explained in this book, which may require some effort on your part at first, will also become automatic for you.

The audience for your college writing is almost exclusively composed of experts. Those experts are in a position above yours (in the sense that professors are above students), they are part of the same organization (college, course) as you, and they usually carefully and actively read what you write (rather than glancing at it and passing it on).

After college all of that changes. Your audience outside college could be experts, laypeople, technicians, or executives—or a mix. They could be above you, below you, or on your level. They could be inside your company (internal) or outside it (external). And they could actively and carefully read your report, just skim it, or just file it. They may be persuaded by it, learn from it, pass judgment on it, or ignore it.

As a college student you may write for an engineering professor, a biology professor, or an accounting professor, but you're always a student writing for a professor. On the job, however, you may be an engineer trying to sell a construction proposal, a systems analyst writing instructions for people who work for you, a quality-control manager announcing a change in company procedures, or a field accountant writing up the results of an audit for official records. You had years and years to get used to the kind of writing required in school, and its kind was just about always the same. But the job-related writing you have to do may well change constantly, frequently allowing you little time to adjust. The effectiveness of your job-related writing often depends directly on whether you have properly judged and adapted to the requirements of each different writing task. Over and over the focal point of that adaptation will be the audience.

Learning to write for different kinds of audiences in different roles requires learning audience analysis and adaptation. Because your college writing—all of your writing as a student—has been for one kind of audience in one role and one position, you probably aren't in the habit of considering these features. To help you sharpen your thinking about the kinds, positions, and roles of audiences in business and industry, the next three sections of this chapter list and explain them in detail. (See also the material on discourse communities in Chapter 17.)

3.1 Four Basic Kinds of Audiences

It is impossible to classify every kind of audience in business and industry, so the best one can do may well be to choose a simple, reasonably efficient scheme and then remember that it may need to be amended in some situations. The classification given here has become almost standard. We will use it throughout this book.

In this classification scheme there are four kinds of audiences: layperson, expert, executive, and technician.

Different Kinds of Audiences

Layperson: A person in whom you can assume no implicit knowledge or interest in your subject. *Examples:* The general public—people reading an account of how DNA works, instructions for installing a garage-door opener, or a critical comparison of two products.

Expert: A person who has substantial knowledge, experience, and (probably) interest in your subject. *Examples:* An engineer reading an engineering report, a marketing professor reading a student's paper, a banker reading a financial statement, or a dietician reading a description of a new clinical diet for cancer patients.

Executive: Someone with decision-making power over your career. *Examples:* A store manager reading a suggested change in inventory procedures, a department head reading the product brochures of various copiers, or the president of the company reading your report.

Technician: The person who actually runs the machine or performs the procedure; a person with some hands-on knowledge of the subject, interested in it in a how-to sense. *Examples:* A secretary trying to figure out a word processor's user's manual, a maintenance technician reading a service manual, or an army private reading a training manual.

Each audience has its own characteristics and requires its own adaptations.

- Because laypeople typically have no built-in interest in your subject, you need to work on motivating them to read what you've written (see "Provide Reader Benefits," Chapter 4, Section 2.4, for more on this.)

- Experts may have a great deal of interest not only in the conclusions you draw but also (and especially) in seeing how you came to those conclusions (see Chapter 14, "Solving Problems," for more on this).

- Executives typically want to see the bottom line first, and only *after* seeing your conclusions and recommendations might they want to see how you arrived at them (see "Executive Summaries," Chapter 7, Section 2).

- Technicians may not be interested at all in how you decided a certain operation needed to be done a certain way, but they will be very interested in exactly how the operation is to be done (see Chapter 13, Section 2).

Figure 1.3   Different Points of View

People in various audiences are looking for the answers to the questions that are most meaningful to them personally, questions which are different for each group.

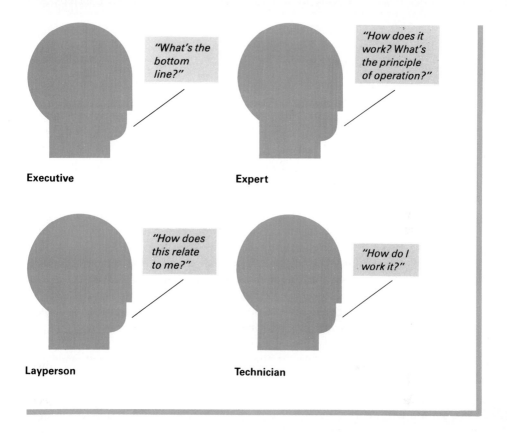

Each kind of reader brings a certain set of expectations to the act of reading (see Fig. 1.3), and for your writing to be successful, you must anticipate and respond to them. Who the reader is, why the reader is reading, what the reader hopes to gain, and how you respond will vary with each kind of audience.

The strengths of this scheme—layperson, executive, expert, and technician—are its simplicity and its comprehensiveness: it is easy to remember, easy to use, and fits most situations. Its weakness is that it names different kinds of characteristics for one kind of audience than for the others. It defines the executive audience in terms of the reader's role and position, the other three audiences by the reader's knowledge and interest. That is, a person can be simultaneously an executive and a layperson, or an executive and an expert. Because your audience may fit more than one of these four types, the scheme needs to be slightly modified, especially to account for complex and multiple audiences.

3.2   Complex Audiences

If your audience is one person who is both an expert and an executive, you have a *complex* audience. An engineer who has become president of an engineering firm is one example. Writing for such a reader is different from writing for the company's vice-president, an MBA who came up through the company's accounting division. If you can recognize those complex qualities in your audience, the adaptation should be no more complicated than for a simple audience.

A particularly common complex audience is the layperson/executive, someone who has no particular knowledge or interest in your subject but has decision-making power over your career. This kind of audience is more and more typical. Chemists in research labs have to justify their expenditures to an executive whose training and expertise are in accounting or management; engineers have to convince lawyers their projects are safe; and managers have to convince stockholders a merger is the sensible thing to do. If your writing class is being taught by an English professor, your audience is most likely a layperson/executive; thus, you must work to build and maintain interest while being aware of the power relationship between you and your reader.

3.3   Multiple Audiences

A *multiple* audience is one composed of several people who fall into different categories. There may be some executives and some technicians, some experts and some laypeople, or any combination of types. Company newsletters are typically written for multiple audiences. So are a company's yearly reports. The instruction booklets for many VCRs are examples of bad writing for multiple audiences: by trying to write for everyone, the writers achieve clarity for no one.

In general there are three ways to adapt your writing to a multiple audience:

- Write two or more separate versions, one for each audience.
- Write for the common denominator—isolate characteristics held in common and work on those.

- Compartmentalize your report—divide it into separate sections and clearly label them so the readers can find the sections they want easily and skip the rest.

Compartmentalization, which is usually the best method of adapting to a multiple audience, is discussed in more detail in Chapter 19, Section 2.6.

4. *Purpose*

What purposes does your writing in college serve? To complete an assignment, to move ahead academically? To show a professor you've mastered some subject area, to be evaluated (and to receive a good evaluation)? What purposes does writing on the job serve? To get jobs (for yourself and your firm), to move ahead on the job, to keep records, to give orders, to make contacts, to gain personal satisfaction, to secure professional advancement—these are only a few of the purposes that apply to writing on the job. And each purpose, each reason for writing, makes its own special demands on how your writing needs to be done. Writing for record keeping (as in a typical quarterly progress report) is about as different from writing to persuade (as in a new product brochure) as night from day. It is the change in *purpose* that keys all these differences—in tone, style, word choice, organization, use of visuals, degree of finish of the final document, and countless others.

Professionals do so many kinds of writing in so many situations that almost any generalization about purpose (such as "there are four purposes for professional writing") immediately has to be amended, made more complicated, and amended again. Having said that, we can then look at a tentative list of purposes for professional writing:

- To record
- To inform
- To teach
- To persuade

Of course, no one piece of writing ever serves just one of those purposes. The same memo that requests authorization from your boss to buy a new personal computer (a persuasive purpose) also informs the boss of your need (an informative purpose) and serves as a record you can refer to a week later when you haven't gotten an answer to your request. If the memo goes into detail about why you need the computer, filling your boss in on facts he or she doesn't otherwise know, it also serves a teaching purpose. Thus, one ordinary memo can serve all four purposes. Despite that, it wouldn't be hard to figure out which purpose was the most important (persuading the boss to give you a computer) and which one was least important (to keep a record of the request, perhaps).

Suppose, on the other hand, you were talking to your boss one day and happened to mention that you could really use a new personal computer in your office, and the boss said, "No problem, just put your request in a memo to make it official, and as soon as I get the memo I'll send through a purchase order." Think about how this record-keeping memo will be different from the other one that has to persuade the boss.

Another good example to consider is the company newsletter. Is it fancy-looking, typeset on shiny paper, with lots of sophisticated visuals? Or is it plain-looking, word-processed, and photocopied on cheap paper, with the only visuals cartoon sketches by the staff? *Purpose* is the primary intangible that explains these differences. The fancy newsletter is probably distributed outside the company as well as inside it and is probably used to represent the company to outsiders (and potential clients). Because it serves a persuasive purpose (whatever else it does), it gets the fancy, Class A treatment (see Chapter 11, Section 1.2, on various classes of documents). The plain newsletter probably serves mainly the purpose of informing people within the company of key events (promotions, picnics, and problems); thus, it gets the plain, Class C treatment.

Besides differentiating among various classes of documents (such as Class A, B, or C) professionals need to differentiate between *routine* and *special* writing tasks. Routine tasks—such as quarterly progress reports, office management memos, and letters requesting information about products or services—cannot be given the close attention and care that special writing tasks—such as performance evaluations, proposals, and reports to shareholders—must have. Once again, *purpose* triggers the differences between the two kinds of tasks. The other intangibles are also involved, but purpose leads the way.

Consider one last example of how purpose shapes a typical piece of professional writing and of how purposes interact in complex ways: Susan Smith is a junior engineer with XYZ Engineering, a rapidly growing firm that specializes in geotechnical engineering—that is, figuring out if the ground below your proposed skyscraper is strong enough to hold it up. Assigned to the team of engineers who are testing the site for the new Wilson Towers high-rise apartment and condominium complex, she also will carry a large part of the burden of writing the final report to the client. She will write some parts of it herself, collect the parts other members of the team write, merge them into one document, shepherd the report through review by senior members of the team (and by her firm's management and its own editors), and keep an eye on the reproduction process. From her company's point of view, what are the purposes of the report, in order of importance?

From the Company's Point of View

1. Susan has to let her firm's client know the information it wants: the building can indeed be built but with certain restrictions as to the type of footings that must be used. So she *informs*.

2. She has to do it in such a way that the client believes what she says and acts on it (otherwise, the report is a failure and Susan is in trouble). Therefore, in the report Susan has to be fully professional, both in the way she handles her technical tasks and in the way she communicates her results. So she has to *persuade*.

3. If the client lacks engineering expertise (maybe the client is a group of investment bankers and brokers), she has to teach, to anticipate questions such as, "Why do we need this type of footings?" and answer them in terms and concepts nonengineers will understand *but* without talking down to the client. So she *teaches*.

4. The report serves as the record of her team's work, the tangible product for which the client paid a hefty sum of money. So she writes to *record* as well.

Thus, her purposes, in order of importance, are to (1) inform, (2) persuade, (3) teach, and (4) record. At least that's what we would guess by looking at the situation through our own eyes and maybe by looking at the report.

From Susan's Point of View

1. As a new employee and as one of a relatively small number of women engineers in the firm (a situation she hopes to help change), Susan is concerned with persuading the other members of the team and the management that she is a top-flight engineer; in this, she's more concerned with persuading people inside the company than outside it.

2. She wants to *inform* her client of the team's findings.

3. She wants to *record* the team's work.

4. She may forget that the report should also *teach* until one of her firm's managers sees a draft of the report and points out that she needs to explain the "why" with the "what."

Thus, we see a different ordering of purposes when we look through Susan's eyes. From the client's point of view, the relative priorities of the various purposes would be different still.

Purpose is a complex aspect of the communication situation, one that merits concern at every level and stage of your writing. You may never be able to make your purpose, your company's purpose, and your reader's purpose exactly the same in every way, but you very much need to consider very carefully the relative importance of these four purposes in each writing situation. Many times, the accuracy of the match between your writing's purpose and your reader's purpose will be a primary indicator of your writing's chances for success.

5. *Message*

During college you write mostly in your area of subject specialization as it is traditionally understood. After college, regardless of your major, you may find yourself writing about nearly anything—personnel problems, company history, new products, management and operations, or the conference you just attended, to name a few topics. Following is a list of some of the main areas you may find yourself dealing with. Remember that in each case any or all of the four purposes may be involved and that any of the four audiences explained in Section 3 of this chapter could be your audience:

Agriculture	Management	Research
Budgets	Medicine	Research and
Business	Operations	development
Computers	Personnel	Sales
Education	Problems	Science
Government	Production	Services
	Products	Technology

Whichever one of these areas you find yourself working in and however your college major fits into that area, you will find that there are certain ways of writing that are acceptable and others that are unacceptable. For example, in government writing people typically have a much higher tolerance for jargon (language that only a particular in-group can understand) than in science writing: the same person who in the first group would aspire to be a "change agent with positive feedback" would in the second group want to be a Principal Investigator.

Of course, each of the main categories of content has almost infinite subdivisions, many of them overlapping with subdivisions of other categories, and as you become more and more of a specialist in your subject, you will inevitably learn more and more about how writing works in that field. It would be impossible to cover here all of the different varieties of writing or even to point out all of the ways in which they differ. A few of the important variables are

- the acceptability of jargon.
- the degree of technicality.
- what constitutes an acceptable proof of a point.
- how tactful the writing needs to be.
- what other concessions are typically made to readers.
- typical structures.
- average sentence length.
- use of visuals.

Each group of readers in each field (each *discourse community;* see Chapter 17) will have its own ways of dealing with those variables. The important point for writers making the transition from college to professional life to remember is that the content of your writing— specifically how well you have learned your field—just doesn't exist apart from your writing, at least not as far as your readers go. If you write poorly, people will always suspect and challenge your subject knowledge; if you write well, your writing will be a major asset in your career.

6. The Writer's Role

"Student" may best sum up your role for the last ten or fifteen years. As your purposes, messages, and audiences change when you become a professional, your role as a writer also changes. After leaving college, each different writing situation you face may require you to size up your role anew. You may be writing a note containing complex instructions to your secretary, a report to your boss explaining a difficult work-related problem, or a letter to a prospective customer. In each case the role you play as writer will need to change, and that change will combine with the particular audience, purpose, and message to shape your final written product.

How many roles can writers in professional life play? Well, you might as well ask how many writers and how many situations there are. Not surprisingly, a list of typical roles looks a little like a combination of the lists of purposes, messages, and audiences. Typical roles include seller, buyer, persuader, reporter, subordinate, evaluator, teacher, consultant, complainer, peacemaker, sharer, even student, and the list goes on and on. In your writing you have probably become so accustomed to playing the role of student that you no longer consider what role you are in as a writer; now that you are moving beyond student life, you need to become accustomed to (1) taking a few minutes to size up the role you play in each situation and then (2) adjusting your writing accordingly. You already have most of the knowledge you need to recognize such roles; we all slide into and out of a variety of them every day. The trick is to apply to your writing the common sense and instinct that guides you through daily encounters.

Figure 1.4   **A Persuasive Passage**

Overview of the Study

With its contribution to trade, its coupling with national security, and its symbolism of U.S. technological strength, the U.S. aerospace industry holds a unique position in the nation's industrial structure (NASA, 1986). However, the U.S. aerospace industry is experiencing profound change created by a combination of domestic policy actions such as airline deregulation, while others result from external trends such as emerging foreign competition (Hannay, 1986).

These circumstances emphasize the need to understand the aerospace knowledge diffusion process with respect to federally funded R&D; to recognize that STI emanating from federally funded aerospace R&D is a valuable strategic resource for innovation, problem solving, and productivity; and to remove the major barriers that restrict or prohibit the ability of U.S. aerospace engineers and scientists to acquire and process the results of federally funded aerospace R&D. However, as Solomon and Tornatzky (1986) point out, "While STI, its transfer and utilization, is crucial to innovation [and competitiveness], linkages between [the] various sectors of the technology infrastructure are weak and/or poorly defined."

These conditions also intensify the need to understand the production, transfer, and utilization of knowledge as a precursor to the rapid diffusion of aerospace technology and as a means of maximizing the aerospace R&D process. Maximizing the aerospace R&D process begins with an understanding of the information-seeking habits and practices of U.S. aerospace engineers and scientists. As Menzel (1966) states:

> The way in which [aerospace] engineers and scientists make use of the information systems at their disposal, the demands that they put on them, the satisfaction achieved by their efforts, and the resultant impact of their future work are among the items of knowledge which are necessary for the wise planning of S&T information systems and policy.

Source: NASA, *NASA/DoD Aerospace Knowledge Diffusion Research Project,* January 1991.

Figure 1.5   A Passage to Record

Breathitt County, Kentucky: Profile

Coal mining is the major industry in Breathitt County, employing 21% of the country's labor force. The county's population declined more than 30% from 1940 to 1970, but in the 1970s the population began to increase.

Almost all of the coal produced in Breathitt County is surface mined, using contour and mountaintop removal methods. There are 900 million tons of strippable coal reserves in the county. Most mining operations are small, and their numbers have declined in recent years due to several economic and practical factors, including the costs of conforming to regulations.

The most common problems associated with surface mining in the county are erosion, flooding due to sediment-choked streams, acid drainage, and disposal of mine wastes. These are chronic problems, because traditionally there was no reclamation of mined land. It will take several years before the effects of new regulations on sediment levels become evident. Because mountain soils are thin, most soil material is likely to be removed during clearing of vegetation, and it is often difficult to find the two inches of topsoil cover required on reclaimed land. There is little usable land in the county, so many landowners would prefer that land be returned to a slope flatter than the original hillsides. In mountaintop removal operations it is permissible for the land to be restored to a level condition, and it is subsequently suitable for alternative uses, such as pasture.

Source: John Seddon and Carl H. Petrich, *Management of Public Impacts in Surface Mining,* ORNL/ TM-7672.

Four examples from actual reports will show you how these elements—purpose, message, audience, and writer's role—interact in any one piece of writing. Figure 1.4 gives an example of writing to persuade. Even though the persuasive quality of this is more low key than you may be used to seeing in advertising, editorials, and other more openly persuasive communication, the passage still has a point it is trying to make. All of the elements—the point or message, the writer's role, the audience, and the persuasive purpose—are closely intertwined. That same intertwining of purpose, message, audience, and writer's role can be seen in the examples given in Figures 1.5, 1.6, and 1.7.

Figure 1.6   A Passage to Inform

Plan Scope

This Plan updates the FY 1991–1995 Five-Year Plan, incorporates (in Section 5) a condensed version of the Draft Applied Research, Development, Demonstration, Testing, and Evaluation (RDDT&E) Plan, and adds Section 6, Transportation. It begins with FY 1990 budget execution and continues through FY 1991 budget request, FY 1992 budget formulation, and outyear cost estimates through FY 1996. The Plan reflects a new Headquarters organization, the Office of Environmental Restoration and Waste Management (EM). This organization, established in November 1989, fulfills a major departmental commitment to create a high-level focal point for the consolidated environmental management of nuclear-related facilities and sites formerly under the separate cognizance of the Assistant Secretaries for Defense Programs and Nuclear Energy and the Director of the Office of Energy Research. Superfund sites at which DOE is considered to be a potentially responsible party continue to be included in the Plan as they are identified.

Source: U.S. Dept. of Energy, *Environmental Restoration and Waste Management Five-Year Plan,* June 1990.

7. The Communication Situation: Key Tangible Elements

There are many concrete and tangible elements involved in writing—punctuation, grammar, spelling, sentence structure, and others. Partly because of the sheer concreteness of these elements, it's easy to focus on them when trying to turn bad writing into good writing. But there's a lot more to good writing than handling these tangible elements correctly, and writing can be "correct" in that sense without being either *good* or *effective*. The difference between correctness and effectiveness is often how well the writer handles writing's key intangible elements. The key intangible elements of any professional communication situation are

- Message: *What* do you need to say?
- Purpose: *Why* do you need to say it?
- Audience: *To whom* are you saying it?
- Writer's role: *What* is your *relationship* to this purpose, message, and audience?

Figure 1.7   A Passage to Teach

VALUE OF THE ENERGY DATA BASE

2.3   *General Approach to Measuring the Value of Information*
       Products and Services

Value of information and information products and services can be
measured from the viewpoint of several participants in the information transfer process, including searchers of secondary products and
services, readers of primary products and services, the organizations
that fund users, and all of society that is the ultimate beneficiary of
energy information. At each level are two kinds of value. The first
value of information is determined by what the consumer is willing
to pay, and the other is derived through use of the information.
Both perspectives depend to a large degree on the extent and purposes of use of information. It is assumed that the more information
is used, the greater is the value. This does not mean that extensive
use of information about one research result necessarily yields
greater value than less-used information concerning another research result. What it does mean is that the value of any valid information is enhanced by factors that increase the amount and extent
of use of the information. The paradigms used throughout this report are totally based on this assumption.

Source: King Research, Inc., *Value of the Energy Data Base*, DOE/OR/11232-1.

Some people—natural communicators—adjust to changes in these
elements intuitively. For most of us, however, the changes require
getting used to; considering these as variables (outside college) rather
than as constants (as in college) takes some effort. The rest of this
book shows you how to make the necessary changes in your writing,
once you realize those changes are necessary.

These four elements are key intangibles in any communication
situation, but there may also be additional factors that become important. Sometimes, for example, the time line you operate on to write
your report may be such that it winds up controlling everything else
you do. If you have only two days to tour a manufacturing facility
and your manager wants your report as soon as you get back, you
may not have time to make as many adjustments in your final report

as you know your purpose, message, audience, and role really require. Or you may be only one writer on a team of writers producing a major report, and you may have to go along with the way the others have decided to do it, even though your own analysis of the situation causes you to disagree with something about their approach. These are only two examples of ways other aspects of the communication situation—such as time problems and office politics—can become as important as these four key intangibles.

The important point to remember is that the time you spend considering these key intangibles for each particular communication situation and then deciding how to treat the situation based on those considerations may be the most important time you spend. The investment in terms of minutes is actually quite small, but the rewards in terms of producing effective professional communication are high.

Finding the answers to the following questions will help you improve your awareness of the particularities of each communication situation you find yourself in and thus strengthen your ability to deal with them:

- What do you need to say?
- Why do you need to say it?
- To whom are you saying it?
- What is your relationship to this purpose, message, and audience?

- *What* do you need to say?
  What are the key points you need to cover? Are they the ones the audience will want you to cover? Is there a natural organization those points fall into? Will that be the best organization for your audience? What level of detail do you need to use? How does the occasion that brings your audience to read this document need to affect what you write? What kinds of presentation techniques (use of statistics, visuals, argumentation, etc.) will your audience respond to most positively? most negatively? How do you need to adapt your vocabulary to this situation?

- *Why* do you need to say it?
  What do you want to happen as a result of presenting this message to this audience? What is there in this communication situation that joins your purpose in writing with that of your audience in reading—that is, what is in this report for them? What will be done with your report (or letter, memo, etc.) once it reaches its audience? Will it be routinely filed away, circulated widely, or argued over? How do all these factors need to shape your document?

- *To whom* are you saying it?
  What is your audience's vocabulary level? How willing are they to hear what you have to say? What is their relationship to you (customer, teacher, peer, public, subordinate, manager, etc.) and yours to them? What is their power relationship to you? Where do they fit into the audience classification scheme presented earlier in this chapter? What does your audience stand to gain from this report? To lose? What is your audience's age, gender, education, primary group membership (accountants, engineers-turned-managers, etc.), and pertinent special interests (personnel, inventory reduction, research and development, etc.)?

- *What* is your *relationship* to this purpose, message, and audience? What kind of person do you need to be in this report? Formal or informal? Reserved or outgoing? Official or unofficial? Demanding or conciliatory? Statistical or rhetorical? Consider carefully who you want to be and who your audience wants you to be. Can you find a compromise voice to adopt?

These questions highlight some of the major considerations that need to go into doing professional writing successfully. *If you were to combine all the considerations of audience, writer's role, message, and purpose into one equation, that equation would have to be different for each report.* Thus, the techniques you need to use to adapt to the situation based on your analysis of it are much more difficult to generalize about than are the analysis techniques. For each situation, you will have to make unique adaptations. The following list shows some of the major kinds of possible adaptations:

- Provide your readers with the right kind and amount of background.
- Avoid specialized terminology when not absolutely necessary.
- If you must use specialized words, define them.
- Use visuals to help clarify your message. If formulas, mathematics, and complicated diagrams are going to obscure rather than clarify, either simplify or avoid them.
- Use plenty of specific examples.
- Meet your reader's standards for thoroughness and accuracy.
- Be sure to answer the obvious questions before they're asked.
- Make sure your reader understands the document's structure. Always remind the reader of what the document has done so far and where it is headed.
- Within reason, make sentences and paragraphs short. The more complicated your content is, the clearer your writing must be. Prefer subject-verb-object word order.

Successful professional communication always occurs within the context of a specific writer, message, purpose, and audience. Successful analysis of the communication situation (and adaptation to it) may not depend on understanding the ideas presented here (or elsewhere). Many people are successful at communicating because they have developed an intuitive sense for one or more of the four key intangibles; often such people cannot articulate exactly how they do what they do. But, for the majority of writers making the difficult transition from writing in college to writing on the job, learning to analyze each new communication situation and adapt to it is essential to effective professional communication.

EXERCISES

1. Compare the two following definitions. How do the assumptions the writer made about the reader's kind and amount of education, the reader's degree of knowledge and interest in the subject, the reader's purpose for reading the definition, and the writer's role differ between the two definitions? What other significant differences are there?

   A. magnet   any body having the property of attracting iron; specifically, a mass of iron or steel having this property artificially imparted and hence called an *artificial magnet;*—called also, according to its shape, a bar magnet, a horseshoe magnet, etc. A magnet usually has two poles of opposite nature, situated near its ends. If a magnetized bar or needle (magnetic needle) is so suspended that it may rotate freely around a vertical axis, as a compass needle, the magnet will assume a direction nearly north and south. The end towards the north is called the north end, or north pole; the end towards the south, the south end or south pole. The earth is a magnet whose poles are not far distant from the geographical poles. . . . If a second magnetized bar or needle is brought near the first, the unlike poles will attract each other while the like poles will repel each other. When a magnet is broken in two, each part is a magnet; hence, each molecule of a magnetizable body, such as iron, is itself a magnet, and magnetization is essentially the arrangement of the molecular magnets in a practically parallel array. Soft iron may be temporarily converted into a magnet, by induction without contact, or by the influence of an electric current (in which case it is called an *electromagnet*).(1)

   B. magnet   a body having the property of attracting iron and producing a magnetic field external to itself; specifically, a mass of iron, steel, or alloy that usually has two poles of opposite nature situated near its ends so that when brought close to a similar body the unlike poles attract

---

A ➤ indicates a case study exercise.

each other while the like poles repel each other, and that in the form of a bar or needle (as a compass needle) suspended so that it may rotate freely around a vertical axis assumes a direction nearly north and south.(2)

2. In what ways are the audiences for the following excerpts the same? How are the audiences different?

   A. Chemist Douglas Covey felt very much at home in his laboratory at the Washington University School of Medicine in St. Louis. The maze of glass tubes, whirling centrifuges, and bubbling flasks seemed to be all he needed to carry on his work, the creating and testing of new drugs.

   Then three years ago he met Garland Marshall, a professor of biophysics at Washington. Marshall told him about a totally new way to confront his molecules: face to face on a computer screen. Covey was skeptical. "Computers have absolutely nothing to do with my work," he said. Today he admits that he was dead wrong. He has become a true convert to computer chemistry.

   Part of Covey's research is now done in front of a computer screen, where he manipulates a mouse and the keyboard as though he were playing some sort of video game. At the flick of a wrist, lines of red, yellow, and green turn and twist before his eyes, each image conveying a bit of information about the electrical charges, structures, and volume of the molecules he may later make in the laboratory.

   Covey is one of many scientists in universities and drug companies across the country who use computers before turning to their test tubes.(3)

   B. Everybody knows what arthritis is. Stiff knuckles, swollen knees, a morning when you can't get out of bed, the poignant and agonized shuffling of people who've made desperate, last-ditch trips to the soothing-and-perhaps-healing waters of the famous bathhouses in Hot Springs, Arkansas, in hopes of a cure.

   A crippler without a clearly defined cause that respects no age, although it has always been looked at as one of the inevitable and irreversible consequences of aging—that's arthritis.

   But there's exciting news emerging from a recently held gathering in Little Rock, just a stone's throw from Hot Springs' Bathhouse Row. More effective treatment, and maybe prevention, of the disease seems possible with trace minerals—copper and zinc in particular.(4)

   C. Any science fiction buff would recognize the creature: Neither plant nor animal, the protoplasm grows in the cool damp darkness of a rotting log. Then, cued by its mysterious inner clock, the blob oozes upward, toward the surface and sunlight, toward the world of open air— there to undergo an astounding transformation.

   That blob actually exists, not in fevered imaginations but in our own woods and gardens. It's the thoroughly terrestrial slime mold, an often lovely organism with an unlovely name. Some five hundred species of this cousin to mushrooms confound zoologists and botanists alike with a life cycle that takes them from beast to beauty to beast again. They first resemble animals that grow into shapeless, slime-covered masses called plasmodia, then change into funguslike spore-bearing "fruiting bodies," or sporangia. Thereafter, they begin the cycle all over.(5)

3. Write three pages to explain any one of the following topics in three different versions—one version (one page) each for three of the audiences specified here:

   Topic 1: Why does a cold glass crack if you put boiling water into it?
   Topic 2: What happens to water after it goes down the kitchen drain?
   Topic 3: How does a color television work?
   Topic 4: How do you deal with an angry customer?
   Topic 5: How does an airplane wing work?
   Audience 1: A curious twelve-year-old (you could be writing part of the script for an educational children's television program or doing an entry for a children's encyclopedia).
   Audience 2: A friend your age (this could come in a personal letter, or as an explanation your friend must understand in order to write a paper on a similar subject).
   Audience 3: A college English professor who will grade your writing and place you in (or out of) Freshman English on the basis of its quality.
   Audience 4: A prospective employer who will use this writing sample to evaluate your writing skills (this would be part of a longer application form you are filling out).

4. Carefully read the following five passages. Each is by the same author, each is on the same subject, and each is for a different audience.(6) As you read the passages, construct answers to these questions:

   • Who is the audience? How much do they know about the subject? Why would they read this piece?
   • Where could the piece appear (give a specific title)?
   • What is the purpose of each piece?
   • What is the writer's role in each piece?
   • What in the writing of each piece caused you to answer each question as you did?

   A. Recent studies have provided reasons to postulate that the primary timer for long-cycle biological rhythms that are closely similar in period to the natural geophysical ones and that persist in so-called constant conditions is, in fact, one of organismic response to subtle geophysical fluctuations which pervade ordinary constant conditions in the laboratory (Brown, 1959, 1960). In such constant laboratory conditions a wide variety of organisms have been demonstrated to display, nearly equally conspicuously, metabolic periodicities of both solar-day and lunar-day frequencies, with their interference derivative the 29.5 synodic month, and in some instances even the year. These metabolic cycles exhibit day-by-day irregularities and distortions which have been established to be highly significantly correlated with aperiodic meteorological and other geophysical changes. These correlations provide strong evidence for the exogenous origin of these biological periodisms themselves, since cycles exist in these meteorological and geophysical factors.

B.   One of the greatest riddles of the universe is the uncanny ability of living things to carry out their normal activities with clocklike precision at a particular time of the day, month, and year. Why do oysters plucked from a Connecticut bay and shipped to a Midwest laboratory continue to time their lives to ocean tides 800 miles away? How do potatoes in hermetically sealed containers predict atmospheric pressure trends two days in advance? What effects do the lunar and solar rhythms have on the life habits of people? Living things clearly possess powerful adaptive capacities—but the explanation of whatever strange and permeative forces are concerned continues to challenge science. Let us consider the phenomena more closely.

C.   Familiar to all are the rhythmic changes in innumerable processes of animals and plants in nature. Examples of phenomena geared to the 24-hour solar day produced by rotation of the earth relative to the sun are sleep movements of plant leaves and petals, spontaneous activity in numerous animals, emergence of flies from their pupal cases, color changes of the skin in crabs, and wakefulness in humans. These periodisms of animals and plants, which adapt them so nicely to their geophysical environment with its rhythmic fluctuations in light, temperature, and ocean tides, appear at first glance to be exclusively simple responses by the organisms to these physical factors. However, it is now known that rhythms of all these natural frequencies may persist in living things even after the organisms have been sealed in under conditions constant with respect to every factor biologists have conceded to be of influence. The presence of such persistent rhythms clearly indicates that organisms possess some means of timing these periods that does not depend directly upon obvious environmental physical rhythms.

D.   A deep-seated, persistent, rhythmic nature, with periods identical with or close to the major natural geophysical ones, appears increasingly to be a universal biological property. Striking published correlations of activity of hermetically sealed organisms with unpredictable weather-associated atmospheric temperature and pressure changes, and with day-to-day irregularities in the variations in primary cosmic and general background radiations, compel the conclusion that some normally uncontrolled, subtle pervasive forces must be effective for living systems. The earth's natural electrostatic field may be one contributing factor.

E. Everyone knows that there are individuals who are able to awaken morning after morning at the same time to within a few minutes. Are they awakened by sensory cues received unconsciously, or is there some "biological clock" that keeps accurate account of the passage of time? Students of the behavior of animals in relation to their environment have long been interested in the biological clock question. Almost every species of animal is dependent upon an ability to carry out some activity at precisely the correct moment. One way to test whether these activities are set off by an internal biological clock, rather than by factors or signals in the environment, is to find out whether the organisms can anticipate the environmental events. The first well-controlled experimental evidence on the question was furnished by the Polish biologist J. S. Szymanski. In experiments conducted from 1914 to 1918 he found that animals exhibited a 24-hour activity cycle even when all external factors known to influence them, such as light and temperature, were kept constant.

5. Inventory your own strengths and weaknesses as a writer in terms of the factors discussed in this chapter. To what extent have you been prepared, by training and experience, to deal with the kinds of writing situations described in this chapter? Write a short report (250 words) based on this inventory.

6. Write a brief analysis of the audience for one of the sample long reports at the end of Chapter 19. Look carefully at all of the evidence the report offers, both explicit and implicit.

➤ 7. You are a summer employee doing an internship in your major field. As one of your first assignments, your supervisor asks you to write a short report (about 250 words) describing the role of writing in your field. The audience is other professionals in the field who may or may not believe writing ability is important. Your task is to make a convincing argument that it is. (In order to paint a convincing picture, you may want to look at trade journals, professional journals, scholarly journals, and newsletters, and perhaps do a couple of short interviews.)

NOTES

1. By permission. From Webster's *New International Dictionary*, Second Edition (p. 1480), © 1957 by Merriam-Webster Inc., publisher of the Merriam-Webster® Dictionaries.

2. By permission. From *Webster's Third New International Dictionary* (p. 1359), © 1986 by Merriam-Webster, Inc., publisher of the Merriam-Webster Dictionaries.
3. Marcia Bartusiak, "Designing Drugs with Computers," *Discover* (August 1981): 47.
4. Bruce Fellman, "Zinc, Copper, and Arthritis," *Prevention* (November 1981).
5. Douglas Lee, "Slime Mold: The Fungus That Walks," *National Geographic* (July 1981).
6. From *Practical Technical Writing* by Ritchie R. Ward. Copyright © 1968 by Ritchie R. Ward. Reprinted by permission of McGraw-Hill.

2. How Working Writers Write and Revise

In general, people write in one of two ways: either they compose very slowly and revise very little, or they compose very quickly and revise slowly. There is more and more evidence that slow composers more frequently have problems with writer's block. This chapter presents a sketch of how professionals write and then examines some typical problems working writers face and how they deal with them. By improving your understanding of the process of writing, you can become a better writer.

1. The Writing Process

Many students typically approach writing a report as something to be done in one sitting: start with several clean sheets of paper, write until you finish, and then make a clean copy. Because the purpose of

Figure 2.1   The Writing Process
Each phase of the process typically involves some looping back on the others.

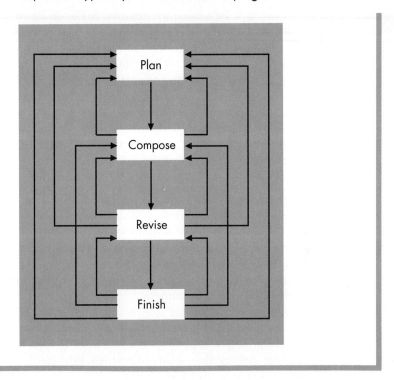

the whole exercise is to achieve a grade, and the teacher evaluates the finished product, it may seem natural for you to try to start at the end, with the finished product. But writing is most accurately and effectively seen as a *process* with several stages or phases. The more effectively you put the process into action, the better your finished product will be. The model of writing presented here puts those stages in the spotlight.

It's convenient to depict writing as a process with four stages: Plan, Compose, Revise, and Finish. But these stages do not really have a 1-2-3-4 sequence. Their relationship is not linear but rather multiply recursive—they frequently loop back on one another (Fig. 2.1).

Unfortunately, it is difficult to discuss all four stages of the writing process simultaneously, despite the fact that research shows they happen that way. When you read the following discussion, remember that in practice the stages can loop back on one another at any time.

1.1 Planning

Planning includes everything from selecting a topic, to formulating ideas about it, to devising a structure for it. Most people who need to write as part of their jobs are not in the position of having to decide what to write about: topics for writing present themselves as an integral part of the work. If your boss wants you to write a report on the tests of the C-2 cooling fan, the raw material of your report is there waiting for you. In such situations, your first task may be to formulate what you want to say about a topic that has already been determined and then to find ways to make the words flow. Here are seven *invention techniques* you can use, listed and discussed from the least structured to the most structured.

Brainstorming. Most invention techniques derive from brainstorming, and most students have done brainstorming of one kind or another, even if the activity was not actually given that name. In essence, brainstorming is free associating in response to an assigned stimulus—such as "things that could be wrong with our C-2 cooling fans"—and recording every idea or notion that comes into your mind as you consider the stimulus. Fundamental to the success of brainstorming is *not* screening out anything and *not* monitoring (during the process) which responses are better or worse. Brainstorming usually works better as a group activity than as a solo one; groups can stir up a level of creative energy that most individuals find hard to match. If you are doing brainstorming by yourself, you may well find that taking the time to write down your ideas slows your creativity too much; if this happens, try using a tape recorder.

If your company's C-2 cooling fan has drawn complaints from customers, obviously there's some laboratory testing to be done. But to be certain that all aspects of the problem are considered, brainstorming may well precede or parallel that testing. The brainstorming group may include a couple of engineers, a couple of technicians, a manager, a secretary, and perhaps one "wild card" such as a graphic artist or technical editor, to help the mix. The obvious points will need to be brought up—maybe a flaw in the manufacturing of the blades or in their mounting causes them to be unbalanced and produce a vibration at operating speeds, or maybe the shafts are not true. But some less-than-obvious points will also come up—maybe the installation instructions or the maintenance instructions are not clear, and the problems are caused by factors external to the fan itself.

This brainstorming will affect not just the process of figuring out what is wrong with the fans but also the eventual report(s) written about that process. Among the results of the brainstorming session, each item later judged to be worthy of serious consideration (check the blades' balance, check the shaft, review the installation and maintenance instructions) will be at least a minor point in the report. The one that was found to be the actual source of the problem—here it was the installation instructions—will get major attention.

Questioning. A slightly more systematic approach to invention is called questioning. In the simplest version, you just answer the journalist's traditional questions about your topic: who, what, when, where, why, and how. *Who* did something; *what* was done; *how, why, when,* and *where?* This very basic technique may have helped more writers get a rough draft going than any other technique. A more elaborate version involves brainstorming your own list of questions about the subject at hand:

- How do we know this problem really exists?
- How can we discover its nature?
- What is the problem's history and future?
- What are we afraid will happen?
- What are the possible courses of action we might take?

Once you have brainstormed a list of 15 or 20 questions, you can start answering them and use the answers to form the beginnings of your report.

Suppose you were part of a group trying to discover ways to reuse an abandoned grade school in one of your city's neighborhoods that was fighting to keep from becoming a slum. You want your report to convince the city council to give your group a $5,000 planning grant to come up with a detailed plan (and, if at all possible, potential funding sources) for the school's reuse. You know that what you will end up recommending will involve in part reusing the structure with its external appearance substantially unaltered (although refurbished) because the building is sufficiently old to be of architectural significance. This kind of reuse is called "adaptive reuse," and federal grants are often available to promote it. So your report must include that aspect, at least.

The questions you and your group brainstorm to answer might include these:

Who built the building?
Who might be interested in reusing it?
What would the reuse be?
What is the building's architectural significance?
When was the building built?
When was it abandoned?
Where is the building located in relationship to other parts of
    the neighborhood and the city? (Is it, for example, on bus
    lines or near a school that is still in use?)
Where might supporters of reusing the building be found?

Why should this building be saved instead of torn down?

Why should a federal planning grant be sought?

How would the inside of the building need to be changed under various possible reuses?

How can other funding be sought?

The answers to questions such as these will be the basis of an early draft of the report.

Verbing. A still more structured process of invention is to use more specific instructions, based on common verbs, to get your writing started. First, let's look at a simple version (often called *cubing* by writing teachers). In this use, the "It" refers to your topic. For a simple beginning, you would write as hard and fast as you can following each of these prompts:

- Describe it
- Compare it
- Associate it
- Analyze it
- Apply it
- Argue for or against it

By the time you finish this *fast freewriting*, you should have several pages on your topic.

The only rule is not to stop writing. If you can't think of anything to write, write, "I can't think of anything to write" over and over again until you can. Once again, try not to screen or monitor what you write. Don't select for good or bad, grammatical or ungrammatical, smooth or clumsy—just *write*.

A more complex version of this technique uses the verbs shown in the following list.*

Some Examples of Verbing Words

Build Up	Display	Simulate
Eliminate	Organize	Test
Work Forward	List	Play
Work Backward	Check	Manipulate
Associate	Diagram	Copy
Classify	Chart	Intercept
Generalize	Verbalize	Transform
Exemplify	Visualize	Translate
Compare	Memorize	Expand
Relate	Recall	Reduce
Commit	Record	Exaggerate

---

* *Source:* James Adams, *Conceptual Blockbusting: A Guide to Better Ideas* (New York: Norton, 1980).

Defer	Retrieve	Understate
Leap In	Search	Adapt
Hold Back	Select	Substitute
Focus	Plan	Combine
Release	Predict	Separate
Force	Assume	Change
Relax	Question	Vary
Dream	Hypothesize	Cycle
Imagine	Guess	Repeat
Purge	Define	Systemize
Incubate	Symbolize	Randomize

Try each verb in the list for a moment or two but do your five-minute freewrites on only the ones that seem to strike sparks with your topic. Suppose you need to write a recommendation report suggesting that the accounting firm you work for begin to switch a bigger part of its business away from corporate accounting and toward public account-ing (such as financial planning and tax preparation) because the profit margin on public accounting is so much higher and there's little com-petition from other large firms. Beyond using the six basic verbs described as *cubing,* which verbs in the list starting on page 34 might you find useful as starters? Let's try these:

• *Work Backward*—Trace how the company got so deeply involved in corporate accounting in the first place. The cause might lie not with any conscious decision but simply because when the company was founded and as it grew, that was where the best opportunities were. But maybe now, when so many more members of the public employ tax and financial planning help than previously, it's time to examine the opportunities in public accounting and how they compare with those in corporate accounting.

• *Hypothesize*—Speculate about where the company wants to be 10 and 20 years from now. Given the increasing competition from other large firms and the resulting decline in profits that can be earned from any corporate account, what will those numbers look like in 10 or 20 years if the same trend continues?

• *Translate*—Try translating the number of hours of your company's time that get spent on corporate accounting and the number of dollars those hours earn into an equal number of hours of time spent in public accounting and the (you hope) larger number of dollars those hours would earn.

Some of the verbs you choose might not produce material you could eventually use, but chances are that some of them will. In that case, you would have the beginnings of a potentially good report.

Capsuling. More structured still is *capsuling:* writing down the gist of what you think you want to say all in one short paragraph (in the case of something that needs to be only a few pages long) or all on one page (in the case of something that needs to be 15, 20, or more pages long). Imagine that your boss has asked you to tell her, right now, in *one minute* (you can stretch it to two), what it is you're working on. What would you say? Try writing that down. Or speak it into a tape recorder and then write it down. Later you can go back over each sentence, each clause, and each phrase, and expand each to the length each might justify.

Maybe your employer has asked you to look into ways that employee training can be improved to make it both more realistic and less time-consuming and to write a short proposal suggesting further research into whatever scheme you think makes the most sense. Perhaps what you've been looking at is the use of expert systems, a subset of artificial intelligence wherein computer-based training modules are developed based on analysis and computer modeling of the practices of humans who are expert at that task. The paragraph-capsule you quickly jot down might look like this:

> Expert systems seem to be the best way to model and so to teach the complicated reasoning and adaptive thought processes involved in responding to orders from customers in our retail sales offices. We can look into purchasing any one of several systems that have already been developed for this kind of training, and some of those systems can be modified to suit our needs for a very small amount of additional money. This training reportedly is both more cost-effective and less time-consuming than the "each one teach one" approach we have become unhappy with. The initial outlay for enhanced computer equipment and relevant software and minor modifications in similar businesses has been paid for by savings gained in training expenses over not more than the next 18 months.

What might you elaborate on when you go back over that paragraph later? Looking here only at the first sentence, you would probably expand on what "expert systems" are, why they "seem to be the best way," why and how modeling is seen as essential for teaching, and what "the complicated reasoning and adaptive thought processes involved in responding to orders from customers" might be. Similar expansions would be based on key words, phrases, or clauses in the rest of the paragraph as well.

Looping. Looping can be used as a second stage of any of the techniques listed so far. To do looping, you have to have something already written. Show that material to a friend or trusted colleague and ask him or her to underline the central idea (the "center of gravity," some would say). Some people are good at picking such lines out for themselves, and others have to learn the knack. The central idea is the thought that seems most interesting, worthwhile, pertinent, insightful, or otherwise worth expanding. Write that line on the top of a new page, and then do fast freewriting with it as stimulus. Identify the center of gravity of that material, and write again. And so on. Many people find that after doing two, three, or four 5- or 10-minute loops, they have found what they want to say.

Working from Notes. The previous five techniques have one result in common: they enable you to begin the more disciplined and directed part of your writing process, the composing, *not* by staring at a blank piece of paper or a blank screen but by sorting through a fairly large number of already written words and pages to find the right ones to put into your writing task. But if you feel paralyzed as a writer when you face a blank sheet of paper, don't begin that way. In general, as soon as a writing task comes to you, begin keeping careful written records. Write everything down, and save it all—from scribbles on dinner napkins to lab results on the computer. Begin writing by gathering up all of this prewriting material, and look for words, phrases, and sentences to use. Then build the skeleton of your first pages from that material.

Outlining. Many writers find that outlining is the most effective way to start writing. Some outlines start as brainstormed lists. If you're working with a committee to try to come up with ways to improve student parking, the results of your brainstorming session may well wind up getting organized into an outline. Some outlines start as careful structural plans. If you're working on a proposal to go to the National Science Foundation (NSF), for example, the structural format of that document may well have been so tightly dictated by NSF that their structure becomes your outline.

In the writing that most professionals do, some structural patterns occur repeatedly. The subjects of other chapters of this book represent a few of the most common structures—proposing, recommending, problem solving, and so on. Deciding which structure is right for your task depends on the communication situation: the topic, your purpose, and the nature and purpose of your audience. The following list shows some more of these typical structures; chances are your outline will reflect one or more of these patterns.

Typical Structures

Chronological   Describing a chemical process, the story of a space flight, or the evolution of a problem.

Pros and Cons   Examining whether to invest in a new plant or operation.

Order of Importance   Presenting goals that need to be met.

Classification   Inventorying a varied stock.

Spatial   Describing the distribution of different types of vegetation.

Analysis   Breaking a problem or situation into its various parts.

Functional Organization   Explaining the different parts of a situation by the way in which each one works.

General to Particular (or vice versa)   Leading a reader from a general background to a specific point, or starting with a specific example and building to an understanding of a generalization or abstraction.

Simple to Complex   Explaining a complex organization by starting with one fixed point of reference and building on it.

Organization by Association   Leading a reader naturally from one part of a subject to another, useful when your subject has lots of disparate elements.

Task-Oriented Organization   Arranging the topic in sequence of what needs to be done.

Outlining as part of invention can be as finished or as rough as you want it to be. Regardless of what your earlier experiences with outlining may have been, when you're doing the outlining for yourself, you do not need to worry about such rules as grammatical parallelism or "a B for every A." But whether you do an elaborate outline using computer software or scribble brief notes on napkins, chances are that at some point in your planning process, you will find outlining to be particularly useful.

Nearly all invention techniques have certain qualities in common. Most often the goal is to help you be less selective about what to write and what not to write, to assure you not to worry about grammar, punctuation, spelling, and style while writing a first draft. Some people find it too restrictive to call such writing "first draft" and prefer to speak instead of *freewriting* or a *zero draft:* you want to write quickly and without detailed planning and don't want criticism to interfere with creation. The goal of all of these methods is to make a start, to get words on paper any way you can, in order to conquer the fear most people feel when confronted by a stack of blank pages.

1.2 Composing

Whether you do a long zero-draft or only an outline, at some point you will be ready to go beyond writing a couple of sentences here or a paragraph or two there and to start composing your document in earnest. That point may be determined by your own feeling of readiness or by the pressure of time. The readiness to begin this process, coupled with some sense of structure and movement toward a goal, characterizes the beginning of the composing stage of this four-part model. Everything before that point is planning.

A number of factors can affect the composing stage adversely, especially *fear* and *fatigue*. Most people write better and more easily when they are not feeling threatened or tired. Unfortunately many students (as well as many professionals) find that the only time they have to write is when they have to work under exactly such adverse conditions. But to the extent that it's in your power, you should try to plan your writing project so that you do not have to work when you are tired or under pressure. This means

- starting well before your deadline.
- doing sufficient research.
- saving style and grammar concerns for the revising stage.

Writing when you are not fatigued means you should plan your use of time for days when you want to write, to allow yourself to write during quality time rather than "totally-exhausted-collapse-at-the-end-of-the-day" time. If you're having difficulty on a writing task, try tackling it first thing in the morning rather than at the end of the day.

Most writers are also sensitive to the *medium* and *environment* of their writing. Some people can only compose longhand, others only at a keyboard. Some people need quiet, private surroundings, whereas others work much better in crowded, noisy offices or libraries. As your career demands more of you as a writer, it will profit you to know more about where and how you work best, so that you can structure your writing time accordingly.

Perhaps the most important generalization about the composing stage is the one behind this statement, made by working writers again and again:

> The first time through I just try to get the subject down on paper, just to see if I've got enough data, and if I can handle what's there. Then after I finish that version I go back and start trying to make it good.

As you become more and more experienced as a writer, you will probably come to demand less and less of your early drafts. Your first written draft is an early stage, not a finished piece; grammar, punctuation, and spelling shouldn't be allowed to deflect or trap your good energy. The focus of your attention during composing should be primarily on content, on capturing and shaping the raw material. *The more aggressively you can approach your first draft and the more you can attack composing it, the better off you will be.*

1.3 Revising

In this model of the writing process, revising holds the real key to quality. Although thorough planning will help you generate the right content, as a writer in a professional field much of the content of your writing will be predetermined. Thus, composing can become a fairly routine process of "writing it up." The burden of making what you write *good* falls on revising. Doing a good job of revising requires three things:

- The *time* to do a good job.
- The *knowledge* to do a good job.
- The *will* to do a good job.

This chapter can teach you how to revise effectively; developing the time and will to do so are up to you.

This chapter focuses on revising for clarity and content more than on revising for grammatical correctness (see Chapter 3 and the Appendix for more), but the two clearly reinforce each other in important ways. It's hard to have consistently clear writing without consistently sound (grammatical) sentence structure.

Figure 2.2 shows a sample section from a technical report. By examining both the first draft, with revision marks on it (Fig. 2.2a), and the final copy (Fig. 2.2b), you can see how a typical report changes from first to final draft. Many of the changes in this report are clarifications of content. If we were to examine more paragraphs from other reports, we would see many other kinds of changes.

Revision Processes. Most writers revise in many different ways and at many different times during the writing process. Five waves (or stages) of revision can be identified—revision for

1. overall purpose and content.
2. overall structure.
3. paragraph structure and content.
4. sentence-level problems.
5. idiosyncratic errors.

Like waves in the ocean, waves of revision follow one another closely and regularly, sometimes overlapping and sometimes intermingling.

1. Revision for Overall Purpose and Content. The first question to ask after you finish your first draft is whether the document serves its purpose; in the broadest sense, is it "right"? If it's a proposal, does it contain the right elements to convince your reader to fund your project? If you're writing a final report, have you included all of the important information?

Remember that *purpose* and *content* in this sense can be seen from two different points of view: that of the writer, and that of the reader.

Figure 2.2a Rough Draft of a Technical Report

3.1 Research and Development.

3.1.1   Slurry Preheater

The coking tendencies of coal slurry feeds were unknown. Rapid preheater coking with attendant flow restriction and loss of heat transfer to the slurry was considered a distinct possibility. Coke buildup *in the tubing of the slurry preheater* was investigated in the one-ton-per-day pilot plant *(Ref. 3.1) to ensure that the preheater designed for the 250 TPD pilot plant (the ECLP) would achieve adequate run lengths*. Two feed coals, Illinois Monterey and Wyoming Wyodak, *with bottom recycle* were tested under two conditions: heating the coal-solvent slurry separate from and with the *hydrogen* treat gas. tests were run under conditions *with various profiles and residence times* that met or exceeded expected commercial-plant design; *specific conditions were not reported*.

Figure 2.2b Final Version, with Notes and Revisions Incorporated

3.1 *Research and Development*

3.1.1 Slurry preheater. Coke buildup in the tubing of the slurry preheater was investigated in the one-ton-per-day pilot plant (Ref. 3.1) to ensure that the preheater designed for the 250-TPD pilot plant (the ECLP) would achieve adequate run lengths. Two feed coals, Illinois Monterey and Wyoming Wyodak, were tested with bottom recycle of vacuum bottoms under two conditions: heating the coal-solvent slurry separate from and with the hydrogen treat gas. Monterey was also tested without bottom recycle. Tests were run with various temperature profiles and residence times that met or exceeded expected commercial-plant design; specific conditions were not reported.

You can't just ask if something is right; you have to ask if it is right for *this* audience. If your purpose in writing a first draft is just to get it all down on paper, the first step in revision is to make sure you've done just that. If you haven't, you should add or move around large sections of the document until, in its large dimensions, it fulfills its purpose and has the right content.

2. Revision for Overall Structure. The second step in this ideal revision process involves checking the large-scale structure of your report. Make sure it has a beginning, a middle, and an end easily discernible to your reader. Suppose you've written a report with five major sections: Why is the second one after the first and before the third, and why is the fourth not last? Your reader needs to see answers to those questions without having to ask them. Is your report's structure as easy for your reader to follow as it is for you? If not, do whatever is necessary to attain that clarity of structure, by making structural changes in your report, adding more structural cues, or both.

A more specific form of this stage of revision requires looking at how each individual paragraph links with the paragraphs before and after it. Can your reader see the connection? Is there a clear reason why paragraph two is after one and before three? This type of structural testing is demanding and important, and if your report cannot meet this test, you need to work with the structure until it does.

3. Revision for Paragraph Structure and Content. In the third wave of revision, you need to examine the structure and content of each paragraph individually. Do the sentences in the paragraph proceed in an orderly sequence—for instance, from general to specific, or from specific to general? Are generalizations supported by specific examples where necessary? If you need to add, delete, substitute, or move sentences, this is the time to do it. Each paragraph should be a structural unit, and each paragraph should make a distinct contribution to the report. Ask yourself, "Does this paragraph help get the report where it needs to go?" If the answer is *no,* you may want to move (or drop) the paragraph.

4. Revision for Problems on the Sentence Level. Sentence-level revision proves again and again to be the best way to improve student writing. It may be that the quality of your *sentences* (their clarity, economy, and straightforwardness, as Chapter 3 explains) tells your reader more about you as a writer than any other feature of your writing. Make these pointers part of your revision habits (also see the Appendix):

Guidelines for Sentence-Level Revising

- Prefer active verbs to passive verbs, and strong verbs to weak verbs. Watch out for strings of passives or "is" verbs.

- Watch your sentence length and complexity: the longer the sentence is, the simpler its structure needs to be.
- Prefer sentence openers to sentence closers, especially when they are critical to the meaning of the rest of the sentence.
- Prefer subject-verb-object (SVO) sentence structure. Use other structures for variety, but let SVO predominate.
- Use words that end in *-tion* sparingly. Too many of them in any one sentence, especially when paired with a weak verb, make the sentence hard to read.
- Check for grammatical errors, such as sentence fragments and comma splices, that make the document difficult to read and understand.
- Be careful not to put long interrupters between subjects and verbs.
- Watch out for abstractions acting abstractly on other abstractions. Try to rephrase them in concrete terms; if you can't, your reader will find them difficult to understand, as well. Can they be dropped without damaging the document?

As you grow as a writer, you will become more aware of your style's strengths and weaknesses and learn which sentence-level revision techniques you need to pay the most attention to. If you really want to develop a clear, economical, straightforward style, practice revision on the sentence level.

5. Revision for Idiosyncratic Errors. Idiosyncratic errors are those that may be unique to you as a writer. Such errors can be large scale (you typically spend too long explaining background and delay getting to the point), pervasive (you tend to overuse "is" base sentences), or of the one-word variety (you confuse *affect* and *effect,* or misspell *occasion*). All of us have errors we commonly make. To grow as a writer you need to keep a list of your typical errors and as one stage in your revision process check for and (and correct) those errors.

Checking for idiosyncratic errors in this way is especially important for people who are chronic misspellers. If you place yourself in that group, it may well be that you only regularly misspell 100 or 200 words. That number of words is easy to list and check for. There still will be words you misspell, but they will mostly be words you hardly ever use, and that unfamiliarity may itself bring them to your attention as words that you should look up in your dictionary. But with a list of your own most frequently misspelled words, you can work effectively toward eliminating those errors. There are also word-processing packages that automatically call your attention to spelling errors (see Chapter 3 for more on spell checkers).

By applying the five waves of revision in sequence to the rough draft of your report, you can improve your writing. The five stages do not require you to *rewrite* your paper five times, only to *go over* your paper five times. Use the five waves of revision in sequence, working from the biggest units of your report to the smallest, from overall rightness down to individual words. The process is like tuning an engine, beginning with large-scale changes and working down to fine tuning. Or you can see the process as one like sculpting: first seeing a shape in a block of marble, then chipping away to achieve the large dimensions, and eventually making only the smallest of alterations.

Perspectives on Revision. One popular explanation for revision describes it as *seeing again* (re-vision), an attempt to look with fresher eyes and fewer preconceived limits. The attempt is to gain a view of the document as a whole, to see whether it fulfills its purpose, to try to see the structure buried beneath the surface, to make alterations that will bring that structure into greater relief. The writer asks, "Will this do what it needs to do? Is it right?" looks for rough stylistic surfaces, ill-matched structures, or vague content, and tries to solve those problems. According to one school of thought, this *revision on the global scale* is something students often fail to do. Studies have shown that students and other inexperienced or struggling writers restrict their attention to individual words and phrases, missing the larger and more important problems that "seeing again" on a large scale could detect and solve. Student comments about revision unite on two points: "I don't know how to," and "I can't tell when I need to." The five-wave process gives you a good start on the "how to" but does not address the problem of "when to." How can you see *when* your work needs revising?

In order to see the need for revising, you must gain a fresh perspective on what you write. The *worst* possible situation is to write your document, revise it, and proofread it all at one sitting. Revision cannot occur effectively in that context. The best way to gain a fresh perspective on what you've written—a perspective that will help you see where you need to revise—is to let time pass between writing and revising. If you finish the first draft Tuesday night, plan on revising it Wednesday. If you can get at least a night's sleep between writing and revising, you will be much better at revision.

Other ways to gain a fresh perspective including the following:

Keyboarding: If you start with a handwritten draft and have the time to keyboard an intermediate draft before the final copy, you may find that the places that need revising are much easier to see on the resulting intermediate draft than they were in the longhand draft. Something about seeing the words arranged by the printer rather than by your hand makes problems easier to spot. Word processing makes this technique especially easy.

Reading: You may find that you can detect problems better by reading your rough draft aloud: that is, your ear may be more critical than your eye. If you try this, it's important to read aloud and try to make the words make sense (the way a good newscaster does) rather than mumbling them.

Taping: If you have access to a tape recorder, try reading your draft as described above while you tape it. Then you can listen really critically on the playback. Do not simultaneously read the printed copy during the tape playback; just close your eyes and listen for whether the report makes sense.

Sharing: All these techniques approximate ways to make yourself into someone else—a different person reading the document for the first time and trying to make sense of it. Professionals who write frequently do this by showing their manuscript to a friend (or an editor), just to see if it makes sense. Of course, if you are writing for a class there will be explicit limits on whether or how you share your paper with other readers. An informal survey of writing teachers shows that most do not mind their students' showing rough drafts to other students (and some even encourage it), with the stipulation that the reader not mark or correct mistakes. The reader responds only to questions such as, "Does the thing make sense to you?" or "Do I need to explain more here?" Check with your teacher about whether you can use this technique and what limitations need to be put on it.

In some fields, such as engineering, this practice of letting someone else's eyes provide the fresh perspective on a writer's work has become institutionalized. When Sue Hobson, an entry-level engineer, completes what she thinks is a readable draft of the portion of her company's analysis of the San Onofrio power plant, she passes it to her immediate supervisor, who reads it only for content, writes a brief note to her about it, and returns it to her. This *review/revise cycle* continues until Sue's supervisor judges the draft ready to show (usually as part of the rest of the report) to upper management. Depending on the firm, it may also be common practice for upper management to bounce the report back to Sue's supervisor for several revisions. (Anyone whose writing comes back *too many* times could be in trouble.)

A good way to employ this technique on your own is to have each document you write that is of any importance reviewed at three levels—self, peer, and supervisory—with a revision cycle to occur after each review. For a student, this might mean:

1. Working through the five stages of revision on your own.

2. Showing the draft to another student to be read for content only ("just to see if it makes sense") and making whatever changes seem appropriate after that reading.

3. Seeking one more reading, this one from someone in a higher position—a writing center tutor, for example, or perhaps your teacher—to be read not for proofreading but for *substance* and making whatever changes seem appropriate after that reading.

You may well not have the time and inclination to do this on everything you write, but documents of any importance deserve at least this much attention. Once you've done five waves of revising and three levels of review, you should be ready to move on to the finishing stage.

The five-wave process of revision presented here does not follow any one writer's particular revision techniques; revision techniques are different for each person and each situation. This process is specifically designed as a way for you to learn to revise, based on experience working with students majoring in a number of different disciplines. Use the five-wave process on enough assignments to be sure you know how to handle each wave, and then modify the process to suit your own needs. If you work with it long enough to become competent at each technique, you will become a better reviser and hence a much better writer.

1.4 Finishing

The finishing stage is similar to the revising stage, except that while you often move or take out words, sentences, or paragraphs, you almost never *add* anything. If you keep finding places where material needs to be added, you're not ready to be finishing. Finishing involves three subprocesses: editing, manuscript production, and "selling."

The editing stage involves rechecking every feature of the manuscript, from overall structure to sentence-level revisions. Now is the time to check style, grammar, spelling, and so on, which you were not concerned about while you were inventing and composing; checking these aspects of your report will help you produce as correct a manuscript as possible. Check also to ensure that the heading levels are correct, that the visuals are properly placed, and that the keyboarding is accurate. Chapter 11, "Writing in a Professional Setting," will present the process of editing in more detail.

Working as your own editor means you work as your own proofreader as well. If you do your own keyboarding, you cannot proofread your report efficiently immediately after you keyboard it. The more time you let pass between keyboarding and proofreading, the better job of proofreading you can do. You can proofread best with a sharp pencil in your hand, actually pointing at each word. The slower you go, the better. The list below is a good reminder of the most common proofreading errors:

Proofreading Errors

1. Missing letters (*th* instead of *the*)
2. Extra letters (*thew* instead of *the*)
3. Transposed letters (*hte* instead of *the*)
4. Wrong letters
5. Missing words
6. Extra words
7. Transposed words
8. Missing spaces
9. Wrong homonyms (there/their, its/it's, to/too, affect/effect, cite/sight/site, etc.)
10. Missing punctuation
11. Wrong punctuation
12. Wrong word
13. Sentence fragments

Some people try to proofread their work while it's still on the screen in front of them, before they print it out. Most find their proofreading ability goes way down when they do so. The best proofreading is still done, pencil in hand and dictionary to the side, looking at hard copy. Similarly, some students like to rely on computer software to do their proofreading for them. While there's nothing wrong with using a computerized spell checker, grammar checker, or style checker, they have profound limitations as far as what they "see" or don't "see." (Chapter 3 discusses them in more detail.) For example, this passage may look acceptable to your computer:

> Bob's architects boss May lose his heir wen he seas the cite four the gnu damn, wear a slough and sluice slice a slurb with a slurry.

Computers are fabulous tools for writers, including the system this text was written on, but they can't do everything a human mind can do.

Manuscript Production. As the documents you write become longer and contain more elements (headings, visuals, cover pages, and so on) producing the final manuscript—the one you will hand in for final evaluation—becomes more challenging. Student writers often fail to allow sufficient time for this step. Visuals and bibliographies, for example, can require time all out of proportion to the number of pages they involve.

A particularly important part of manuscript production is the *final check:* the writer puts the finished manuscript alongside the immediately prior copy (the "dead copy") and compares them to make sure that everything is as it should be. Certainly it's important to go over the text of the document one last time, but it's equally important to check all of the nonverbal elements and copy that isn't body—lists, illustrations, formulas, tables, titles, and so on. Whatever process you're using to produce the final copy, chances are that many of the errors introduced in production are of a sort that affects such nontextual elements. And do make sure the page numbering is right!

"Selling" Your Work. As a student, you might be reluctant to try to create a situation in which you submit your report not as just one more paper in the stack the teacher collects in class but as something that you hand over in your teacher's office at the end of a brief chat about the project. But in professional life, that's exactly what you want to do. Try to avoid situations in which you just put something in the company mail; instead, create situations that allow you to hand deliver any major piece of writing you've done. Never underestimate the value of a short person-to-person method of delivery even if what you say during that brief encounter sounds pretty much like the summary on the first page of your report.

Another part of selling your work involves keeping a good copy of it—not just the next-to-the-last draft but one with all the visuals, cover, and everything else just as they are in the one you submitted. Perhaps someone else in your firm will want to see it, or you need to bring it to a meeting with you in order to remind people about it, or you need it for your own records. If you put very much work at all into something you write, you'd be foolish indeed not to save a good copy of it for future use.

Once again, the four stages of the writing process presented here are artificial divisions in what is for most writers a very intuitive and naturally unified process. But if you focus on each stage more than you have before and become more conscious of what you are doing when you write, you will become a better writer.

2. Eight Problems Writers Face and How to Deal with Them

Just as all writers go through some version of the writing process, so all writers sometimes have problems. Here is a list of typical problems writers have and how they cope with them.

2.1   What If I Don't Have Any Inspiration?

It's always tempting to wait to begin until the winds of inspiration blow, but working writers can't afford to. If you want to finish your project, write on a schedule—a certain number of words or pages or hours per day on a regular basis—until you finish. You don't *wait* for inspiration; you *work* for it.

2.2   How Faithful to My Preplanning Must I Be?

Most writers make some kinds of plans before they write, but they keep those plans flexible as they proceed. They view writing as a process *that generates some of its own best ideas* at the same time that they are writing down other, previously conceived ideas. Working writers stay alert for these new ideas and trust their intuitions about when to let those spontaneously generated ideas help shape the document in process.

2.3   How Do I Deal with Procrastination?

Writers often put off writing by straightening their desk, vacuuming the floor, or using a million other creative avoidance mechanisms. Working writers use this as part of their preparation for writing. Just as athletes limber up before exercising, so you can use a physical ritual (drinking a soda, sharpening a pencil) as a habit that tells your mind that it's time to write. Go along with starting-up rituals, plan them into the way you use your time—but then write.

2.4   How Fast Should I Work?

You can't hope to work to standards if you try to do it all at once, especially at the last minute. Working writers work in stages, with the individual stages written either slowly with slight revision, or rapidly with great revision, but not at the last minute. Again, there's reason to believe that doing a draft of each stage quickly, and planning on doing lots of revision, is the best way to proceed.

2.5   When Should I Revise?

Most writers revise *while* they write, and then revise again *after* they write. However, don't fall into the trap of doing so much revising while you're composing that you lose your momentum. Most writers learn to produce at least two drafts of anything they write, and there may well be more than two phases of revision involved in producing those two drafts.

2.6   How Do I Deal with Deadlines?

All writers must cope with deadlines, and few turn in their work much before the deadline. There are three main differences between amateur writers and working writers:

- Working writers set their own deadlines, independent of externally imposed ones.

- Working writers extend their deadlines in both directions—not just pushing the deadline as *late* as possible but also expanding the time by beginning as *early* as possible. As soon as you receive an assignment, start on it in some way, if only by jotting down ideas on a note pad. Increase your writing time in *both* directions.

- Working writers realize that often a *pretty good* document turned in on time is better than a *perfect* document turned in late. (See item 2.7.)

2.7   How Do I Know When to Stop?

That's what the deadline is for. There is a small but noticeable number of people who simply cannot let go of writing tasks; they keep tinkering in search of perfection no matter what the deadline is. Often those people have gotten so frustrated by the time they enter college that they try to avoid writing entirely. The key word for dealing with this problem is *satisficing*, a combination of "satisfy" and "suffice"; it means knowing when something isn't perfect but will do. Certainly no one admires the auto mechanic who gives up on a tune-up before the motor runs quite exactly right or the eye doctor who says, "What do you want to see that well for anyway?" We focus less, however, on people who push their drive to perfection to self-destructive levels in situations where running the risk of self-destruction (or professional suicide) isn't merited. If you find yourself unwilling to let go of a piece of writing even though the deadline is upon you and everyone you show it to says it's good enough already, remember the key word: Maybe you aren't *satisfied*, but maybe all you really need is to be *satisficed*.

2.8   How Do I Deal with Other Writing Problems?

All writers have problems with their writing at one time or another. Probably the most talked about problem is "writer's block." Working writers who have this problem usually also have developed routine ways of dealing with it, such as working on another part of the piece, writing on something else, talking the subject out into a tape recorder or to a friend, or going back and rewriting the last few pages up to the problem area. Working writers also have external ways of getting help; they turn to friends, family, other writers, or editors. When they need help, working writers seek it out, rather than just withdrawing into a blue funk and not producing.

Learning more about other people's writing processes (and about your own) can help you become a more effective writer. Not only do other people have the same kinds of problems you do; they've also worked out solutions to those problems. Writing can feel like a very solitary occupation; being able to profit from the experience of other writers is a pleasant reassurance that other people have not only shared problems similar to yours but have also overcome them.

EXERCISES

1. For practice with the invention stage, try the following exercise: Pick an everyday object (my favorite is a glass cola bottle) and write down as many questions about it as you can think of in ten minutes. Brainstorm the list of questions. Don't try for good questions, and don't rule out bad ones; just try for quantity. Then do cubing (see section 1.1) with the bottle as your topic, giving yourself five minutes of writing time for each verb. Finally, pick the cube passage that strikes you as most promising, and do three five-minute loops starting from it.

2. One way to make writing easier is to practice composing faster. For an exercise in fast composing, choose a simple process you know well (changing a tire, registering for classes, swimming freestyle, etc.) and see how full a description of it you can write in five minutes. Try this with three different processes. Once again, go for quantity of words and ideas, not quality.

3. The best way to learn revising is to do it. Take one of your earlier writing assignments from this class and recopy it, writing on every other line and only on the left half of the page. (With a word processor, you can simply change the line spacing in the file and print out a fresh copy.) Now work through it with each of the five waves of revision. If you have access to colored pencils, use a different color for each wave. Finally, make a fair copy of the results, and hand both copies in to your teacher.

4. Another way to learn revising is to do it on someone else's writing. For this exercise, trade one of your earlier writing assignments with a classmate. Then proceed as in Exercise 3: Recopy the document, leaving lots of room for revision; put it through the five waves, preferably with a different color for each wave; make a fair copy; and turn in each version.

5. On the next writing assignment you do, try the five waves of revision. As you do it, keep track of the kinds of revisions you do. You may find it

handy to classify them as substitution, deletion, addition, and reordering. Write a short analysis of the strengths and weaknesses of your own revision process, based on the results of that study.

6. Find five students in other writing classes and survey them about their writing processes, the problems they face as writers, and how they deal with those problems. Write a brief report summarizing the results of your survey and analyzing their implications. Attach to it as appendixes lists of the specific questions you use. Specify the audience you are writing to—other students, people training to become writing teachers, the general public, or other (specify).

7. Write a report in which you analyze your own writing process. The report should be presented in four drafts:

   1. the planning draft, including whatever initial notes you work from.
   2. the first composing draft (the first full draft you write).
   3. the revising draft—either a middle-level draft or changes marked in a different color on the first draft.
   4. the final draft, with finishing completed.

   Try to achieve insight into your writing process, and try to show clear and methodical use of the four stages. Specify your audience: your teacher, the public, a graduate student in English doing doctoral research on the writing process, or other (specify).

8. Choose a famous author (Ernest Hemingway, Alice Walker, Carl Sagan, Ishmael Reed, Annie Dillard, Richard Rodriguez, etc.), and spend two hours in the library trying to find out about his or her composing process, what the problems were, and how the writer overcame them. A reference librarian can help you find published interviews with your author. Write a brief report on what you found, in the same four different versions described in Exercise 7. Specify your audience.

9. Choose several journals in a professional field that people with your college major usually enter (see page 422 for help with this). Search back five years in each to see whether there have been articles published on the writing problems people in that profession have. Choose one of the two following contexts, and write a short (300–500 words) report describing what you found.

   Context A: The faculty adviser in your major field is reluctant to accept the credit hours from this writing course as hours that should count toward your major. You're writing this report to convince her that writing ability is a necessary part of any student's preparation for professional life in this field—not just something that's "nice to have."

   Context B: You're the newest employee in a firm that has 40 professionals (engineers, accountants—or whatever your field is) and one editor. You've observed that the professionals spend large amounts of their time in writing-related activities, especially revising, editing, and manuscript production. They tend to bring their work to the professional editor only at the last minute and only for a final check because he is so overworked. Thus, 40 $80,000-a-year people are doing work that could be done by adding perhaps four more $25,000-a-year people, thereby making the quality of the written work better, the 40 happier in their work (because they're spending more of their time doing what they wanted to do when they entered their profession), and the company richer (because at least some of the time the new editors would save those 40 people would be spent by the 40 drumming up new business and doing better professional work on existing business). Thus you're trying to make a case to the director of human resources to hire four additional editors, and you're using this report, "The Importance of Writing in Our Company, and the Problems Professionals in This Field Have with Writing," as part of your argument.

3. *Style and Readability*

Ask a group of professionals from a variety of fields to describe briefly what they look for in the way material they read is written, and the answer will probably be "a readable style." Ask the same people what they mean by that, or how it can be achieved, and you might get as many different answers as there are people in the group. Many different descriptions of style exist, from *high, medium,* and *low* to *tough, sweet,* and *stuffy.* And many different measures of readability have been suggested, from syllables per word to words per clause and unfamiliar words per 100-word unit. Yet there is no agreement among communication professionals about what constitutes a fair measure of readability, and any survey of the "best" writers will produce a wide variety of styles. All absolute statements about what constitutes a readable style have at least one shortcoming: *no piece of writing (especially professional writing) can really be judged good or bad, readable or unreadable, apart from consideration of the communication situation in which it occurs.* We can make sound judgments about the quality of a piece of writing only by considering the audience, purpose, message, and writer's role.

With that limitation clearly in mind, what can be said about style and readability that will be of concrete use to people who are trying to improve their writing? In fact, there are certain features of writing—some qualitative and some quantitative—that do crop up repeatedly in discussions of readability or good professional style. Those features are the building blocks of style and readability presented in this chapter.

1. What Factors Characterize a Good Professional Writing Style?

Three qualities characterize good writing in any field:

1. *Clarity:* Your writing has one and only one meaning, and that meaning is readily apparent to the reader.
2. *Economy:* You don't use more words when fewer will do.
3. *Straightforwardness:* Most of your sentences are written in subject-verb-object word order.

1.1 Clarity

In your writing as a professional, try to convey one and only one meaning. Writing you do on the job is not meant to be expressive or open to interpretation. This writing also needs to move along in an orderly way and give the reader a clear sense of direction. Incorporating these qualities into your writing can help you establish a clear style.

Many times a sentence is not clear because the author has not reread it with an eye to whether it could mean two or more completely different things. Such *ambiguity* often happens because important details (those contributing to precision and exactness) have been inadvertently left out of the sentence. The following sentences show some typical violations of clarity:

1. *Example:* "In its notice of proposed rulemaking, the Federal Trade Commission made it clear that they wished to limit advertising messages to children too young to understand their content."
   Discussion: The sentence can have two possible meanings that run counter to each other. Does the FTC want to limit the number of ads to which children are exposed, or does it want the ads to go only to children too young to understand them? Adding just one more detail can make the meaning clear: ". . . they wished

to limit the number of advertising messages that are aimed at children too young to understand their content."

2. *Example:* The Industrial Distribution program here was established in 1956 and had its first graduate in 1958. The total enrollment is now 450 students, and last semester it graduated 105 seniors. At the Placement Center, the graduates had a choice of three job offers."

Discussion: This paragraph seems to be saying that the program sent out 105 seniors to compete for a total of three jobs. The third sentence should read "At the Placement Center each of the graduates had a choice of three job offers." Careless wording of the sentence and the omission of two key words ("each of") changed the meaning of the paragraph drastically from what the writer intended.

Examples 1 and 2 were unclear because the writers left important details out of their sentences and then failed to reread the passages with an eye toward possible misunderstanding on the reader's part. A good guideline for that kind of rereading is this motto of successful writers and editors in all fields: "Anything that *can* be misunderstood *will* be."

Another common reason for loss of clarity is the careless use of abstract words in your writing, as the following examples demonstrate:

3. *Example:* "A major factor in determining acceptance or rejection of a loan centers around the concept of risk."

Discussion: The sentence makes a reasonably simple thought needlessly hard to grasp because it uses abstractions instead of specifics—a *factor centers around a concept.* It might profitably be rewritten as, "Loan officers determine acceptance or rejection of a loan application in part according to the degree of risk involved."

Some writers make careless abstraction a habit, using words like *concept, factor, aspect, facet,* and *centers around* over and over. The person who wrote the sentence in Example 3 followed it with this one:

4. *Example:* "This risk refers to the general picture that is portrayed by all the demand sectors for loans."

Discussion: Now the concept the factor centered around is referring to a picture portrayed by sectors! To clarify that sentence would require introducing some tangible, concrete content, beginning with some people to perform the action in the sentence: "By 'risk,' people in the loan business mean . . ."

Here are more examples of the loss of clarity through careless and needless abstraction and the lack of specifics:

5. *Example:* "Most of my communication skills lie in the verbal facets."

6. *Example:* "The art aspect of photogrammetry relates to the imagination and past experiences of the person working with the photographs."
   Discussion: Aspects and *facets* are two of the most overused and meaningless words around today. Often they combine with an equally popular but useless verb phrase, *relates to.*

7. *Example:* "The effect of decentralization in agriculture concerns itself with pricing efficiency and how accurately market news is reported."
   Discussion: Because the writer gives us an *effect* that *concerns itself,* it's hard to determine what the writer actually means to say, which is that the effect of decentralization in agriculture *depends on* pricing efficiency and the accuracy of market news reporting.

8. *Example:* "This area can be broken down into two questions."
   Discussion: As with the other examples, improving this unclear sentence requires introducing some specifics, including someone to do the breaking down and specific identification of the area: "The project engineer then asks two questions: Is the petroleum recoverable? What will recovery cost?"

There are many more ways to write unclear sentences than any book could list. You can find more examples and ways to improve them in Section 3 of the Appendix, *A Brief Review of Sentence Grammar.*

1.2   Economy

When you're trying to improve your style, ask with each sentence you write whether you can say the same thing more clearly. But readers in business or industry demand more than clarity; they also want economy. Without making your writing too dense, try to say the same thing in fewer words. Choose words to minimize the energy your reader will need just to decode your message. The next four examples show sentences that need revision for economy. (Because violations of economy are also violations of clarity, there is little point in debating whether this or that is a violation of clarity or a violation of economy. The important thing is to recognize that the sentence has a problem and to take steps to fix it.)

9. *Example:* "The area of communication I am interested in learning more about is along the line of technical reports used in industry and government."
   Discussion: This kind of writing characterizes the style of nearly any writer's first drafts. The meaning evolves in the writer's consciousness as the sentence proceeds out onto the page. The completed sentence then has a structure that may not be the most economical way to transmit that thought to the reader. The example needs to be revised to take the clutter out of its structure and make its meaning clear: "I want to learn more about technical reports . . ."

10. *Example:* "Effective communication is the key to the success of any project, whether in school, business, or between husband and wife."
    Discussion: By using grammatical parallelism— putting similar items into similar grammatical structures—the series of items at the end of the sentence can be shortened: "whether in school, business, or marriage." Look for chances to use parallelism when you have items in a series. You can tighten up the sentence even more by eliminating the verb *is*. One standard way of doing that is to change the basic form from "X is dependent on Y" to "X depends on Y." Another standard form for revising "is" verbs also works here: try flip-flopping the sentence, rotating it 180 degrees around the verb. The sentence that originally read "a . . . b" becomes "b . . . a," which requires changing the verb. Here, instead of "Effective communication . . . success of any project," we use: "Any successful project begins with effective communication, whether in school, business, or marriage." A 20-word sentence with a weak verb becomes a 13-word sentence with a strong verb, and the writing takes on the kind of crispness and clarity that characterizes the best professional writing.

Many times when you revise a particular sentence, you use all of the methods discussed here: getting rid of extra words to reveal structure, revising "is" verbs, and being sensitive to places in which your first draft reflects the order of your thinking as it progressed rather than the order through which the reader's thinking needs to progress.

There will also be times when you will have to divide one sentence into two or more. This typically occurs in sentences that try to say too much:

11. *Example:* "Legal liability is a difficult question consisting of many parts, with the most important being the parties to which the accountant has legal responsibility for unaudited replacement

cost information and how much work the auditor can do without increasing his or her liability."

Discussion: To improve this sentence, first establish a generalization and then support it with two examples: "Courts look at many factors to determine legal liability. The two most important are (1) the people to whom the accountant has legal responsibility, and (2) the amount of work the auditor can do without increasing his or her liability."

One last pattern for revising is especially helpful. Most sentences that begin with an "It . . . that" can be made more economical by deleting the entire construction.

12. *Example:* "It is a common occurrence with me that people tend to mistake my name for something else."

    Discussion: Deleting the entire "It . . . that" construction makes the sentence much more economical. All that needs to be saved out of the deleted part is one word to capture the thought's strength or frequency: "People often tend to mistake my name for something else." Depending on how strong a statement the writer originally intended to make, the *tend to* might also be deleted.

As with violations of clarity, there are more uneconomical ways to write sentences than any book can cover. Remember the patterns discussed here and ask yourself this question: "Can I say the same thing more clearly and in fewer words?"

1.3   Straightforwardness

One of the most important qualities of effective writing on the job is straightforwardness. Within each sentence, are your thoughts written in an order that makes sense, or do you carelessly require your reader to loop back to the beginning to understand the sentence? Violations of straightforwardness, like violations of economy, often occur in first drafts, when the writer is simultaneously writing down thoughts and formulating new ones. Sometimes a whole sentence can wind up backward, as in the following example:

13. *Example:* "Soils are dynamic bodies having properties that reflect the integrated effects of climate and biological activity on the parent material at the earth's surface, which is modified by the topography of the landscape and the passage of time."

    Discussion: The sentence is too long, with too much in it. First we need to separate out its ideas:

    • Soils are dynamic bodies.

- Those bodies have properties that reflect the integrated effects of climate and biological activity on the parent material at the earth's surface.
- The earth's surface is modified by the topography of the landscape and the passage of time.

Next we must ask what the logical order of the thoughts is. The whole sequence should be reversed here, to go from the general to the specific. The reader needs to be led from the most familiar concept—the changes in the earth's surface—to the least familiar—the dynamic properties of soils (which is also the point of the passage). The principle of leading your reader from the known to the unknown is crucial to establishing the straightforwardness of sentences, paragraphs, and reports. The quality of straightforwardness is established according to the *reader's* point of view: "As changing topography of the landscape and the passage of time modify the earth's surface, the parent material of soils at the earth's surface is affected by climate and biological activity. Those effects make soils *dynamic* bodies."

Just as there are recognizable patterns to use when you are trying to make your sentences more economical, so there are patterns that make sentences more straightforward. The next sentence shows a pattern you should recognize as not being straightforward, and you should use another pattern to fix it:

14. *Example:* "Explorations of the importance, physiology, economies, equipment, media, method, requirements during growth, what is happening now, and what the future holds for aseptic methods of propagation are included in the following report."

    Discussion: This sentence, like Example 13, is backward. It conforms to a recognizable pattern, called *passive voice.* There are two basic voice patterns: active ("X does Y") and passive ("Y is done by X"). In this example, the sheer number of things that are having something done to them (explorations, physiology, equipment, etc.), combined with their being listed before the thing being done (they are included in the following report), makes the sentence difficult to read and understand. The thrust of the sentence is backward. To become an effective writer, you need to recognize such sentences and reverse their order: "The following report includes . . ."

Examples 13 and 14 show sentences whose whole structures need to be turned around. Another kind of sentence that puts last the information that needs to come first is the "roundabout" sentence. It contains a phrase or group of words at the end that need to be at the beginning, while the rest of the sentence's order stays the same. Reordering is necessary when the phrase or group of words in question states the conditions under which the rest of the sentence occurs, as in the following example:

15. *Example:* "Initial attention pertaining to waveform analysis precedes the amplitude modulation discussion in order to provide background information."
    Discussion: The "in order to . . ." needs to be first because it gives the preconditions for the rest of the sentence, the purpose behind it. Apart from those words, the rest of the sentence is also unclear; it needs an agent, a person to perform the action: "In order to provide background information, we will discuss waveform analysis before discussing amplitude modulation."

Another typical kind of roundabout sentence promises the reader to do one thing but then breaks the promise, as Example 16 shows:

16. *Example:* "The four main purposes of air traffic control are to prevent collisions between aircraft and ground obstructions; to provide for a fast, orderly flow of traffic; and to provide advice and information useful in planning and executing flights."
    Discussion: If you say *four,* your readers will think it odd if you list only three. The clarity problem can be solved by the addition of a few key words: ". . . to prevent collisions between aircraft, to warn of ground obstructions, to provide for . . ."

One last kind of roundabout sentence truly baffles readers. This is the sentence with a head, a tail, and another head. It is actually one sentence that has overrun another:

17. *Example:* "Important fluid levels (gas, oil, etc.) should be checked, body condition, that the controls are functioning properly, and that the engine is running up to specifications are included in the preflight check."
    Discussion: Here the writer forgot he or she had nearly finished one sentence and proceeded to finish another one. The sentence can be easily sorted out once one realizes it's mainly about the preflight check: "The preflight check includes . . ."

You develop the habit of clear, economical, straightforward writing by building those qualities into your writing during *revision.* As you reread each sentence, watch for opportunities to enhance its clarity, economy, and straightforwardness. The practice sentences in the Exercises section at the end of the chapter will let you measure your ability to use revising to build these qualities into your writing.

2. What Stylistic Features Change from Situation to Situation and from Field to Field?

Clarity, economy, and straightforwardness are positive characteristics of any professional's writing. There are also factors that change from situation to situation and from field to field. In general, those changes are matters that can be productively understood as adaptations based on a particular purpose, message, role for the writer, and—especially—audience. Two features of particular concern are objectivity and advocacy.

2.1 Audience Adaptation Techniques

Ensuring that your writing is clear, economical, and straightforward means *revising* it specifically for those qualities. When you write only for yourself (for example, in a private journal or in the first draft of a report) you need not worry very much about those qualities. Creating a readable style becomes important only when you plan to show your writing to someone else. In that sense, clarity, economy, and straightforwardness are audience adaptation techniques which lead to the production of a readable style.

Changes in style mainly have to do with the way you express yourself. But making changes only in style—without changing content—will not necessarily produce writing that is readable for a specific audience. Without changing the fundamental truth behind what you're writing, there are three basic ways to adapt the content of your report to your specific audience; you can adjust

- vocabulary.
- concepts.
- kind and amount of detail.

The way you handle each of these features will subtly tell your reader whether he or she is your real audience. The key issue is whether you can resist the natural tendency to place your own needs, wants, and purposes as a writer above those of your readers. On the one hand, you naturally want to use the words that come into your mind first, to take knowledge of certain concepts for granted, to go into complete detail whenever you want to, and thus to fulfill your own purposes for writing, even if it means accidentally frustrating the reader's purposes. Your reader, on the other hand, naturally wants the vocabulary, concepts, and kind and amount of detail to be adjusted and customized just for him or her. Communication is difficult when reader and writer each clutch personal needs and purposes so tightly; the document's purpose can be fulfilled only when there has been a successful negotiation between the reader's purpose and the writer's purpose. Most of that negotiation takes place during the writing process, and the signs of its success include effective style and content adjusted to that specific audience.

Vocabulary. The most obvious way to adjust to your reader is by controlling the vocabulary you use. How much highly specialized language can you use? Consider this example:

> Aerodynamic tradition derives hydrofoil craft drag as a sum of drag coefficients to produce the drag polar which provides a convenient nondimensional display of the most pertinent design features affecting craft performance. Coefficients for the drag polar are based largely upon empirical data. When consideration is limited to craft with fully submerged hydrofoils, and hydrofoil wave drag is represented in drag polar form, the total drag for such craft can be expressed analytically with a precision which matches that of the laboratory.(1)

A hydrodynamics expert can understand that passage fairly easily. For anyone else, the specialized terminology makes comprehension difficult or impossible. Many professionals will tell you they wish their colleagues would use a less specialized vocabulary, and even experts often choose not to read a piece because its language is too dense. The question is, "If you really have something to say, why don't you say it so I can understand it easily?"

Specialized technical language becomes undesirable *jargon* when it interferes needlessly with the reader's process of reading and understanding. Identifying jargon is a little like identifying weeds: any plant growing in the wrong place can be considered a weed, and any technical term (or element of an otherwise restricted vocabulary) in the wrong place can be called jargon. Most problems with vocabulary in your writing will involve determining when you can and cannot use technical terminology. Whether you are being precise and professional or dull and pretentious depends partly on the words you use, partly on your purpose, and largely on your audience.

There are all kinds of jargon; every in-group has its own language. Newcomers to any professional field feel strong pressure to use specialized language to show that they are, in fact, insiders. Yet in every professional field, there are people at the very top who speak and write in effective, plain English. Those people are admired and respected (and at the top of the profession) in part because of their abilities with language.

One of the interesting and rewarding things you can do to improve your writing is to become more critical of writing by people in your field. As you read professional journals, ask yourself how much of the specialized vocabulary you see is really necessary. Find out who the good writers in your field are, read their writing, and try to imitate it. Remember that in writing the test is effect, not what you prove about your membership in a particular group. Be careful not to use technical vocabulary inappropriately.

Concepts. Avoiding jargon is mostly a matter of being aware of what it is and how it blocks effective communication. A more subtle problem is using concepts your reader does not understand. Problems with concepts are harder to detect in your own writing than are problems of vocabulary, but they are equally disruptive to effective communication.

Whether your reader will really understand a concept is difficult for you to judge. Consider the following definition of plagiarism, written by a university lawyer at the request of a professor. The idea was to hand out to students a definition of plagiarism that would stand up in court. Here is the definition as the lawyer wrote it:

> Plagiarism definition: Any student who copies, reproduces, or in any manner presents the written work of another or others with the intent to cause any person to believe that such work is a product of the student's own mind and effort; or any student with knowledge that any work is that of another who submits same in a form which a reasonably prudent student would know is likely to induce a teacher or other person to believe that the submission is the product of the student's own mind or effort shall be guilty of Scholastic Dishonesty as that term is used in Section 34 of the University Regulations for 1992–1993.

The attorney is writing for an audience that mirrors him or her in terms of knowledge, interest, and experience in reading legalese. Although courts and lawyers may understand that definition, students and faculty (outside of law school) generally do not. Is it the legalese that makes the definition able to hold up in court, or can the definition be rewritten for laypeople in such a way as to make it clear to them and still hold up in court? If you didn't discuss this definition with a lawyer, you probably would not realize that there are two concepts buried in the definition that would be critical in its courtroom use: *intent* and *reasonable prudence.* Any translation of the lawyer's definition into lay terms needs to capture those concepts if it is to hold up in court. Here's how one writer dealt with rewriting the definition for a student audience while keeping the concepts intact:

> Any student who either:
> *Copies, reproduces, or in any other manner presents the written work of another person or other people with the intent to cause any person to believe that such work is a product of the student's own mind and effort,*

Or who:

Uses what is known to be the work of another by submitting it in a form which a reasonably prudent student would know is likely to induce a teacher or other person to believe the submitted work is the product of the student's own mind or effort,

Shall be guilty of *scholastic dishonesty* as that term is used in Section 34 of University Regulations for 1992–1993.

The problems with the lawyer's definition of plagiarism are common to all kinds of writing professionals do. Under many vocabulary problems there lurk concept problems. The accountant who suggests that her client try "amortizing good will" has more than vocabulary to explain. And the computer-science technician who tries to explain the advantages of magnetic bubble memories to his employer cannot rest after explaining "magnetic" and "bubble" and "memory."

Concepts can be difficult to explain without long digressions. Imagine trying to explain "government by laws, not by people" to a European count from the Middle Ages, or "limited warfare" to a Crusader. Closer to home, ask a management professor to explain "management by objectives," or ask a wildlife and fisheries science professor to explain "maximum sustainable yield." But despite the difficulty of explaining key concepts to your readers, the ability to do so is one of the things that separates effective writing from ineffective, and readable writing from unreadable.

Kind and Amount of Detail. Adjusting the kind and amount of detail you include may be the form of audience adaptation you are most familiar with already. In everyday conversation we automatically adjust the detail we include. You see your listener's attention start to drift away, and you immediately hurry on to your point by cutting out details. Adaptation by paring detail is so easy some computers do it. In "Help" files or other settings, you can request a very basic text:

Efflorescence. When a substance evolves moisture upon exposure to the atmosphere, the phenomenon is known as efflorescence. If the substance has a higher water vapor pressure than that of the atmosphere at the given temperature, water vapor is evolved from the substance.(2)

Or you can request more detail:

Efflorescence. When a substance evolves moisture upon exposure to the atmosphere, the substance is said to be efflorescent, and the phenomenon is known as efflorescence. If the substance has a higher water vapor pressure than corresponds to that of the atmosphere at the given temperature, water vapor is evolved from the substance until the water vapor pressure of the substance equals the water vapor pressure of the surrounding atmosphere.

Substances that are ordinarily efflorescent are sodium sulfate decahydrate, sodium carbonate decahydrate, magnesium sulfate heptahydrate, and ferrous sulfate heptahydrate. When the saturated solution of a substance in water has a water vapor pressure greater than that of the surrounding atmosphere, evaporation of the water from solution takes place.

See *Deliquescence* for the converse phenomenon.(3)

You can determine how much detail to include for any particular reader by balancing out two considerations:

1. How much (or how little) does your reader want or need to know?
2. How much (or how little) do you have to say to do justice to the subject?

When you feel those two considerations as forces pulling against each other, then you are becoming aware of audience analysis and adaptation as it must occur in order for you to become an effective writer.

Sometimes you want to do more and the reader wants you to do less; sometimes you want to do less and the reader wants you to do more. Unlike conversations, your writing cannot adjust as the reader goes along. You will have to determine the correct amount of detail the first time. That requires taking your reader's needs and wants into consideration mentally while you are writing or revising. It's not like playing chess against yourself in your head but more like playing chess against someone else in your head. If you can do this successfully, the kind and amount of detail you use will tell your reader that he or she is at the center of your concerns as a writer. *Reader-centered writing is readable writing.* Its opposite, *writer-centered writing,* may be good for diaries, journals, rough drafts, personal letters, or avant-garde novels, but it is not effective in professional work. It sends the message that "I (the writer) am more important than you (the reader) are." For a full discussion of reader- and writer-centered writing, see Chapter 18, Section 4.1.

2.2   Objectivity and Advocacy

Two characteristics of writing style change often enough from field to field and from situation to situation to be especially important: objectivity and advocacy. *Objectivity* in reference to writing style means that the writing wears the pose or takes the stance of being disinterested. It means, for example, that a piece of writing can present two possible solutions to a problem without the writer's seeming personally in favor of one over the other. It means the ability to describe procedures or situations in such a way that anyone reading the description will think, "Yes, I see it that way too." *Advocacy* in reference to writing is in some ways an opposite to objectivity (in other ways, "subjectivity" is). Discussing options with a tone of advocacy means doing so in such a way that people perceive the writer as clearly favoring one particular option. Some fields prize objectivity in all writing situations; others prize advocacy. Most want objectivity in some situations and advocacy in others.

Two different versions of the same passage illustrate these two characteristics. As you read the versions, ask yourself which one illustrates advocacy and which one illustrates objectivity. Here's the situation: Russel Hart, a senior accounting major doing an internship at Mitchell Accounting Company, has been asked by his supervisor to write a short report comparing two software packages, both of which claim to improve people's writing. The question he is supposed to answer is, "Are these any good at all, and, if so, should we purchase one of these for our employees—especially our accountants—to use?" The question Russel is wrestling with concerns which version of the report to submit. Here's the one he wrote first.

Version A

An Analysis of PunctuPro and Xprtwrtr:
Grammar and Style Checking Software for
IBM Personal Computers

PunctuPro and Xprtwrtr have been the two most popular software packages for grammar and style checking on IBM computers for several years. Although the two products are comparable on nearly all key issues, a close comparison of the two (including looking at the results of published reviews by independent journals) shows that the additional minor features of PunctuPro make it the program our office should purchase.

Both software packages detect a high number of the most common grammatical errors, both offer very easy to understand statistics on readability, and both are relatively inexpensive (around $100; multiuser site licenses are also available at about $500). Both let the user specify the desired style (PunctuPro: General, Business, Technical, Fiction, Informal; Xprtwrtr: General, Business, Report or Article, Manual, Proposal, Fiction). Both also flag racist and sexist usages. Both programs provide document summaries that present readability coefficients, stylistic traits, and other relevant features both numerically and graphically. Each allows the user to turn off any part of the program the user does not want to apply to a particular document and to insert sensitivity to any of a number of other features (words ending in *-tion*, for example).

The key difference is that while PunctuPro allows the user the choice of either viewing an entire document before changing anything or making changes as the text proceeds, Xprtwrtr requires the user to make any desired changes as each item comes up. Xprtwrtr assesses each document's readability according to only one readability measure (Gunning's Fog index), while PunctuPro assesses each document's readability according to three widely used readability measures (Gunning, Flesch, and Flesch-Kincaid). While Xprtwrtr does compare each document's readability to that of three other presumably well-known standards (the Declaration of Independence, the IRS directions for filling out the basic income tax short form, and an Edgar Allan Poe short story), a feature PunctuPro does not match, frankly the usefulness of such a comparison is difficult to imagine.

Neither program is sensitive to the context of the writing; neither can sense when to break the rules. Tests by both *PC Magazine* and *PC Week* showed that both missed instances of subject-verb number disagreement, misplaced punctuation, and sentence fragments. Both programs did, however, catch all the spelling errors and flag all the uses of passive voice and sentences with "is" verbs in the passages on which they were tested.

Either program will probably improve some aspects of the writing our accountants produce. Given the relatively small cost involved, purchase of a site license with enough individual copies for each employee thus makes sense. PunctuPro's additional features—the ability to make changes at any time or withhold changes until an entire document has been viewed and the use of three different readability indexes—make it the recommended choice.

After Russel had finished Version A, his roommate (an engineering student) suggested he make the changes that resulted in Version B. "This one is more professional," his roommate said.

Version B

An Analysis of PunctuPro and Xprtwrtr:
Grammar and Style Checking Software for
IBM Personal Computers

PunctuPro and Xprtwrtr have been the two most popular software packages for grammar and style checking on IBM computers for several years. Both detect a high number of the
most common errors, both offer very easy-to-understand statistics on readability, and both are relatively inexpensive (around $100; multiuser site licenses are available
at about $500). Both let the user specify the desired
style (PunctuPro: General, Business, Technical, Fiction,
Informal; Xprtwrtr: General, Business, Report or Article,
Manual, Proposal, Fiction). Both also flag racist and sexist usages. Both programs provide document summaries that
present readability coefficients, stylistic traits, and
other relevant features both numerically and graphically.
Each allows the user to turn off any part of the program
the user does not want to apply to a particular document
and to insert sensitivity to any of a number of other features (words ending in *-tion,* for example).

PunctuPro allows the user the choice of viewing an entire document before changing anything or making changes
as each item comes up; Xprtwrtr requires the user to make
any desired changes as each item comes up. Xprtwrtr assesses each document's readability according to only one
readability measure (Gunning's Fog index) but compares
each document's readability to three other (presumably)
well-known standards: the Declaration of Independence, the
IRS directions for filling out the basic income tax short
form, and an Edgar Allan Poe short story. PunctuPro assesses each document's readability according to three readability measures (Gunning, Flesch, and Flesch-Kincaid) but
does not offer a comparison to other documents.

Neither program is sensitive to the context of the writing; neither can sense when to break the rules. Tests by
both *PC Magazine* and *PC Week* showed that both missed instances of subject-verb number disagreement, misplaced
punctuation, and sentence fragments. Both programs did,
however, catch all the spelling errors and flag all the
uses of passive voice and sentences with "is" verbs in the
passages on which they were tested.

Either program will probably improve some aspects of
the writing our accountants produce. Given the relatively
small cost involved, purchase of a site license with

```
enough individual copies for each employee thus makes
sense. Although the two products are comparable on nearly
all key issues, the additional minor features of PunctuPro
(in terms of letting the user make changes in the text at
any time and representing the text's readability in terms
of several well-known readability measures) make it the
program of choice.
```

Version A lets its readers know from the start where the writer stands on this issue; Version B withholds judgment until the end. There's not much point in asking which one is a better piece of writing in any abstract sense; they're almost exactly identical. But one might be better in the situation Russel is in. Some would say one is more like the writing accountants generally do and the other more like the writing engineers do. In fact, there are plenty of situations when an accountant needs to use the pose of objectivity rather than advocacy and plenty when engineers need to use the pose of advocacy. *The difference between the tone of advocacy and the tone of objectivity is important in professional writing; only each writer in each situation can fairly judge which tone is desirable.*

Use of First-Person Pronouns. Often when writers struggle over the use of first-person pronouns, the problem is really one of advocacy versus objectivity. For example, Russel could just as easily have rewritten Version A this way (only the first paragraph is shown here; similar changes would be made throughout):

> PunctuPro and Xprtwrtr have been the two most popular software packages for grammar and style checking on IBM computers for several years. Although I find the two products are comparable on nearly all key issues, a close comparison of the two (including looking at the results of published reviews by independent journals) shows that the additional minor features of PunctuPro make it the program I believe our office should purchase.

Then Russel might have worried whether it's acceptable to use first-person pronouns in the writing accountants do (generally it isn't). But the real issue behind that worry is whether the tone of advocacy indicated by the first-person pronoun is appropriate. In chemistry, engineering, and agriculture, objectivity is usually the goal for writing's tone; thus, the first person is frowned upon because it adds the elements of advocacy and subjectivity (the two opposites of objectivity) to the writing. In business, government, and human ecology, advocacy is much more generally tolerated or expected, so *sometimes* first person is more acceptable.

The bottom line on these key issues is that the writer has to make the right decision about which tone is preferable (and, if advocacy is the choice, whether first person is then appropriate) according to the situation. Usually the standard for making the decision is set by the community of all writers in that field—what scholars today call the *discourse community*. For more on discourse communities, see Chapter 17, Section 3.1.

3. What Troubleshooting Methods Are There for Style and Readability in Professional Writing?

The concern for such factors as clarity, economy, and straightforwardness and for such issues as advocacy versus objectivity is properly something for writers to apply during *revising*. There are other ways to improve writing style that are perhaps best saved for an even later stage of the writing process, *finishing*. Such troubleshooting methods are the subject of this section.

3.1   How Not to Write Gobbledygook

We've all seen gobbledygook—scientific, or governmental, or legal, or technical, or any other kind of professional writing in which the use of jargon, generally poor word choice, and overly long sentence structures makes the writing unreadable. The response of most readers to such writing is "Even if I could understand this, it's just not worth the effort." Different kinds of gobbledygook have different names—legalese, medicalese, Pentagonese, and so on. Here's an example of several kinds of Pentagonese rolled into one:

> Any fiscal planning subsystem requires considerable promulgation and synthesis at the division level to ensure efficacy of the relevant data base. Furthermore, the policy of redundant standardization performance implies anticipation of future growth dependence and preliminary evolution testing. Subsequent configuration finalization must function in concurrence with postulated hardware interrelationships and full utilization of integral criteria qualifications. While the purview of this office includes dissemination and characterization of functionally interwoven criteria subsystems, pertinent

regulatory guidance, particularly the applicable concomitant sections of CFR 10.58.193, precludes furtherance of fiscal analysis considerations without high-tier functionality assurance. It is therefore necessary that triplicate submittals be availabilized for conceptual checking of utilization commonality.(4)

One way to discuss the passage's lack of readability would be to look at it in terms of clarity, economy, and straightforwardness, as discussed earlier in this chapter. Another way would be to look at it in terms of its audience—is there an audience who would read that and try to understand it? But there are some other criteria for readability that are not as general as audience or style in a broad sense. Although these criteria are generalizations, they provide several good rules of thumb for writers who want to avoid writing such gobbledygook as the passage just cited. These criteria remind you to check for sentence length, stacked nouns or adjectives, and abstract verbs.

Shorten Your Sentence Length. Frequently a sentence that is unreadable is unreadable in part because it is too long. For writers who are not professionals and are dealing with complex subject matter, it seems that the longer the sentence is, the greater is the chance that its structure will break down. Just what constitutes a sentence that is too long is a debatable point, but many writing specialists use a length of 17 words as a guideline. That is, while it is certainly possible to find very clear sentences longer than 17 words, or to find very unclear sentences shorter than 17 words, as a rule of thumb it's a good idea to look twice at any sentence you write that is longer than 17 words and make sure its structure really carries its weight. Generally that means using a subject-verb-(optional) object structure. (See the Appendix for more on this.)

Unpile Your Noun and Adjective Stacks. Some writers develop a habit of piling up modifiers in their sentences. This produces strings of words like "fully operationalized bidirectional real-time multiplexer" or "cantilever truss reinforcing gradient sloping beams." It takes only a couple of phrases like that to tempt the reader to stop reading. When each sentence has two or three such groupings, you get a paragraph of solid, impenetrable gobbledygook. Such phrases are characterized by the use of too many modifiers for any one phrase and the frequent use of nouns as modifiers. Avoiding the use of noun and adjective stacks will make your writing more readable.

Minimize Your Abstract Verbs. One other key element contributes to the making of gobbledygook. The same writers who use overly long sentences and fill them with noun and adjective stacks also usually use weak and abstract verbs. Such verbs are forms of *to be, to have,* and other verbs such as *relates to, revolves around,* and *centers around.* Avoiding such verbs when they are not necessary to your sentence's meaning will help make your writing more readable.

3.2   Using Computerized Style Checkers

A number of computer software packages claim to provide different kinds of readability measures of the text files they are run on and to improve the user's writing. Certainly most of these programs do detect some writing errors—spelling errors and subject-verb disagreement, in particular. But when the editors of *PC Magazine* compared the performance of three popular software packages to that of a human editor on a poorly written business memo they created just for this test, they found the computer software inconsistent at best (Fig. 3.1).

As the report about "PunctuPro" and "Xprtwrtr" in Section 2.2 pointed out, the problem with such software goes beyond its sheer inability to detect some problems (such as confusion among *sight, cite,* and *site;* run-on sentences; or misplaced modifiers, just to name a few). The problem is that no practical way has been found to make the computer sensitive enough to the context in which the writer is working to be able to tell the writer when to bend the rules. And bending the rules seems to be an inherent characteristic of good writing in English.

A further problem exists with such software, and that is the notion of readability on which the various popular readability measures (used with variation by these software packages) as "readable at the eighth-grade level" are based. For example, to find the Fog index of a piece of writing, one selects a sample that is 100 words long and divides the total number of words by the number of sentences to find the average sentence length. Then that number is added to the number of words of three syllables or more in the passage, with the sum multiplied by 0.4. The result, the Fog index, is said to represent the number of years of schooling a person would need in order to read the passage with ease and understanding. The problem is that the formula is not sensitive to the context the passage occurs in. What might be readable with a certain degree of ease by an engineer sitting in the office might pose real problems for a power plant operator surrounded by flashing red lights and alarms. But the Fog index bases an estimate of readability on the passage itself, not on its purpose, the nature of its readers, the nature of its content, and especially not the situation the passage is being read in. The passage from a physics textbook that absolutely baffles you at midnight may look much more readable at 9 A.M., after eight hours of sleep, a shower, a bowl of

Figure 3.1   Analysis of Computer Software for Checking Grammar

	Program			
Grammatical Element	*Correct Grammar*	*Grammatik*	*RightWriter*	*Glen Becker*
Passive voice	Yes	Yes	Yes	Yes
Spelling errors:				
personel	Yes	Yes	Yes	Yes
unwillingly	Yes	Yes	Yes	Yes
Minitores	Yes	Yes	Yes	Yes
Subject/verb disagreement	Yes	Yes	No	Yes
Missing word	No	No	No	Yes
Misplaced comma	No	No	No	Yes
Misplaced hyphen	No	Yes	No	Yes
Misplaced modifier	No	No	No	Yes
Redundancies	No	Yes	Yes	Yes
Double word	No	Yes	Yes*	Yes
Cliché	No	No	No	Yes
Run-on sentence	No	No	No	Yes
Fragment	Yes	No	No	Yes
Sexist language	No	Yes	Yes*	Yes
Inconsistent verb tense	No	No	No	Yes
Wrong word usage:				
affective	No	Yes	Yes*	Yes
on	No	No	No	Yes
there/their	Yes	Yes	No	Yes
that/which	No	No	No	Yes
who/whom	No	Yes	No	Yes

*Noted the word but did not specify the nature of the problem.
Source: PC Magazine, December 11, 1990.

cereal, and two cups of coffee. But the readability measure stays the same.

Computerized style and readability checkers, like the readability measures they include, have serious limitations. Nevertheless, they can be useful. Although spell checkers cannot catch every possible error (recall the passage "Bobs architects boss May lose his heir . . ." in this chapter), they do catch quite a few, and they are great for eliminating many typos, as well. Programs that call writers' attention to the overuse of passive sentences or overly long sentences are very beneficial *so long as users realize the programs' limitations.* Let these programs help where they can, but do not rely solely on their readability judgments. Let your own sense of the situational context, along with responses from other readers, weigh more heavily in your final assessment of your writing than the computer software's assessment. Finally, don't rely on computerized grammar and style checkers to catch every error. They can't do it.

3.3 Eliminating Biased Language

There are a number of forms of language usage that people in American society have come to view as biased, that is, unduly favoring one group of people over another (or disfavoring one group or another). Such language may discriminate against people on the basis of age, gender, race, ethnic group, or physical abilities. Typical examples of biased language include usages that (however inadvertently) characterize older people as being forgetful, or women as being "charming," that identify one person's race but not another's ("Jack Sellars, a bright, young, black attorney on our staff, will accompany Thomas Watson, another of our attorneys, to the hearings"), that stereotype ethnic groups ("another one of those overachieving Vietnamese-Americans"), or that imply people's physical abilities necessarily correspond with their professional abilities (the blank labeled "health" on an application for managerial employment). In most professional settings today, use of such language is discouraged; in some settings (such as government contracts), it is strictly forbidden.

As a writer, you may be producing biased language without knowing it. During the finishing stage of your writing process, an increased awareness on your part of what biased language is and how to eliminate it can help you deal effectively with this important issue. Here are three simple guidelines to follow that will help you eliminate biased usages from your writing:

- Be careful about stereotyping.
- Pay attention to the gender of pronouns.
- Eliminate unnecessary references to gender, race, age, ethnic group, or physical abilities.

Be careful about stereotyping. Is it appropriate, in a letter of recommendation, to refer to a woman as "delightful"? No (unless, perhaps, it's for a modeling job). Such language stereotypes women as people whose quality as prospective employees should be measured in part by their ability to keep everyone around them comfortable. How about, "State legislators and their wives will attend Saturday's game as the university's guests"? State representative Anne Dorrell will resent that, as will her husband. Stereotyping can be visual as well as verbal. How many times have you seen a film in which the secretary brings a note into the boss's office, and the secretary is *always* a woman, and the boss is *always* a man? Such stereotyping may still be common in Hollywood productions, but it is unacceptable in professional communications of any sort. Bosses aren't always men, secretaries aren't always women, nurses aren't always women (a stereotype male nurses resent as much as female nurses do), and construction workers aren't always men. Make sure that your professional communications don't reinforce such stereotypes; they're offensive at least, and illegal at worst.

Pay attention to the gender of pronouns. More than 50 percent of the people in today's workplace are women; to use only masculine pronouns to describe workers is discriminatory and untruthful. If you are writing text for a brochure describing what your accounting firm does for its clients, you cannot say, "When our accountant comes to your office, the first thing he does is . . ." Nor can you write about the number of "man-hours" a task requires or that "each stockholder will cast his vote." There are a number of ways around such usages:

- Rewrite passages to use *plural constructions* where possible: "When our accountants come to your office, the first thing they do . . ." or "Stockholders will cast their votes . . ."

- If singular pronoun references cannot be avoided, consider alternating between masculine and feminine as you move from one passage to the next. If your instruction manual for operating the robot welder refers to the operator as "he" in the first section, switch to "she" in the second section. You can even use fictitious individuals, with names, to avoid pronouns almost entirely, with a woman for your example in one section and a man in the next. A safety manual could explain a floor supervisor's role in ensuring workplace safety by talking about Jack Smith (a fictitious example) in one section and then talking about an assistant manager's role by discussing the responsibilities of Mary Anne Epstein (another fictitious person).

- As an absolute last resort, if it comes up only once, you can use "he/she" or "he or she"—or "s/he" or "she or he"—but most people find those usages awkward reading.

Eliminate unnecessary references to gender, race, age, ethnic group, or physical abilities. Using "Dear Sir" as the only possible salutation unnecessarily implies all letter readers are male. That's fine if you know the reader is male. If the reader is female, it's not. To avoid the problem some companies now routinely recommend all letters use "Dear [recipient's name]:" as the salutation. Other companies always use "Dear Mr. Smith" or "Dear Ms. Jones." In some instances, you can simply use the person's job title: "Dear Credit Manager." And in others, you can dispense with the salutation entirely and use a *subject line* instead. (Chapter 4 shows an example of this.)

Company titles also can unnecessarily be biased. "Chairman," "foreman," "night watchman," and "stewardess," for example, all have desirable alternatives: "chair" or "manager," "supervisor," "the night watch," and "flight attendant."

There are many other usages that contain unnecessary references to gender, age, and so on and that thus stereotype or discriminate. During the Gulf War in 1991, it was interesting to listen to television newscasters struggling to avoid referring to "our servicemen on duty in the Persian Gulf" or the equally biased "our brave boys in the army, navy, and air force." Your writing needs to be free of such biased usages, and it can be if you check for them during the finishing stage of each document's writing process.

This chapter has given you many different ways to make your writing style more readable. Whether you focus your attention on clarity, economy, and straightforwardness or on controlling sentence length, avoiding noun and adjective stacks, replacing abstract verbs, and eliminating biased language, the goal is to produce more readable writing. A popular misconception about professional writing would have you believe that it's okay for such writing to be really hard to read because the subject is really complex (or really technical). The principle here is something like "the harder something is to do, the less well you have to do it." Operating under a principle like that, human beings would never have made the ceiling of the Sistine Chapel a timeless masterpiece, or broken the four-minute mile, or eliminated polio, or walked on the moon. Reject that principle and realize that the tougher the topic you are writing about, the more important a readable style becomes. *As your topic becomes more difficult or more complex, your writing should become more readable, not less.*

EXERCISES

1. Find a technical paper you wrote before you took this class. Revise it for clarity, economy, and straightforwardness. Turn in both copies.

2. Revise these sentences for clarity:

   (a) My educational background basically centers around a B.S. degree in business administration as an accounting major.

   (b) I spent two quarters studying the government securities market and three quarters of computer (BASIC, Pascal, and COBOL) which will allow me to handle your data processing.

3. Revise these sentences for economy:

   (a) It is my hope that this letter will allow you to have a better understanding of my writing.

   (b) The focal point of my interest is presently directed to achieving a major in marketing with a double minor in computer science and biological science.

   (c) To elaborate, it should suffice to say that this is the only formal technical letter I have ever written.

   (d) It has been determined through research that the vaccination of adult cattle will nearly eliminate clinical disease.

   (e) The reason why I am taking this course is because it is required for me to graduate from the university.

4. Revise these sentences for straightforwardness:

   (a) The position described in your advertisement of Sunday, October 6, 1992, in the *Washington Post* interested me very much.

   (b) In response to your job opening notice concerning geotechnical engineers positions advertised in *Civil Engineering* magazine September 1992 issue, I am submitting my job application letter.

   (c) In response to the letter you wrote Dr. Randall, I am applying for the position of programmer/analyst described in the aforementioned letter.

   (d) By the time you finish reading this letter, statements related to past life, present thoughts, and future hopes should be clear.

5. Revise these sentences according to all three principles:

   (a) If the insured exceeds the benefit limit due to a certain sickness or injury the insurance company will still pay additional money if you have a totally different sickness or injury from the one which caused you to exceed the benefit limit.

   (b) Factors that cause a shift in demand curves are a change in the number of buyers, this is caused by population growth or an extension of the

market, changes in the income of people and their tastes and preference for their product contribute greatly.

6. Choose any two of the following passages and revise them for clarity, economy, and straightforwardness to fit a layperson/executive audience. Eliminate any biased language.

   A.   The aim of the present work may be bracketed by a series of disjunc-
   tions. In the first place, the question is not whether knowledge exists,
   but what precisely is its nature. Secondly, while the content of the
   known cannot be disregarded, still it is to be treated only in the schematic and incomplete fashion needed to provide a discriminant or de-
   terminants of cognitive acts. Thirdly, the aim is not to set forth a list
   of the abstract properties of human knowledge but to assist the reader
   in effecting a personal appropriation of the concrete dynamic structure
   immanent and recurrently operative in his own cognitional activities.
   Fourthly, such an appropriation can occur only gradually, and so there
   will be offered, not a sudden account of the whole of the structure, but
   a slow assembly of its elements, relations, alternatives, and implications.
   Fifthly, the order of the assembly is governed, not by abstract considerations of logical or metaphysical priority, but by concrete motives of
   pedagogical efficacy.(5)

   B.   Henry County is hereby authorized to incur indebtedness to the extent
   of not exceeding $700,000 in aggregate principal amount, and to issue
   its bonds in evidence of the indebtedness so incurred, for the combined
   purpose of constructing and equipping a new courthouse and a new
   jail in said county. Said bonds may be issued only after the question of
   the issuance thereof shall have been submitted to the qualified electors
   of said county at an election called for that purpose by the governing
   body of said county and a majority of said qualified electors voting at
   said election shall have voted in favor of the issuance of said bonds,
   which election shall be called, held, conducted, canvassed and may be
   contested in the manner provided by the then existing laws of the State
   with respect to elections on the issuance of bonds by counties, provided,
   however, that if a majority of the qualified electors of said county participating in the election on the adoption of this amendment shall vote
   for the adoption thereof then the approval of this amendment expressed by the vote in said county in favor of its adoption shall of itself
   authorize the issuance of the bonds, and in that event no additional
   election by the electors of said county shall be required to authorize
   the issuance of said bonds. In the event the majority vote in said county
   on the adoption of the amendment is against the adoption hereof, or
   in the event the majority vote at any election held in said county pursuant to the provisions of this amendment after its adoption is not
   in favor of the issuance of the bonds proposed at such election, the

governing body of said county may from time to time call other elections hereunder on the issuance of said bonds, but not more than one such election shall be held during any period of twelve consecutive months. The power to become indebted and to issue bonds in evidence of such indebtedness shall be in addition to all other powers which the said county may have under the constitution and laws of the State, and any bonds issued pursuant to this amendment shall not be chargeable against the amount of indebtedness which said county may incur under the constitution and laws in effect prior to the adoption of this amendment.(6)

C. Should purchaser fail to pay said indebtedness or any part thereof when due, or breach this contract, or should seller feel itself or chattels insecure, or if any execution or writ be levied on chattels or any purchaser's property, or a receiver thereof is appointed, or if a petition under the Bankruptcy Act or any Amendment thereof should be filed by or against purchaser, the entire unpaid balance shall at once become due and payable at seller's election, and seller may, without notice or demand, by process of law or otherwise, take possession of chattels wherever located, and retain all moneys paid thereon for the reasonable use of chattels and purchaser will pay for necessary repairs because of damage thereto.(7)

D. Any person, firm, corporation or association or agent or employee thereof, who, with intent to sell, purchase, or in any wise dispose of, or to contract with reference to merchandise, real estate, service, employment, or anything offered by such person, firm, corporation or association, or agent or employee thereof, directly or indirectly, to the public for sale, purchase, distribution, or the hire of personal services or with intent to increase the consumption of or to contract with reference to any merchandise, real estate, securities, service, or employment or to induce the public in any manner to enter into any obligation relating thereto, or to acquire title thereto, or an interest therein, or to make any loan, makes, publishes, disseminates, circulates, or places before the public, or causes, directly or indirectly, to be made, published, disseminated, circulated, or placed before the public, in this state, in a newspaper, magazine, or other publication, or in the form of a book, notice, circular, pamphlet, letter, handbill, poster, bill, sign, placard, card, label, or over any radio or television station or other medium of wireless communication, or in any other way similar or dissimilar to the foregoing, an advertisement, announcement, or statement of any sort regarding merchandise, securities, service, or employment, who includes any false or misleading statements or assertions, shall be guilty of a misdemeanor.(8)

E. For the preceding year or two, man has been witnessing a dynamic and explosive war of the newly developed schematic noise reduction systems, but the struggle is unceasing against a background of impending digitalization. From the digital point of view, binary playback is inevitable and the skirmishes over analog noise reduction merely a rearguard action to forestall obsolescence. The analogist, whose attitude is distinctly favorable, assumes that the three C's will anchor the digital Hun at bay for an equally elongated period: cost, complexity, and compatibility. And then there's vested interest. With untold billions of analog discs and tapes actively engaged in the world's household hi-fidelity systems, the conventional medium will, at ultimate, demise a lingering death of attrition many years after the successful launch of a home digital medium.(9)

7. Select a 100–150-word sample from one of the extracts from long reports in Chapters 11–15 or from the student report at the end of Chapter 19, and rewrite it for a lower-level audience. Begin by specifying the audience you are rewriting it for and the reasons they have for reading the passage.

8. You have applied for an entry-level job in your field of choice and now are one of the final three candidates. Your prospective employer has now asked each of the three final candidates to write a brief letter explaining their work-related communication skills (both writing and speaking). Write the letter, and remember to support any claims or generalizations you make. (In a short note on a cover sheet, explain to your teacher what profession you're planning to enter.)

9. Revise the following memo to improve its style and readability.

MEMO

February 12, 1992

TO: Salesmen and Office Personnel in the Woodson Building

FROM: B. J. Williams, Assistant to the President

Upon initiation March 1st and for a duration of approximately two or more months workmen will commence refurbishment of floors one through four (1–4) of the Woodson Building, 1717 E. Lee St. In order to facilitate accomplishment of minimization of disruption or crippling of work effort and wastage of man-hours resultant from this activity, immediate effectuation of the following policies and procedures is hereby announced:

1. Parking in the lot adjacent the building will be restricted by space allocation designation for workmens' vehicles and the four outermost spaces will be reserved for foremen of the construction crew so employees should make other arrangements for parking during that time frame and consider implementation of vehicular co-transportation. Parking slip payment vouchers from the pay lot at the corner of 17th and Lee Streets will receive reimbursement at a rate of fifty (50) percent, but all such parking reimbursement voucherization must be initialized by each unit's supervisor or his secretary. In the evenings, women who are afraid to walk to the pay lot will be escorted by the security guard in between his regular rounds.

2. Due to plumbing repairs and general refurbishment/replacement of old equipment scheduled initially for the restroom facilities on the East and South sides of the building followed by the West and North sides, employees should make themselves aware of alternate locations for utilization in other parts of the building during this time frame. At this point in time, each unit's administrative secretary should post a floor plan for her area identifying alternate locations.

3. Beginning on four and proceeding downwards one floor at a time in order, offices on each floor have been scheduled for new carpeting, painting, and replacement/refurbishment of old equipment. Floating space has been made available on five for those who have been temporarily displaced by implementation of this refurbishment/replacement.

4. The Staff Lounge and Lunch Room on two will also be closed for the duration of the above stated renovation. As per above, alternate dining locations will have been posted by each secretary for each of her units.

On behalf of all the members of the Board, and especially Chairman Adams, I want to thank you in advance for taking the inconvenience of this necessary temporary displacement like the great group of guys and gals you are.

NOTES

1. H. Raymond Wright, Jr., and Frank W. Otto, "Hydrofoil Craft Drag Polar," *Journal of Hydronautics* (October 1980): 111.
2. and 3. This example was adopted from *Van Nostrand's Scientific Encyclopedia*, 5th ed. (New York: Van Nostrand Reinhold, 1976), pp. 864–865. Reprinted by permission.
4. Used as an example of bad writing in Bill Minkler, "Let Me Speak to Your Computer," *Nuclear News* (November 1980): 240.
5. Father Bernard Lonergan, *Insight.*
6. Amendment 237 to the Alabama State Constitution.
7. Part of a standard sales contract.
8. *Printer's Ink* model statute, 1945.
9. Robert Long, "Tape and Tape Equipment—A Time of Change," in *High Fidelity* (Sept., 1981).

PART TWO

Letters and Memos

The letters you write in your job can make or break your day, week, month—or career. Many people in the working world dread writing letters and memos, put off doing them, take time with them they can ill afford, and (as often as not) wind up doing a poor job. If you talk with a few competent, trained professionals in any field about their problems with letters and memos, often you will find beneath that smooth professional surface a mass of anxieties and fuzzy ideas about how to write business communications. Yet a professional in any field who cannot write effective letters or memos is like a tennis pro without a backhand.

What makes a letter or memo effective? There are two main factors:

- It must do its job—fulfill its purpose—for both reader and writer.
- It must be efficient for the writer—not take time or energy (or cause stress) out of proportion to its importance.

The three chapters in Part Two show you how to write effective letters and memos—correspondence that you produce efficiently and that gets results.

The organization of Part Two is a miniature of this book as a whole. There are too many different kinds of letters or memos to give specific attention to each individually: letters of inquiry, complaint, adjustment, goodwill, application, rejection, information, reminder, and intent, just to name a few. Therefore, these chapters first present principles that apply to all types of letters or memos: Key concepts such as *solicited versus unsolicited, "you" attitude,* and *goodwill* are ex-

4. Forms and Principles

5. Patterns

6. Job Applications and Résumés

plained in detail. Then the needs shared by all correspondence writers receive specific attention; how to ask and how to say no; how to decide between a direct or an indirect approach; how to send bad news when you must; and how these patterns respond to human psychological needs. Finally, one most important kind of letter, the job application with résumé, is discussed in detail, beginning with the important obvious (and not-so-obvious) roles writing plays in the job search, including detailed instructions and examples for writing effective application letters, explaining three different kinds of résumés, and concluding with guidelines for succeeding in interviews. The goal throughout is to teach you the principles you need to be able to adapt the basic patterns to your own situations.

4. *Forms and Principles*

This chapter provides a working knowledge of elements common to all kinds of professional letters: both external elements (forms) and internal elements (principles). Custom dictates how you deal with some of them. For instance, if you double-space your letter instead of single-spacing it, or you write it in longhand instead of keyboarding it, your letter may strike such an odd note when first opened that the message you intended to send may never really be considered. Whatever the message you are trying to send, the message the reader gets may be that you aren't very professional.

Although some elements of correspondence seem to be dictated by custom and require only imitation of good models, others require considerable thought, vary from one person to another and from one situation to another, and create considerable debate among professionals. Chief among these may be the issue of negative messages: how to say something the receiver may not want to hear. For example, how do you write letters telling customers that their accounts are past due and do it so that it increases the probability that they will pay up?

Anyone who has ever gotten an "account past due" letter knows the first impulse that races through the receiver's brain: "Well if they're going to get tough about it, they can just wait another week or two." Maybe the only thing professionals agree on about negative messages is that there is no single right way to send them. Despite the philosophical nature of such concerns, the person who has learned how to transmit negative messages effectively—how to move the reader in a positive direction—will find that business and industry will reward that ability. Furthermore, the ability to use letters persuasively—to gain people's cooperation and goodwill even under adverse circumstances—is the mark of a fully effective communicator, a very special kind of person.

1. *Proper Form*

Maybe the best way to explain the importance of using proper form in correspondence is to compare it to having the right clothing and equipment for various sports. If you invite your friends over to play volleyball, and if they are good volleyball players, they may not be impressed by your plastic volleyball and flimsy backyard badminton net (you need a leather ball and a very strong, very tight net made especially for volleyball). It may seem unfair to be judged by appearances, but the leather ball and special net perform much better. Similarly, in professional letters there is usually an important function behind the appearance.

For example, what important functions could something like the placement and content of an inside address have? There are at least three:

- To ensure that the letter goes to exactly the right place and the right person, even if the envelope has been thrown away or lost.
- To ensure that the person who receives the letter realizes you know his or her position in the firm.
- To establish a certain level of formality or familiarity.

Behind each of the following matters of form there are several good reasons. But remember, even if you cannot think of the reasons, the power of custom, of presenting a professional (hence safe and familiar) appearance, should not be dismissed lightly.

1.1 Basic Elements for Letters

The basic elements of form in letters include the margins, typography, inside heading, inside address, subject line, salutation, body, closing, and supplement line(s). The following paragraphs explain each key point in detail. Figure 4.1 demonstrates one good form (unblocked) with each part labeled. There are a number of other popular forms for letters: the block form (Fig. 4.2), the military form

Figure 4.1   The Unblocked Form
This traditional letter form is still widely used.

Heading —
Department of English
The University of Tennessee
Knoxville, TN 37996
September 28, 1992

John Q. Public
5000 Oak Ridge Highway — Inside address
Knoxville, TN 37914

Subject: Demonstration of Unblocked Form

Dear Student: — Salutation

    This letter shows you the unblocked letter form which I recom-
mend for use in your business correspondence. While there are
many other possible styles, this one is easy to learn.

.
.
.

— Body

    This is probably the most conservative letter form. Once you
learn the elements of this form, the other forms are simply matters
of arrangement. Good luck in your writing class.

Closing and
signature —
Sincerely,

Michael L. Keene

Michael L. Keene

mlk/js — Supplement line

Figure 4.2   The Block Form
Everything begins at the left margin.

312 Left-Hand Canyon Road
Boulder, Colorado 80301
August 17, 1992

William Devereaux
Attorney
1512 Oak Terrace
Des Moines, Iowa 50321

Dear Mr. Devereaux:

As we discussed on the telephone yesterday, this letter confirms my
acceptance of the Johnsons' offer for my house on

me here in Boulder. Thank you again for your help. Please send me
a copy of the closing papers as soon as the deal is finished.

Sincerely,
Kevin J. O'Farrel
Kevin J. O'Farrel

(modified; Fig. 4.3), the simplified form (Fig. 4.4), and a typical letterhead form (Fig. 4.5). Your decision about which to use should be
based on which form you are comfortable with and which form you
think will appeal to the reader.

Margins. Consistently use margins of 1 to 1½ inches. This "white
space" makes your letter more attractive visually, and it gives your
reader a chance to make notes in the margins.

Figure 4.3   The Military Form
This form is very orderly but too severe for some people.

Acme Testing Laboratories
Region 6 Headquarters
128 Classen Blvd.
Atlanta, GA 30320

Date: April 10, 1992

To: Wilson Parnell
          District Supervisor
          Acme Testing Labs
          1432 East Magnolia
          Norman, OK 73069

From: E. C. Wyatt
          Regional Coordinator

Subject: Shipment of Test Samples

Purpose. This letter is to confirm that I have received your request
for a change in the way we ship the test samples from the lab here
in Atlanta to you. As you suggest, beginning with the May shipment
we will ship by Rapid Express instead of Acme.

Action. We will ship by Rapid Express, following our usual policy
of shipping on the first of every week. I hope the change really does
result in the cost savings you estimate. If there are problems with
these next shipments, just let me know and we will find another
shipper.

Thanks for the good suggestion.

                                      Sincerely,

                                      E. C. Wyatt

                                      E. C. Wyatt
                                      Regional Coordinator

Figure 4.4 Simplified Form
A growing number of people use some variation of this for all but their most formal business letters.

1212 North Avenue
Minneapolis, MN 55401

February 15, 1992

Wiscossett Town Board
Wiscossett, WI 54841

REQUEST FOR REAL ESTATE INFORMATION

My wife and I have often driven through your town on our summer vacations and now are considering buying property.

Thank you for your assistance. I hope we will hear from you soon, and that we might in the future be among your neighbors.

Tom Parsons
Tom Parsons

Typography. Your letter needs to be printed "letter quality" (that is, with fully formed characters as opposed to dots) on good paper (at least 20 percent rag content). The lines of print should be in single-spaced blocks, with extra line spaces left between blocks (for example, between heading and inside address, between body paragraphs, and so on). Although some people think a letter or memo looks more official if it's printed with the right side justified (that is, the right edge of the blocks of text will be aligned rather than ragged), there is convincing evidence that right justification produces a less-readable page than its alternative, "ragged right."

Heading. If you do not use letterhead paper, place your address and the date in a single-spaced block. If you abbreviate your state, use the Postal Service's official abbreviation. The heading should begin 1 to

1½ inches down from the top of the page and far enough in from the right edge of the page for the right margin to be 1½ inches. If you want to emphasize the date, skip one line between it and the rest of the heading. Most letterhead paper needs only a date inserted in the appropriate place (Fig. 4.5).

Inside Address. The inside address, which tells the person opening the letter exactly who is to read it, should be at least double-spaced down from the heading, in a single-spaced block beginning at the left margin. It should include the reader's name (and title, if any), corporate affiliation, and address.

Subject Line. Many writers use a subject line between the inside address and the salutation to inform the reader of the point of the letter immediately. The advantage is that a reader, seeing the subject at a first glance, can decide whether to read the whole letter now, put it off until later, or route it to someone else. If you choose to use a subject line, it should be short, important words should be capitalized, and it should always be phrased positively.

Salutation. At least two spaces below the last line of the inside address (or the subject line, if you use one), you should greet your reader. Custom dictates, but that custom has changed in the last ten years. Begin with "Dear Mr. [Ms.] Smith," followed by a colon. If you know the name but not the gender of the recipient, you may choose to use "Dear [name]" (as in "Dear C. J. Smith") or to omit the salutation entirely (as in Fig. 4.4) and use only a subject line. If you do not know the recipient's name at all, you can use either a job title ("Dear Credit Manager") or a subject line. Omitting a salutation entirely and using just a subject line for routine correspondence is increasingly popular because it avoids these problems and decisions. If you do choose to use only a subject line, you may want to omit the closing ("Sincerely") as well.

Body. The body of your letter should begin on the second line below the salutation (or subject line). In the unblocked form, each paragraph should be indented. Normally your letter will be only one page long, which gives you enough space for three or four short paragraphs and the closing. In these letters, one-sentence paragraphs are fairly common, especially in the last paragraph.

Closing. Begin the closing at least two spaces below the last line of your last paragraph, far enough to the left in the unblocked form that the longest element in the closing ends at the margin. In most instances, the traditional "Sincerely" is the right choice. Depending on the situa-

Figure 4.5 A Typical Letterhead
The centered letterhead combines well with an unblocked letter form. A left-side letterhead would work better with block or simplified form.

Thomson Heating, Inc.
1551 Laurel Lane
Duluth, Minnesota 54840
Phone (613) 841-4880

January 13, 1992

Gary Powers
Personnel Manager
Wallace Motors
Duluth, MN 54839

Dear Gary,

Thanks for sending the people from Ace Labs to us for their new heating and cooling equipment. We visited their plant last week, and it looks like we'll be able to do quite a bit of work for them.

I hope we have the chance to return the favor soon. In the meantime, if I can do anything at all for you, just give me a call.

Sincerely,

J. Thomson

Jim Thomson
Owner

tion, you may wish to use "Thank you." Leave enough room for your signature (three or four lines); then type your name. It is often a good idea to include your job title below your typed signature. Be sure to sign your name. In simplified forms, there is no closing line.

Supplement Line. In the lower left corner of the page, at least two lines below the typed signature and at least an inch up from the bottom, you may want to use another line (or lines), usually called a supplement line. Any or all of the following may be used. You may have typist's initials ("MLK/js"—typed for Michael L. Keene by John Smith). You may add an enclosure notation ("Enclosure") or a list of enclosures ("Enclosures: Final Cost Estimates"). There may be copy notations ("xc" has come to stand for photocopy; "cc" means carbon copy), and finally there may be a personalizing note or a postscript.

You would not normally want to use a "PS" in a professional letter; you don't want to look like someone who has such last-minute changes of thought. If you think of something else important after the letter has been printed, you should probably redo it and incorporate the new material.

There is one very important exception to not using closing notes. Many letters are recognized by the people to whom they are sent as form letters, with only names, titles, dates, and such inserted as they are needed. The letter shown in Figure 4.6 is one such stock letter. While you may often use such a letter, fully aware the reader will recognize it as a form letter, you can ease your reader's feelings by adding a personalizing note at the bottom, as Figure 4.7 shows.

Second and Succeeding Pages. Sometimes you cannot say everything on one page. In those instances, place the following information 1 inch (six lines) from the top of the second (and any succeeding) page(s): At the left margin, the name of the person, then the page number (centered), and the date against the right margin. Letters of more than one page should be paperclipped together, not stapled. Figure 4.8 shows a typical continuation page.

1.2 Basic Elements for Memos

The nature of your audience tells you whether to use a letter or a memo: memos are in-house (sent to a person in your own firm or department), and letters are sent outside the firm. In some professional environments, be cautious about sending memos too many steps up the organizational chain of command. Obviously, a shipping room clerk would be ill advised to send a memo to the Chair of the Board. Similarly, a student petitioning the dean to count FORTRAN as a foreign language should send a letter, not a memo. The example

Figure 4.6   A Stock Letter
Only the name is personalized.

C onsolidated Communications
318 Webb Road, Madison, MI 48071

November 9, 1992

Dear Jane Smith:

As you may be aware, we have recently acquired controlling interest in ABC Printing. Their facilities and personnel will be added to ours, and our customers will be better served by this addition.

While this should in no way affect your position with our company, we wanted you to hear the good news directly from us. The addition of ABC Printing should give us many more opportunities to make Consolidated Communications the leader in the field.

Sincerely,

Donna Banrow

Donna Banrow
Vice-President

db/js

---

in Figure 4.9 (p. 97) shows one good memo form; there are many others.

Your choice of words and phrases creates as big an impact in memos as it does in letters. Choose the way you approach your reader carefully, and avoid the temptation to dash off hasty, abrupt commands or complaints in memos. All the principles of correspondence discussed in the next section apply to memos as well as to letters.

1.3   Record Keeping

Whether your letter or memo seems important or not, you should always keep a copy, even if it is only on disk. Again and again this

Figure 4.7   A Personalizing Note
Such notes become more and more common as stock letters are used more and more.

.
.
.

addition of ABC Printing should give us many more opportunities to make Consolidated Communications the leader in the field.

Sincerely,

Donna Banrow

Donna Banrow
Vice-President

db/js

Forgive the form letter, Jane — there are just too many people to tell. Why don't you come by next week and we'll talk about some ideas I have for your next project?

Figure 4.8   A Continuation Page
Second and succeeding pages begin this way.

Jane Smith   2   November 13, 1992

and that is why the shipment as ordered never came through. I suggest that in the future we accompany the shipments with invoices clearly stating the type and number of parts shipped. Although this new procedure requires more work and tracking, I'm sure that the

.
.
.

Figure 4.9   A Typical Memo Form
Memo forms vary widely from company to company.

June 15, 1992

TO: Employee Members of "Basic Plus" Health Plan
FROM: Denise Walker, Program Administrator
SUBJECT: Extended Benefits for Families

From June 15 until July 15 you can increase the coverage on your immediate families to include dental care with no initial enrollment fee. For the average member, this means that for an extra $8/mo. your spouse and children living at home can have the same schedule of benefits for their dental care which you now enjoy.

Should you wish to take advantage of this limited offer, contact my office at Extension 5411. Remember, the offer expires July 15, 1992. Prompt action on your part will assure you of full dental coverage for your immediate family at a very low cost.

will pay off for you—as customers lose original estimates, as you try to remember exactly what you said to prospective employers, or as someone questions the exact date your letter was sent. Too many people begin keeping copies only after several bad experiences caused by the lack of copies. You will have an advantage if you begin keeping copies now.

2.   *Principles*

If the matter of proper form seems almost totally controlled by custom and if that frustrates you because it leaves little room for your own creative thinking, take heart. The principles of correspondence are almost endlessly debatable. Depending on the situation, each principle applies in a different way and to a different degree. The principles are presented here in sequence from the least debatable to the most debatable. (As you might expect, the most debatable ones are often the most important.)

2.1   Distinguished between Solicited and Unsolicited

From the reader's point of view, all correspondence can be placed in one of two groups: *solicited* or *unsolicited.* Your reader either has or has not asked to hear from you. In terms of effect on the reader, this distinction is vitally important. Most readers at least mildly resent unsolicited correspondence, and in the face of that initial negative message your task as a writer is considerably tougher. Any letter that can follow its salutation with the three magic words, "As you requested," makes an initial good impression on the reader; everything else that letter does is easier.

Think of the way you read your own mail. Many of us first read the letters from friends and people we know, the letters we've somehow expected. Then later, when everything important has been done, we read the unsolicited mail. Thus, if you want your letters to be as effective as possible, you want them to be solicited letters. You begin the body of the solicited letter with "As you requested" or "As we discussed on Thursday, August 12," or some such *contact phrase.* This phrase reminds the reader of earlier contact with you and helps create a favorable predisposition toward the rest of your letter.

Although your reader will mentally categorize your letter as either solicited or unsolicited, as the writer you actually have some freedom to move otherwise unsolicited letters into the solicited category.

While there will be times when your letter must be unsolicited, there are other times it is really in a gray area, and you can move it over into the solicited category. It is worth the effort to avoid sending unsolicited letters, perhaps by making an initial telephone call or by securing an introduction and reference from a third person who knows you both. Then you would have the contact phrases "As we discussed" or "Jim Smith in your Chicago office suggested that I contact you concerning . . ." In these and many other ways, you can work toward moving your otherwise unsolicited correspondence to the solicited category.

If you really must write unsolicited correspondence, remember that all the elements of the letter are precarious to handle because it is unsolicited (unless it is a "good news" letter, as discussed in Chapter 5, Section 2.1). In unsolicited correspondence, each principle is trickier to deal with. Reader benefits must be made clearer sooner; making your own purpose clear is more delicate; ensuring a return message is harder; *you* attitude makes a bigger difference; negative messages seem to be flashing red lights; maintaining a positive emphasis and goodwill takes more work; and the whole letter needs to be more persuasive.

2.2 Remember What to Do First and Last

Nothing aggravates people who receive letters more than letters that beat around the bush. *State your letter's purpose* plainly in the first paragraph—if possible, in the first sentence. Often you need to tell your reader not only why you are writing but also what you hope the reader will do in response.

Suppose you want to write for more information about a new word-processing software package you have seen advertised. You might begin your letter:

> I would like more information about the new PhraseMaster word-processing package.

Your letter would probably bring you a more satisfactory response if you detail specifically what you want the reader to do:

> I would like more information about the new PhraseMaster word-processing software. Would you please send me whatever brochures you have available, specifically including information on increasing readability through . . .

Adding the second sentence that lets the reader know exactly what you expect the next step to be increases the chances that the response you get will be the right one.

In a longer letter, you would put last the line(s) that lets the reader know exactly what needs to be done next. This principle—*the action closing*—is as important as putting the statement of purpose first. Even if a whole paragraph earlier in the letter detailed expected future actions, the last one or two lines restate or summarize briefly what you are requesting.

The action closing, like all of the other principles discussed here, can take many different forms. One of the most interesting—because it shows how precise the applications of these principles can become—is the *resale* ending. The concept of resale is borrowed from the field of executive selling. Any experienced computer (or yacht) salesperson will try to end each encounter with a customer with resale—a phrase or line that both lets the customer know exactly what will come next and paves the way for future communication, providing the contact phrase that will make the follow-up communication solicited instead of unsolicited. In the simple letter we are discussing here, "I hope to hear from you soon" would conform to the principle of resale, but it would be better to give that line some impact: "Our office is considering purchasing a new line of equipment, and your product is one we want to consider very carefully. I hope to hear from you soon." This time the reason behind the resale is explicit.

A computer salesperson, writing an answer to a request for information, might add, "I will be in your city next week; why don't I call your secretary and arrange for us to go over these details in person? I hope to see you soon." Or a yacht salesperson, having just sold a 40-foot sloop, would say something like: "Be sure to let us know if you have any problems. I'll call you in two weeks just to make sure everything is satisfactory." There are several reasons for calling on the client after the sale:

- To make sure this sale will bring the client back for more.
- To sell accessories.
- To sell yourself to friends and co-workers of your client through that contact.

To make your letters as effective as possible, open with a statement of purpose and close by making clear what you expect to happen next, using resale as appropriate to keep that communication channel open for future messages.

2.3   Use "You" Attitude

You can explicitly mold your letter to satisfy both your needs and those of your readers if you employ a statement of purpose and action closing in your correspondence. Another important principle of letter writing is to use "you" attitude. Although at its simplest, "you" attitude simply means using the personal pronoun *you* in business letters, effective "you" attitude includes anything a writer does to demonstrate that the reader's needs, wants, questions, and concerns are matters of importance.

Maybe a teacher once told you a good letter uses at least as many *you*'s as *I*'s, or that you should try not to begin every sentence with *I*. "You" attitude is that principle, but extended to more than use of the word *you*. To illustrate this concept, consider the two versions of a simple courtesy letter shown in Figures 4.10 and 4.11. (Courtesy or goodwill letters are sent to reinforce and strengthen relationships; see Section 2.7.) The first version has the opposite of "you" attitude, which some call "me" attitude. Another example of "me" attitude can be found in exercise 6 on page 134 of Chapter 5. Notice that within the context of appreciation, this first letter actually does nothing for the reader, except perhaps to show how self-centered the writer is (Fig. 4.10). Figure 4.11 shows the same letter, rewritten to demonstrate "you" attitude. As you might expect, "you" attitude can be overdone. You, as the writer, must carefully judge how much attention to your reader is enough.

Figure 4.10 "Me" Attitude
This makes the writer seem self-centered and needlessly ignores the reader.

> Dear Susan,
>
>   I want to tell you how pleased I was with your presentation yesterday. I have seen countless such presentations in my years with the company, and I can truthfully say that it got me excited about the new sales campaign more than any in a long time. Since the presentation, several other people have mentioned to me that they agree with me. Congratulations!
>
> *Jim*

Figure 4.11 "You" Attitude
This makes it clear that the writer puts the reader's feelings over his own.

> Dear Susan,
>
>   You should be very proud of the good sales presentation you made yesterday. Yours was more impressive than any this company has seen in a long time. Several other people have also commented on how well you did. Congratulations!
>
> *Jim*

2.4   Provide Reader Benefits

"What's in it for me?" is a question you can count on your reader asking: "How does this concern me?" "Why should I take the time to read this?" The experienced writer explicitly answers such questions before they're asked. The more unsolicited your letter is, the more important it is to answer those questions before your reader has time to ask them. A good business communication article begins this way: "Keep this article handy. It will be useful whenever you have to present statistical information in writing."

Whenever your correspondence asks someone to do something, you need to explain why it is to your reader's advantage to do it—how it benefits the reader and why it is important, as in a letter of request or a persuasive letter. People usually take action faster if they understand how it benefits them or why it is important. Thus, instead of saying "Please mail us all forms pertaining to your third-quarter expenses as soon as possible," you could say: "In order to make sure you receive your reimbursement promptly, please mail us all forms."

Reader benefits can be written in either or both of these ways:

1. I can help you in the following way(s).
2. I identify with you in the following way(s).

The first option requires no particular explanation here; it is the stuff most reader benefits are made of. The second version of reader benefits can be very effective. Figure 4.12 presents a classic negative-message letter, with its impact softened by both kinds of reader benefits.

Coming up with reader benefits usually requires only a moment's thought and an extra phrase on the writer's part, but their importance in terms of effect on readers cannot be overstated. Reader benefits tell your reader you care about him or her as a human being, a creature who deserves and responds to considerate treatment. Skillful use of "you" attitude and reader benefits gives your letters and memos an orientation toward the reader and thus a tone of humanity that will always make them more effective.

2.5 Be Careful with Negative Messages

So far, we have seen several examples of ways to make letters and memos more positive through "you" attitude and reader benefits, but there are times when you simply cannot be totally positive: you may have to turn down a request, notify someone of a failure, or terminate a professional relationship. Any situation in which you send a message the receiver will not want to hear is called a *negative-message situation*. One of the largest bodies of literature in the field of business communication concerns identifying negative messages and determining how they can best be handled. Because at times you cannot escape sending negative messages, we will discuss three different types: in this chapter, obvious and hidden messages; in the next chapter, those that require special structures for entire letters.

Obvious Negative Messages. When you know your reader will not be pleased with your message, take pains to minimize that unhappiness. Nothing is gained, and a great deal can be lost, by needlessly making

Figure 4.12   Two Kinds of Reader Benefits
The editor identifies with Mrs. White in the first line of the third paragraph and
then tries to help her.

XYZ Publishers

1512 Commerce Street, New York, New York 10012

September 12, 1992

Brenda White
1802 Lee Street
Oklahoma City, OK 73124

Dear Mrs. White:

Thank you for sending us your manuscript, *The Search for Amelia*, to consider for publication. You obviously know a great deal about your subject, and you present your knowledge quite clearly.

After due consideration, our editorial board feels we should not publish your book at this time. We simply do not feel that Amelia Earhart's disappearance generates the kind of public interest that can make publication profitable for us. What you set out to do is done very well, but you add nothing to generate new interest in the subject, and there is not enough interest in it currently to generate sales.

I remember at the start of my own career how unhappy I could be when I was told "It's well written, but not right for us." Let me encourage you to persevere, with this and other projects, but since you've told me you're a first-time author, let me give you two pieces of advice. First, before you invest the time and energy to write a book-length manuscript, do some preliminary groundwork, such as letters of inquiry, to see whether anyone will be interested in publishing what you might say on that topic. Second, you might consider hiring an agent, who would help your writing find its way to the right kinds of publishers.

I.D. with reader

Offer of help
(or alternatives)

Good luck with your career. I hope you will consider XYZ Publishers again in the future.

Sincerely,

John Wolfe

John Wolfe
Editor

people unhappy. There are several ways to soften blows. Here are three different principles for handling negative messages:

1. A telephone call is usually the worst way to send a negative message. Give the message in person if possible; as a second choice use a letter.
2. Always explain the reason for the negative message. If there is a positive side (or a reader benefit), be sure to emphasize it.
   Example: Not "We cannot ship your order," but "Due to the postal strike, we cannot ship your order . . ."
   Example: Not "We have chosen not to hire you," but "Because of your lack of full-time experience, we have chosen not to . . ."
3. To the extent that you can, cushion the negative message by putting neutral or good messages before and after it.

There is much more to dealing with negative messages than this (see Chapter 5, Section 1.5), but before digging deeper into the subject, let us look at a classic negative message instance—the rejection letter—and see how it is handled by two professionals (see Figs. 4.12 and 4.13). Chapter 5 discusses such letters in more detail.

Hidden Negative Messages. Some negative messages are inadvertent on the writer's part and can be very harmful. Consider the following, taken from a brochure advertising an executive management institute.

> For the first few years of our institute's existence we arranged programs for individuals as well as groups, but now we have turned our attention to serving groups exclusively so that we can offer this unique experience to more people each year.

Those lines contain an example of a hidden negative message: the hint of past financial struggle. Although some readers may not notice it, others certainly will; they will wonder if the institute was not financially successful serving only individuals. Had the pamphlet's authors been watching for possible negative messages, they might have written:

> In order to bring this unique experience to as many people each season as we can, we serve groups exclusively.

Now the hidden (and needless) negative message, the hint of possible past financial problems, is gone.

So far we have considered negative messages as isolated elements in otherwise positive letters. Such negative messages come in all varieties and are likely to creep into your writing when you least expect them. In almost every instance, you can avoid them if you sensitize

Figure 4.13 A Classic Negative-Message Letter

Notice the ways in which the negative message (rejection of an application for a job) is softened by "you" attitude, goodwill, and reader benefits.

Agro-Tech, Inc.
P.O. Box 20598
Pocatello, ID 83201

March 15, 1992

Terry Bell
4239 Brazos Blvd.
Hansen, ID 83334

Dear Mr. Bell:

Thank you for sending us your application for the agricultural engineering position we advertised. We are always happy to receive applications from A&M graduates. Your credentials were especially good in terms of your education and agricultural background.

We received a number of applications from well-qualified candidates. Several of them had full-time engineering experience as well as agricultural backgrounds and excellent educations, and we have hired one of those experienced people. Therefore we cannot offer you a job at this time. We will, however, place your application on file, and automatically consider you for any position that opens up in the next six months. Of course you should feel free to re-apply for any opening we advertise that you feel is appropriate.

Positions such as the one you applied for are difficult to get right out of college. May I suggest that you look for an entry-level position with a local equipment company, and that you try us again when you have a year or two of work beyond college to support your application?

Thank you again for considering Agro-Tech. Best of luck in your job search.

Sincerely,

Jerry Bowen

Jerry Bowen
Personnel Manager

yourself to the possibility of their presence. But you may have to write a letter someday whose very essence is unavoidably negative. In the next chapter you will see how to write negative-message letters effectively—how to say No without needlessly losing customers, clients, and friends.

2.6   Use Positive Emphasis

Many statements can be phrased either negatively ("Your widgets will not be shipped until Thursday") or positively ("Your widgets will be shipped Thursday"). Although the strict meaning of such statements stays the same either way, their effects on your reader may be quite different. Maintaining a positive emphasis in your letters encourages readers to make positive responses. To state the same idea another way: Whether the cup is half empty or half full may only be a matter of attitude, but if the person you are writing to is thirsty, it's better to say the cup is half full. The following example illustrates negative and positive emphasis:

> *Wrong (negative emphasis):* You will be charged for each check unless you maintain a $300 minimum balance.

> *Right (positive emphasis):* Checking is free with a balance of $300 or more.

2.7   Show Goodwill

Applying goodwill in letters means considering your reader's needs and wants in ways that go beyond your own and then responding to them. Although you may feel that saying "Your widgets will be shipped Thursday" does all you need to do, your reader will be much happier if you add what the reader really wants and needs to know: "Your widgets will be shipped Thursday, and you will receive them by Rapid Express Saturday morning." Here we clearly see the frequent split between the writer's needs (to notify the reader the widgets have been sent) and the reader's needs (to know when they will arrive). Effective correspondence *balances* the reader's needs and the writer's, paying attention to both, and thus accomplishing its purpose.

> *Wrong (no goodwill):* We have received your application.
> *Right (goodwill added):* We have received your application, and you can expect a reply in two weeks.

The following list summarizes the principles presented in this section.

Solicited Versus Unsolicited

"As you requested . . ."

Reader Benefits

"Keep this article handy—it will show you how . . ."

"You" Attitude

"Congratulations on your . . ."

Goodwill

Wrong: "Your order has been shipped as you requested."
Right: "The widgets you ordered have been shipped as you requested and should arrive in about 10 days."

Negative Messages

Wrong: "Your application has been turned down."
Right: "Thank you for sending us your application. Choosing among so many qualified applicants is a difficult process, and choices are often made because one candidate offers something others do not, rather than because one candidate is better than the others. While we have chosen not to consider your application further at this time, we wish you the very best of luck in the future. We hope that you will continue to think of XYZ Corp. as future openings become available."

Positive Emphasis

Wrong: "You will be assessed a service charge . . ."
Right: "Checking is free when . . ."

Action Closing

Wrong: "I would like to request that . . ."
Right: "Please call me at . . ."

3. *Professional Goals and Human Goals*

Striking a balance between your needs as writer and someone else's needs as reader means paying a great deal of attention to human goals. Communication always goes from one human being to another, and all human beings have joys, fears, hopes, and frustrations that your writing must deal with. Ignoring the human qualities of your audience sends a clear negative message of a sort that often causes readers to take their business elsewhere.

A narrow focus on goals produces this kind of letter:

Dear Customer:

We have received your order, and will ship it as soon as possible.

There are times when such a curt note is all you need. This may be your only transaction with that person, or you may have to rush, or there may be a follow-up letter later. But it really doesn't take more time to be a little more human. This is especially true if your letter is a form letter, as routine correspondence usually is. Form letters are the least human letters of all, when they could easily be the most human. The same computer that printed out the letter above could just as easily turn out this one:

Dear Mr. Adams:

Thank you for your order. We're happy you've chosen to use our Superba Freeze-Dried Trail Mixes.

We received your order on the 15th, and shipped the Trail Mixes on the 16th. As you requested, the shipment was made via Rapid Express.

Should you have any questions, please call us at 1-800-555-1234. We look forward to your continued business.

Thank you,

John Yary

John Yary
Catalog Sales

All the principles described in this chapter share this one characteristic: they help you see beyond your own needs to those of your readers. Thus, they show you how to go beyond the narrow professional goals of your letters to satisfy the human goals. The two are not opposites, for human goals wrap around professional goals. Satisfying both professional goals and human goals, making each reinforce the other, is a secret not just for successful communication but for success in general.

EXERCISES

1. Revise the following sentences to eliminate negative messages:

   (a) If we don't receive your payment by the 1st, you'll be billed an additional 5 percent late fee.

   (b) I'm answering your last letter much later than I should.

   (c) Unless you send us this information, we cannot issue your policy.

   (d) Please don't hesitate to call if you have trouble understanding these instructions.

   (e) I'm sorry I won't be here when you come to town, and so we won't be able to meet, but I hope we won't miss each other next time.

   (f) Until you have received our authorization you are not to begin to get bids.

   (g) If you fail to report for work as a result of any unauthorized job action, your position as a state employee will be jeopardized.

   (h) If you took the test, as you claim, we have failed to receive your score.

   (i) I wasn't in class yesterday. Did I miss anything important?

   (j) It is true that my major area of study deals with large industries, but I will gladly work in a smaller company.

2. Revise the following sentences for positive emphasis:

   (a) My major is General Industrial Technology, in which I have no particular area of specialization.

   (b) My work experience has been limited to part-time jobs.

   (c) During my five years of industrial experiences I have shuffled from job to job, each with increasing responsibility.

   (d) I am interested in getting started in the engineering field and would like to start with your company.

   (e) While I missed your recruiter on campus last week, I would still like to arrange an interview somehow.

   (f) Our records show you did not return two of the twenty books you checked out.

   (g) Four of the five samples you sent failed to pass the minimum standards.

   (h) I hope to graduate this May.

   (i) The project seems to have a good chance of success.

   (j) We see no reason for you not to sign this contract.

3. Analyze the following memo's use of (or failure to use) the principles discussed in this chapter. Using your analysis, rewrite the memo so that it stands a better chance of accomplishing its goal (preventing a strike) by making better use of those principles.

OFFICE OF THE GOVERNOR

Memorandum To: All State Employees

I want you to know of my deep concern about reports of a possible strike by state employees if a new labor contract is not agreed to by August 21. Management is determined to reach agreement on a new contract that is fair to state employees and responsible to all the state's taxpayers. Irresponsible talk of a strike doesn't help collective bargaining for a new contract; it ignores both the law against strikes by state employees, and the law that provides for a fair contract through binding arbitration if the parties fail to reach an agreement in bargaining.

All state employees should know just how seriously they may hurt their future if they participate in a strike.

- They would be commiting a crime—a felony under the state criminal code.
- They would be forfeiting the right to hold a state job—the law says they may not hold positions in the government.
- They would not only lose their pay, but would lose their paid health and life insurance coverage and other fringe benefits.

We intend to abide by the law and to enforce it in every practical way.

We have a deep interest in the well-being of all state employees and their families and I want you to be personally aware of the grave consequences of participation in a strike. I hope this entire issue will be resolved as the parties continue to work toward reaching a fair contract by August 21. That remains our uppermost objective.

Clayton James
Governor

4. Revise the following memo according to the principles presented in this chapter.

STATE UNIVERSITY
Computer Center
MEMORANDUM

TO: Dean Laura Bolt
College of Arts and Sciences

FROM: Steve B. Dickson, Director
Computer Center

SUBJECT: Funds for Instructional Use of Computer

Reference memorandum from Vice-President Knebel on this subject dated September 2, 1991 in which you were allocated funds in the interest of encouraging greater instructional use of the computer, and my memorandum on this subject dated September 4, 1991, in which you were allocated supplemental funds from the CC to stimulate use of Dr. Knebel's allocation earlier in the fiscal year.

We have reviewed your accounts, and our records indicate that you did not spend at least one-half of the Account 82122 funds provided by Dr. Knebel. Accordingly, the CC Supplemental funds allocated on September 4, 1991 are withdrawn. It is emphasized that this action does not in any way affect the allocation of funds from Dr. Knebel.

In the interest of stimulating use of your Account 82122 funds prior to the peak summer load, the Computer Center will again allocate supplemental funds for the Spring Semester, 1992. Your CC Spring Semester supplemental allocation is an amount equal to 30% of one half of the total amount allocated to you by Dr. Knebel.

82122 Allocation for 91-92	Amount of CC Fall Supplement Used	YTD Expenditures	Amount of Spring Supplement Available
$7000	-0-	$1366	$1050

Balance Available
for Use by
05/20/92*

$6684

Any unused portion of the CC Supplement will be withdrawn on May 20, 1992.

*Please note that the CC supplementary allocations are in no way intended to limit the original allocation of funds from Dr. Knebel's office. It only provides an added bonus for spending the funds early.

SBD/js
xc: Dr. A. Knebel

5. The following memo is one administrator's response to the memo in exercise 4. Write a brief analysis of its content. Will it be likely to increase the use of computer funds? Rewrite the memo to increase its likelihood of increasing the use of computer funds.

STATE UNIVERSITY
COLLEGE OF ARTS AND SCIENCES

Office of the Dean   January 16, 1992
MEMORANDUM

TO: Department Heads in Arts and Sciences
FROM: Laura J. Bolt, Dean
SUBJECT: Computer Funds for Instructional Purposes

Attached is a memo from Steve Dickson indicating that we have again failed to spend computer funds that were available to us. Please forgive me for being so terse, but I have difficulty in being sympathetic to complaints that we have insufficient funds for instructional purposes when each year we fail to take advantage of the available funding.

Attachment

xc: Vice-President Knebel

6. Using one of the letter forms presented in this chapter, write a formal letter of introduction in which you present yourself to your instructor. Here are some of the elements that you might include in such a letter:

   - What is the purpose of your letter?
   - What is your area of academic specialization? Within this area, what do you like to work on most? If you had a chance to study something in this area, what would you choose?
   - What are your plans for a job? What kinds of communication will be important to that job? Be as thorough as you can.
   - What kinds of communication will be important in your private life? In your home? At parties? At meetings?
   - What is your communication background? Include your experience in freshman English, speech, debate, and dramatics.
   - What areas of communication would you like to work on most in this course? Why?
   - Why did you take this course? What is your honest attitude toward it?
   - Anything else you think it is important for your instructor to know about you.

   Caution: Do not use this list as a step-by-step guide to writing your letter, which will probably produce "canned" writing (like "canned" laughter). If everyone does this, all the letters in your class will be the same, reflecting the list but not your individuality. With the exception of the first item (purpose), the ordering of these items is not meant to be significant.

   Remember to analyze your audience for this assignment. Most students assume that readers know much more about what the students are writing than they actually do. Be especially careful about that on this assignment. Write your letter for your instructor's *real* level of knowledge and interest, not the level you imagine.

7. Despite its best efforts, once in a while the Postal Service does some minor damage to a letter. When that happens, the customer gets the letter in a plastic bag, printed with a note like the one below. Read it carefully, and then come to class prepared to discuss how it uses the principles discussed in this chapter. Pay particular attention to ways it could be made better.

   > Dear Postal Customer:
   >
   > The enclosed article was damaged in handling by the Postal Service.
   >
   > We realize that your mail is important to you and you have every right to expect it to be delivered intact and in good condition. The Postal Service makes every effort to properly handle the mail entrusted to it; but due to the large volume, occasional damage does occur.

When a post office handles approximately one million pieces of mail daily, it is imperative that mechanical methods be used to maintain production and insure prompt delivery of the mails. It is also an actuality that modern production methods do not permit personal attention if mail is insecurely enveloped or bulky contents are enclosed. When this occurs and our machinery is jammed, it often causes damage to other mail that was properly prepared.

We are constantly striving to improve our processing methods to assure that an occurrence such as the enclosed can be eliminated. We appreciate your concern over the handling of your mail and sincerely regret the inconvenience you have experienced.

Sectional Center Manager/Postmaster

5. *Patterns*

L etters and memos serve a variety of purposes in professional life: to get jobs, to agree on the conditions of a deal, to modify the terms of an agreement, and to request or receive payment, among others. This chapter presents two levels of discussion and examples of letters and memos:

- The first structural decision to make, and in some cases the only decision necessary, is whether you as writer of the letter are sending a *direct* or an *indirect message.* In some situations that may be all you need to know in order to write a good letter or memo.

- If what you need to do isn't apparent after you decide whether the situation is one that calls for direct or indirect communication, or if you have trouble making that decision at all, then you will profit from Section 2 of this chapter, on types of letters and memos. It examines in detail letters designed to share information or good news, to make simple requests, to be persuasive, to send bad news, or to create or maintain goodwill.

Additionally, the chapter presents two slightly more philosophical discussions: some thoughts on the proper uses (and common misuses) of the kinds of patterns presented in Section 2 and a review of the kinds of psychological needs all human beings have as readers (as a reminder that *audience* is every bit as big a factor in letters and memos as it is in any other kind of writing).

1. *Direct or Indirect?*

In simple situations, you can easily select the large-scale pattern of any letter or memo: does it need to be direct, or indirect? In a direct pattern, the writer gets right to the point (you'll see this, for example, in the job application letters presented in Chapter 6). In an indirect pattern, the writer holds off divulging the point of the letter until a later (maybe second or third) paragraph (you saw this in the negative message letters in Chapter 4, Figs. 4.12 and 4.13). The direct pattern is typically used when the writer anticipates no possibility of resistance or a negative response on the part of the reader. The indirect pattern is used when there is a possibility or a likelihood of a negative response or some resistance on the reader's part.

Often one type of letter (as they are discussed in Section 2 of this chapter) can appear with either pattern, depending on the situation. Consider the following situation: Chris Ward is asking her boss to purchase a new computer for her office. In Situation A, the direct message situation, she anticipates no resistance from her boss. She has already spoken to her boss several times about the inadequacy of her computer, and the last couple of times he has said, "When the time comes, we'll get you a better one." Yesterday she told him the time had come, and he said, "Put the request in writing, and I'll sign it." In Situation B, the indirect message situation, she has been told by her boss, "Money's tight around here right now. Put your request on paper and do your best to justify it, and I'll see what I can do. But I'm not promising anything." Thus, in Situation B she does anticipate the possibility of resistance. Here's the body of her letter, written two different ways.

Situation A. Direct Message—No Resistance Anticipated

SUBJECT: Request for Computer Upgrade

The recent increase in budgets and budget-related materials that route through this office is placing more demands on my office's computer than it can handle. This letter is a request for authorization to upgrade the computer system from the present IBM PC (a System 2 Model 50) to a Horizon System 2.

As you may recall, when we acquired the Wilson accounts, we also acquired the budgets for three new subsidiaries, all of them produced substantially in this office. Our computer, purchased five years ago, has a 20-meg hard drive that lacks the memory for the task. In particular, the interactive spreadsheet programs required to keep those budgets up to date place great demands on the computer, which it handles slowly, if at all. The new accounts will require networking capabilities that our computer could handle only with an expensive upgrade.

With your authorization, I will request Purchasing to cut a check for the cost of a new Horizon System 2 with a faster microprocessor and a 120-meg hard drive. We can get the Horizon 2 for $1,000 less than its price of $5,000 by trading in the IBM, so the check will be for $4,000.

This necessary computer upgrade will allow my office to handle all current accounts faster and to produce, if needed, daily updates on each one. In addition, this computer will also easily handle the next three years' projected growth in our bookkeeping activity. All I need is your authorization to proceed.

Situation B. Indirect Message—Some Resistance Anticipated

SUBJECT: Request for Computer Upgrade

The recent increase in budgets and budget-related materials that route through this office is placing more demands on my office's computer than it can handle. As you may recall, when we acquired the Wilson accounts, we also acquired the budgets for three new subsidiaries, all of them produced substantially in this office. Our computer, purchased five years ago, has a 20-meg hard drive that lacks the memory for the task. In particular, the interactive spreadsheet programs required to keep those budgets up to date place great demands on the computer, which it handles slowly, if at all. In addition, properly maintaining the new accounts will require networking capabilities that our computer could handle only with an expensive upgrade.

Replacing our computer with a bigger, faster model will allow my office to handle all current accounts faster and to produce, if needed, daily updates on each one (a function totally beyond the current PC). Thus, this letter is a request for authorization to upgrade from the present IBM PC (a System 2 Model 50) to a Horizon

System 2. With your authorization, I will request Purchasing to cut a check for the cost of a new Horizon System 2 with a faster microprocessor and a 120-meg hard drive. We can get the Horizon 2 for $1,000 less than its price of $5,000 by trading in the IBM, so the check will be for $4,000.

In addition to handling all the current accounts quickly and efficiently, this new computer will easily handle the next three years' projected growth in our bookkeeping activity. All I need is your authorization to proceed.

Both letters are letters of request, but one uses the direct pattern and one the indirect. Both are right for the situation that the writer, Chris Ward, is in. If you can size up your situation as a writer as clearly as Chris sized up hers, you can usually make a good decision about whether to use a direct or an indirect pattern. And in many letter- or memo-writing cases, that's the only real structural decision you need to make. If you can't make a clear distinction about whether to use a direct or indirect pattern, look to the more specific patterns presented in the following section for one that more clearly meets your needs.

2. Types of Letters or Memos

Here you will find discussions and examples of letters in five broad categories that cover most types of professional correspondence. We will look in detail at patterns to use when you are

1. sharing information or good news.
2. making simple requests.
3. being persuasive.
4. sending bad news.
5. maintaining goodwill.

These patterns provide specific applications of the principles discussed in Chapter 4. Using these patterns (and the principles discussed earlier), you should be able to construct the kind of letter or memo you need for any situation.

2.1 Writing Informative or Good News Letters or Memos

Assuming that your reader's attitude toward the information will be neutral or positive, one good pattern for an informative letter or memo, the kind usually written in response to a request for information, is given in the following list.

The Informative Pattern

Paragraph 1: Provide the requested information and/or state the policy or good news in question.

Paragraph 2: Explain how the information specifically touches on your reader's request and/or how the policy or good news applies to the current situation.

Paragraph 3: Present any negative factors (usually areas in which the information or policy limits the reader's actions); try to show the motivation for these factors; retain positive attitude and positive emphasis as much as possible.

Paragraph 4: Restate reader benefits briefly; add "goodwill" ending.

Rationale. Placing the good news first or responding directly to your reader's request in the first paragraph establishes your contact immediately. This way you can help make the reader favorably disposed to the rest of your message. The detailed explanation in paragraph 2 is the meat of the letter, but do not get so caught up in presenting details that you neglect to explain the why's—especially if paragraph 3 will contain any negative messages. If your reader understands why the policy affects him or her in this way, everyone will get along better.

When you have negative messages that are *incidental* to a positive letter, be sure to keep them in perspective for the reader by placing them after the detailed policy explanation and before the reader benefits. Remember to show why these negative elements are necessary; the more negative messages you send, the more important following up with reader benefits is. If nothing else, reiterate the positive aspects of paragraph 2. For example, list several things the policy will allow the reader to do.

> **Caution:** Do not try to disguise a negative message that far outweighs the new information or good news by burying the negative message within this pattern. Readers invariably detect—and resent—this ploy.

Your "goodwill" ending should convey to the reader the extent of your concern. Try to use that human concern to balance out the

Figure 5.1   An Informative Letter
Notice the author anticipates unasked questions by providing the name of the nearest dealer.

Acme Technologies, Inc.
31 Skyline Circle, Redmond, WA 98073

April 9, 1992

Dear Ms. Eubanks:

Thank you for your request for information on our new Acme TravelWriter. I have enclosed brochures that should answer all your questions.

The TravelWriter weighs 1.5 pounds, opens for easy use on a lap or a desk, has a 1.2 meg memory, and a 60-character, 24-line amber screen. It uses either 120-volt AC current or an optional 12-volt DC battery pack (described in the enclosed brochures). The keyboard is standard, its size equal to that of most portable typewriters. Its built-in printer operates in its dot-matrix mode at 100cps and in near-letter-quality mode at 50cps. An optional add-on FAX unit is also available. The unit comes in its own carrying case, especially designed to protect it from normal travel bumps and knocks.

Software available with the TravelWriter includes two word processing programs and five spread sheet programs. We expect to have additional programs available within the year. Recommended retail price of the TravelWriter is $350.

I hope this information leads you to give our product serious consideration. We believe there is nothing on the market that can come close to it for lightness of weight, high durability, and low price. The dealer nearest to you is Wilson Office Equipment, 1511 East Main Street, phone 555-1423. If you visit their store, you can see the TravelWriter and try it out for yourself.

Thank you for your inquiry.

Sincerely,

Arlene Phelps

Arlene Phelps
Customer Service Representative

tendency in such a letter to sound too rule oriented. Be sure to invite the reader to ask questions if they come up; you might suggest a telephone call to save time for your reader in the future. Figures 5.1 and 5.2 show applications of this informative pattern.

2.2   Making Simple Requests

For this discussion, assume you have a request you expect your reader to grant readily. The following outline shows one good way to make such a request.

> ### The Simple Request Pattern
> *Paragraph 1:* Request for information, services, etc.
> *Paragraph 2:* Show why you need the information or service, and/or how you will use it.
> *Paragraph 3:* State the specific action you want the reader to take.
> *Paragraph 4:* Add reader benefits (if possible) and "goodwill" ending.

Rationale. If you are reasonably certain your reader will grant your request readily, you can save time by making the request right at the beginning. Then, in the second paragraph, specify why you need the information (or services) in order to justify your request and to ensure that your reader will send you precisely the materials you need. Your third paragraph needs to make clear exactly what action you want your reader to take and explain any limitations and restrictions. (For example, if there is a time limit involved, explain the necessity for it.) Your "goodwill" closing and statement of reader benefits (if there are any) are especially important in this kind of letter or memo, because the rest of it concentrates more on what the reader can do for you than on what you can do for the reader. Figure 5.3 shows specific applications of the simple request pattern.

Most job-application letters are a combination of the request and the persuasive letter patterns. Because they are so important, they are covered in a separate chapter (Chapter 6).

2.3   Being Persuasive

Persuasive letters or memos include a variety of kinds of correspondence, all of which have in common your expectation that the reader may initially disagree with your request or at least will not be immediately inclined to go along with it. You might be trying to persuade

Figure 5.2   A "Good News" Letter
Notice this letter contains no negative elements, and it goes out of its way to restore goodwill. The one possible negative message ("We regret we cannot . . .") is handled very carefully.

World's Inn

436 Harris Avenue
Alden, MN 56009

September 8, 1992

Dear Mr. Adcock:

Thank you for calling the error on your recent statement to our attention. As you suspected, we had indeed billed you twice for one night's lodging.

We have adjusted your bill to $58, the posted charge for one room, one night, one person. Since you sent the $58 with your last letter, your account with us is now marked paid in full, as the enclosed statement shows. We regret we cannot follow your suggestion of making your stay free simply because of our billing error.

We hope you will accept our apology for this error, and that you will consider staying with us again during your next visit to St. Louis. We hope you will also accept the enclosed Special Guest certificate, which brings you one night's stay at any World's Inn for half price. It is our way of saying we appreciate your continued business.

Thank you again for staying at World's Inn.

Sincerely,

Martin Estéves

Martin Estéves
Customer Relations

one of your local elected officials to change a policy or a vote or a corporate executive to schedule an appointment for you to demonstrate your product line. The basic pattern is given in the following display.

The Persuasive Pattern

Paragraph 1: Catch the reader's interest; establish mutual goal(s).

Paragraph 2: Define the problem you both share—the one that the reader's granting your request will solve.

Paragraph 3: Explain the solution. Show how any negative elements (cost, time, etc.) are outweighed by the advantages of the solution you propose. Present it not as *your* solution but rather as *the* solution.

Paragraph 4: Reader benefits.

Paragraph 5: State the specific action you want the reader to take.

Rationale. Try to soften your reader's opposition by presenting the reasons for your request in the context of solving a mutual problem. This requires *establishing a common ground* between you and your reader. You can try to arouse the reader's interest in something *you* are interested in (the more difficult option) or relate your concern to something you know your reader is already interested in (the easier option). If you're writing to management, stress increased productivity; if your audience is labor, stress better working conditions. In this kind of letter, you must choose your approach carefully, striving for the right statement of the problem—the one with which your reader will identify and that will point your reader in the direction of the solution you propose.

After you have established the problem as something you are both interested in solving, define it in greater depth in paragraph 2. Be objective and detailed, using concrete examples to make your presentation vivid; you are trying to stir the reader to take action. Then in paragraph 3 you can show that, given the need for a solution, the one you propose is best.

Before you present your own proposal, briefly show why any obvious alternative solutions are unsatisfactory. Then present the solution you propose, but present it as *the* solution, not as uniquely yours. Show your reader that any clear-thinking person, once aware of the facts, must come to this same solution. Part of doing this includes answering the objections you know your reader will come up with. Do your homework ahead of time, and build into this part of the letter answers to foreseeable objections. For example, where will the money, or personnel, or equipment for this solution come from?

Figure 5.3   A Simple Request
Notice that the second paragraph tells the *why* of the request, and the third
tells the *what*.

Peter Ledger
21 Mayfern Rd
San Diego, CA 92101

June 11, 1992

Subject: Request for Information on Acme TravelWriter

Your ad in last week's *Computer Times* for a laptop word processor has prompted me to write and request more information.

I am looking for a unit I can take on business trips and use wherever I stay. It needs to have battery power as an option, as well. It
needs to be capable of sophisticated word processing and simple
spread sheets, and it needs to be PC compatible. But the qualities
that most attracted me about your computer are the built-in modem,
the built-in printer, and the optional FAX.

Please send me a description of your laptop word processor's
capabilities. I would especially like information on its keyboard,
compatibility, and what kinds of software packages are available for
it. Obviously, I would also like to know its recommended retail
price.

If this product is as good as it looks, my firm may well order
quite a few. I hope to hear from you soon.

Peter Ledger

Peter Ledger

The reader benefits for this kind of letter should reinforce the common ground built in paragraphs 1 and 2. Explain the benefits that will come to those who solve this problem. Your last paragraph should specifically state what you want the reader to do; make sure that besides being moved to action, your reader will also know exactly the right action to take. Figure 5.4 gives an example of this persuasive pattern.

2.4 Sending Bad News

There will be times when you will know that your reader's attitude toward your message will be unfavorable. You may have to say No to a request, end a professional relationship, or raise a price. Remember that it is usually better to transmit such a message in person; the receiver will usually appreciate your dealing with the situation in person. But of course there will be times when you will need to use a letter. One good pattern for a "bad news" letter or memo is

The "Bad News" Pattern

Paragraph 1: Establish goodwill. Then present positive aspects of your previous relationship with the reader.

Paragraph 2: Present reasons for the negative message. Then present the message.

Paragraph 3: Explain any positive aspects of the situation. Suggest alternatives. Reestablish goodwill if possible.

Rationale. It may seem to you that such a letter is a charade that does not fool anyone. In fact, it is not designed to fool anyone. Rather, its aim is to try to cushion a blow in a human way and to maintain as much goodwill as possible under the circumstances. You try to write the letter in a reasonably positive manner not to deceive the reader into thinking it actually contains good news but to show your concern for the reader's human feelings by buffering the negative message. Most experts agree you shouldn't introduce the negative message with "However" or "Unfortunately" or "I'm sorry, but . . ." because those phrases ring too false. (Although you should try to phrase the message as positively as you can, it is still a negative message.)

Compare the "bad news" letters in Figures 5.5 and 5.6. Figure 5.5 satisfies only business goals; while Figure 5.6, which follows the acceptable pattern for conveying bad news, satisfies business goals *and* human goals. Which would *you* rather receive? You might also take another look at Figures 4.12 and 4.13.

Notice it is *our* problem and *our* solution and that the letter is relatively infor mal (Jack and Bill went to school together).

Environmental Testing Laboratories

121 Macon Road
Lackawanna, NY 14218

November 6, 1992

Dear Jack,

After we talked on the phone yesterday, it came to me that there is a way to deal with our situation that we haven't considered, and it may well be the solution we both need.

I understand how frustrated you can get when your crews have to wait on the results of our tests. But we cannot hurry the tests, or take shortcuts on these standards.

Let me suggest this: usually we receive your samples around noon on one day, and give you the results 48 hours later. Thus when we get the samples on Tuesday noon, we give you results on Thursday noon. But if we were to get your samples between 8:00 and 8:30 a.m., we could probably get you results by 5 p.m. the next day. And there would be a good chance we could get them to you at least part of the time considerably earlier, around noon. You see, we handle our work on a first-in, first-out basis, and we get most of it in around 9-10 a.m. By bringing in your samples earlier, you could get ahead of the pack, and get your samples back much earlier. That may well be the best solution.

This way you get the results you need, and we can get our own work started (and finished) earlier.

If this seems reasonable to you, give me a call and I'll alert our people in the lab that you're bringing the samples in earlier and ex- pecting them to get the results out earlier. Fair enough?

Bill

Figure 5.5 A Poorly Done "Bad News" Letter
This letter may satisfy the writer's business goals, but it is destructive of the important human relationship between the writer and the job applicant.

2737 Avenue of the Americas
New York, New York 10012

July 1, 1992

Dear Applicant:

   Thank you for your application. Unfortunately, we have already chosen the person for that job. Good luck in your continued job search.

                                            Personnel Manager

2.5 Maintaining Goodwill

The term "courtesy" (or "goodwill") letters names a widely varied group of letters including thank-you notes, congratulations, sympathy notes, season's greetings, and letters of welcome. They all share these characteristics:

- These letters are never letters that *must* be sent.
- These letters almost never *require* answers.
- These letters always have as their major goal to create and build goodwill.
- Readers especially appreciate these letters *because* they are not necessary.

You send a goodwill note to strengthen your relationship with a person or a company. You should not try to do business in the letter; building and reinforcing the goodwill is sufficient. Figures 5.7 and· 5.8 show you *how* to do these letters, but the most important element is *why* to do them: they epitomize correspondence that has as its primary goal the satisfying of human needs.

Figure 5.6   A Well-Done "Bad News" Letter

Notice the use of "you" attitude, positive emphasis, and reader benefits to cushion the negative message.

Acme *Software* **Corporation**

24 Technology Parkway, Ventura, CA 93001

February 16, 1992

Dear Mr. Morgan:

Thank you for your letter of January 15, 1992, regarding employment with our firm. We are delighted that so many people of your education and experience seek to work in our industry, and particularly with us at Acme.

The large number of qualified applicants made our hiring decision depend upon a wide range of variables in addition to education and experience. It is always difficult to choose from among such good candidates, and this occasion proved no exception. While we have selected another candidate to fill the particular position for which you applied, we do anticipate openings of a similar type in the future, and I would encourage you to re-submit your application. Keep in mind that knowledge and experience with computers can be a determining factor for employment in this field, and anything you can do to enhance your own qualifications in this area would increase your potential in the job market.

Congratulations on your graduation from State Tech University, and good luck in your continued job search.

Sincerely,

Marla Trager

Marla Trager
Personnel Representative

Figure 5.7   A Simple Courtesy or Goodwill Letter
Such letters are done solely to build goodwill; this one happens to be a thank-you note, but there are many other possible varieties.

Mooreland University
Ballard, WV 24918

March 4, 1992

Dear Vicki,

Thank you for the good job you did on the library tours for my professional writing students last week. I appreciate the care and preparation you obviously put into it, and I admire your skill at presenting the material.

The students thought the introduction to the AIRS system and the OCLC was especially interesting and useful.

If I can be of any assistance to you, be sure to call on me. Thank you again for doing the library tour so well.

Sincerely,

Sylvia Leonard

Sylvia Leonard
Professor of English

Figure 5.8   **Another Sample Goodwill Letter**
Here a graduating senior has been on a plant visit to tour the facilities of a prospective employer. The writer very carefully does not plead her case for the job in this letter; she focuses only on building goodwill (which she hopes will advance her chances for the job).

> Thelma Martos
> 34 Dietz Street, Apt. 3
> Akron, OH 44301
> May 2, 1992
>
> Dear Ms. Jenkins:
>
> Thank you for the hospitality you and your staff at In-
> tergrid showed me during my plant visit last week. The ar-
> rangements your office made for me put me at ease from the
> moment I arrived.
>
> More than ever, I am certain that Intergrid and I would be
> a good match. I know you must have many qualified candi-
> dates to consider, but whatever your hiring choice, I
> value the time I spent touring your facilities and meeting
> your staff.
>
> I look forward to hearing from you soon.
>
>
> Sincerely,
>
> *Thelma Martos*
>
> Thelma Martos

3. *Use of Patterns*

The patterns, rationales, and examples offered here demonstrate how to deal with typical professional correspondence situations. If you want to grow as a writer, apply the patterns with flexibility, using the principles explained in the previous chapter to tailor each letter or memo to its situation. By combining principles, patterns, and audience analysis and adaptation, you should be able to write effective letters and memos.

Remember the *myth of the 100-percent model:* no single pattern unthinkingly applied will quite work anywhere. You must use your own intelligence and your own understanding of the specific writing situation (and of the principles behind these patterns). Blindly following any prefabricated plan in such a human situation is as foolish as having no plan at all.

4. *Human Psychological Needs*

Writing successful letters and memos means paying strict attention to your reader's needs. In many ways, each reader will have specific needs that shape each letter you write. But there are certain qualities and feelings all readers seek—and certain qualities and feelings all readers seek to avoid. The pressures of professional life can tempt writers to be ruthless about ignoring readers' needs: the business goals of the letter may seem to dictate its content and structure. It's easy for any of us as writers to think only about our needs and not about those of our readers. But successful communication builds on shared needs, and the reader's share will always include the qualities described in Sections 4.1 and 4.2.

4.1 Qualities All Audiences Seek

All audiences, and especially all readers, seek to draw three feelings from a piece of writing: identity, stimulus, and security.

Identity. If you treat your reader as a unique individual you will be reinforcing his or her sense of identity, a pleasurable experience for the reader. Think of the times a new acquaintance breaks into a smile simply because you remembered his or her name; on the other side, think of the times you have been offended by not being treated as an individual. In general, the more your writing shows that it is tailored to fit exactly the person who is reading it, the better it will be received because you are recognizing and reinforcing that person's identity.

Stimulus. No kind of writing is intrinsically boring. Some kinds of writing have more chances to be exciting than other kinds, but the responsibility of not boring the reader, like the responsibility for clarity, rests with the writer. In professional correspondence, not boring your reader requires asking yourself how much and what kind of detail to include and how to make your content easier to understand. This may mean putting the "bottom line" of a report at the beginning, or using a telephone call and follow-up letter rather than a long, complicated letter that might be misunderstood. Certainly it means paying attention to sentence length and clarity, paragraph length, and overall length of the letter. In most letters and memos, it also means identifying the purpose in the first paragraph.

Security. No reader wants to feel threatened by a piece of writing, and this is especially true when it comes to letters and memos. For letter writers, having proper concern for not threatening the reader's security means being careful not to include needless negative messages (and handling necessary negative messages carefully).

4.2   Qualities All Audiences Seek to Avoid

The opposites of the feelings all readers seek are those they seek to avoid: anonymity, boredom, and anxiety.

Anonymity. "Dear Occupant" mail makes readers feel anonymous. Canned laughter makes television viewers feel the show is anonymous; canned speeches make the speaker, the occasion, and the audience anonymous. Being dealt with as a number, not a name, makes anyone feel anonymous. To keep your reader from feeling anonymous, build concern for your audience into your letter.

Boredom. Writing for business and industry requires placing a premium on the reader's time and energy, especially in letters. The people who see your letters face a three-way choice: read it, route it, or skip it. If you write your letter carefully and organize it efficiently, your reader is more likely to choose to read it.

Anxiety. Remember the things that make you anxious when *you* are the reader: material that you cannot understand, that is not organized efficiently, or that indicates the writer's needs are more important than yours. If you as writer pay attention to only your own goals, readers may well express their anxiety by consistently responding negatively to your letters. However, if you build shared goals with your readers, you can increase the probability they will respond in the way you desire.

EXERCISES

1. For this exercise, write a variety of letters as either the student or the instructor in an upper-level writing class. The class carries a requirement for a twenty-page term paper. The following letters all concern negotiations between student and instructor on the topic for that term paper:

   (a) As the student, write a letter in which you ask your instructor to approve a major report on a topic you suggest in the letter. For this assignment, you only need to choose a topic you know enough about to write these letters. The letter is simple; you don't expect any problems getting the topic approved.

   (b) As the instructor, write a letter responding positively to the student's letter, spelling out the term paper's requirements (get these from your instructor).

   (c) As the instructor, write a letter turning down the student's request. Possible reasons for turning down the request include the following: the library doesn't have enough information; other students who tried that topic couldn't do it; there isn't enough time; the topic is too simple (or too broad); or there's no conceivable reason why anyone would be interested in such a paper. Be sure to offer alternatives.

   (d) Now, as the student, write a letter to the instructor, responding to the letter in part (c) (above), trying to get the instructor to accept some modification of the topic you originally proposed, a modification that satisfies the objections in part (c).

2. The following letters involve a student (you) and a dean (Dean Smith). You request that the college's requirements for successful completion of two years' study of one foreign language be modified to allow you to substitute one year each of French and German. The dean refuses, you persuade, and offer the dean a compromise.

   (a) As the student, write a letter requesting that you be allowed this substitution. Use your own major field of study—Dean Smith is dean of the college that department is part of.

   (b) As the dean, refuse the request. Offer as your main reason the fact that in an era of declining educational standards, your university is determined to continue to turn out students with well-rounded educational backgrounds—and that means knowing foreign languages.

   (c) As the student, write a persuasive letter trying to change the dean's mind. Feel free to offer a compromise.

A ➤ indicates a case study exercise.

➤ 3. You are the customer service manager for Beta Boots, a major manufacturer of all sorts of boots. The recent popularity of Western-style boots caused Beta to manufacture 50,000 pairs hastily, and many of the people who have bought them have written in complaining that the boot's sole comes unglued from the rest of the boot. You have learned that it costs an average of $7.50 for a consumer to have them repaired locally. Your company has decided to offer each person who complains $3.75 toward repair: by the time the boots fall apart, half of their useful life is gone, so the company will pay half of the repair bill.

   (a)   Write a form letter to customers who have complained, attempting to persuade them to accept the $3.75 and let the issue end there. Obviously, you are also trying to salvage as much of the reputation of Beta Boots as you can.

   (b)   Write a letter to a very angry customer. This person (Mr. Samuelson) not only refused the $3.75 check but also mailed back the boots. He wants reimbursement for the $3.50 it cost to mail the boots, the $38.50 the boots originally cost, plus an additional $10 for "sheer aggravation." Your boss has authorized you to offer him a new pair of boots (an improved version of the originals) and the $3.50 postage but has made it very clear that he will *not* comply with the customer's other demands. Write a letter persuading Mr. Samuelson to accept the deal. Be very careful (Samuelson was furious) but very persuasive (you want the issue to end here).

4. You are a graduating college senior, and the chairperson of your department has invited you to apply for graduate school and to accept a half-time teaching assistantship in that department. Write a letter turning down the invitation but keeping open the possibility that at this time next year, after you've gotten a year's work experience, you may want to accept the invitation. Phrase your letter so as to keep the invitation open without making any kind of a firm commitment to apply next year.

➤ 5. You are the personnel manager of Acme Engineering Consultants, and David Powers has applied for a job as a materials testing supervisor, a position you recently advertised. He worked for you part time for the last half of his junior year and the first half of his senior year in college, but now you need to deny his request for a full-time job after he graduates. Your confidential reason for not hiring him is that he was only average in everything he did for you—always needing more supervision and guidance than anyone thought necessary. He was acceptable as a part-time employee, but you don't want him working for you on a full-time basis. The situation is complicated because you just hired a classmate of his for the job, and you know he will discover that. Write him a rejection letter that will maintain at least a reasonable amount of his goodwill.

6. Revise the following letter, eliminating the needless "I" attitude and replacing it with "you" attitude wherever possible.

```
                                    128 W. 5th Street
                                    Cedar Rapids, Iowa 53108
                                    April 10, 1992

        President
        P.O. Box 90
        Dubuque, Iowa 53110

        Subject: Job Application

           I believe I have the qualifications for the job of-
        fered by your firm in the April 8th Dubuque News and
        would appreciate strong consideration for the position.

           I have the business skills that would be required for
        your firm and myself to excel. My academic background at
        the University of Iowa is a Bachelor of Science in Ac-
        counting degree with a strong achievement level in fi-
        nance.

           I think of myself as an energetic person with a
        strong desire to accomplish and do things correctly.
        Working together well with other people is the best way
        to solve problems and usually creates the optimum re-
        sults.

           Being a part of a creative company is one of the high-
        est goals I plan to attain in my career. I would be
        proud to be a part of your firm. Would you please call
        me for an interview?


                                    Sincerely,
```

6. *Job Applications and Résumés*

n the year before you finish college, you begin writing one kind of letter that is very important to you. Writing an *effective* job-application letter can be more than satisfying and time efficient: the letter can bring you professional rewards for years to come. Writing a poor job-application letter can be both personally frustrating and professionally damaging. (So can a poor graduate school, medical school, or law school application letter.) If the letter fails to get you a position you are otherwise qualified for, that poorly written letter has also cost you money. If someone else gets a job that you have better qualifications for, then your letter may well have cost you a year's salary. This may be your first real experience with the dollar value of good communication skills, especially of good writing. There are many other instances, but few are focused so clearly on just one short letter.

This chapter provides you with the knowledge and strategies you will need to write an effective job-application letter—a letter that not only communicates your individual qualifications but also gets you to the next phase of the hiring process. Besides the job-application letter, you will also learn to assemble an effective biographical résumé (also called a vita or data sheet) in one-page and multipage formats, and a one-page qualifications sheet (or functional résumé). Job-application letters, résumés, and qualification sheets provide concrete examples of the key principles of business communication. Because these documents are so important in your life right now (or soon will be), you need the precise focus on your audience those principles give you.

Although most college graduates use application letters at some point in their job searches, there are other approaches to finding a job. The percentage of jobs secured through personal contacts (many varieties of knowing the right person) runs from 50 to 90 percent. Some people go through interviews only, while others do interviews before letters. Some people use specialized types of résumés to go with particular interviews or letters. And some people use less direct methods of getting a job.

To make the most of your job search, keep your mind open to these other possibilities. But you will probably still find yourself at the point where nearly all of us have been: facing the single biggest writing problem of your senior year, the job-application letter.

1. *Using Writing in the Job Search*

Although the principal focus of this chapter is on the job-application letter and résumé, there is more to the role of writing in the job search than those two documents. You can use writing to your

---

Note: This chapter deals only with solicited applications—those written in response to a notice of a job vacancy. Unsolicited applications are rarely successful and are difficult to generalize about.

advantage throughout your job search, and there are important reasons to do so. One important reason is the paper trail that writing creates. Even more important, the ability to communicate—and beyond that, the willingness to communicate—are qualities employers especially look for.

1.1   What Employers Are Looking For

One generalization that is true across all fields of study and employment is that employers are looking for three qualities (in whatever order) in the college graduates they hire:

* aptitude.
* attitude.
* communication ability.

Aptitude means your level of knowledge and expertise in your field. If you're an accounting student, you've got to know accounting in order to get a job, and an engineering student must know engineering. Prospective employers measure your knowledge of your field directly by your course grades, academic transcripts, and letters of reference.

Attitude means what kind of a worker you are, what kind of person you will be to work with. Are you reliable? Do you need continuing supervision? Are you a self-motivated overachiever? Are you friendly? Different fields and different positions require different kinds of attitudes from entry-level employees. An entry-level employee in hospital management or sales may well be a little more outgoing and assertive than, say, a computer programmer. Employers find attitude a little harder to measure directly than aptitude. That's what interviews and plant or office visits (and, to some extent, letters of reference) are for—so people can meet you face to face and make a well-founded primary judgment about your attitude.

Communication ability is essential in any field and in any position; the higher you go into management in any field, even the more essential it gets. Writing and speaking ability is in some ways the hardest to measure directly of these three qualities all employers look for. A prospective employer will make a judgment about your speaking ability—can you discuss key aspects of your field clearly?—during the interview. But a prospective employer who wants to make a judgment about your writing faces a harder task. For employers, measuring your written communication ability involves answers to two questions:

1. When you need to use writing, do you handle it well?
2. When you have a chance to use writing to help accomplish your professional goals, do you seize the opportunity—or avoid it?

Where do employers go to find evidence of how well you write? And beyond that, where can employers find evidence of your willingness to use writing to accomplish your professional goals?

- Your freshman composition grades are probably several years old by now and may well not reflect much about how well you write in your major field. They certainly don't speak to your willingness to use writing.

- Your grade in a course like the one for which you're using this book is important in terms of measuring how well you write, but it does not say much about how willingly you write.

- Your job-application letter, and the other documents and correspondence that accompany the job search, will speak more loudly than anything else about your ability as a writer and your willingness as a writer.

1.2 Using Writing to Give Them What They Want

The way you handle the writing associated with your job search will answer more clearly than anything else the two questions employers need answered about your writing. The first, about the quality of your writing, is too obvious a point to say much more about here; in one sense, that's what this whole book is about. The second, about your willingness to write when the situation *invites* but does not *demand* it, needs some elaboration.

From your boss's point of view, it's less than ideal to have an employee who is capable of being a good writer but routinely dodges situations that may require writing. Your job-application letter and résumé, like your course grades, tell your future boss nothing about how willing you are to write. If a report is due, will you do it? On time? Up to the required standards? Willingly? Or do you have to be dragged into doing it? If a follow-up letter to a prospective client will help solidify a deal, will you do it? On time? Willingly? Or do you have to be reminded and reminded and reminded? If your company's only tangible product is written documents—whether they are proposals, engineering reports, or audits—how forthcoming you show yourself to be about writing is a critical part of your job qualifications.

How do you send a message that says, "I not only have the ability to write, I'm confident about my writing and use it freely to my advantage"? One answer lies in the correspondence of the job search above and beyond the job-application letter and résumé. Consider the stages in the job-application process represented in the following flowchart. (The left side begins with a job-application letter; the right side begins with an application for an on-campus interview.)

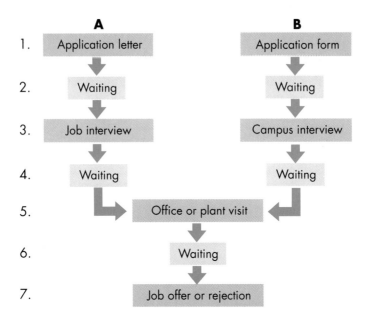

Let's look at the role of written communication and the opportunities for written communication for each step and each column.

1A. Application Letter. The rest of this chapter is about how to write effective application letters, so there's no need to dwell on them here, except to say to write them. Usually the goal of the application letter is an interview at the company's office or plant. It's not unusual for a graduating student to have occasional contacts during the senior year with people who say something like, "Be sure to let me know when you go on the job market." It's important to follow up that kind of a lead; oddly, some students don't do it.

1B. Application Form. Many colleges and universities have a placement office that schedules representatives of firms that are hiring to come to campus and interview there. Usually these interviews are preliminary screening interviews followed, for a few selected students, by a visit to the company's offices or plant. Interviewers coming to campus usually screen these forms and select the students whom they choose to interview. There *may* be an opportunity for you to use writing to your advantage here. Some placement offices prohibit students from having any contact with interviewers prior to the interview; others don't. If your campus placement office doesn't, you may be able to send a letter to the interviewer introducing yourself and emphasizing your interest in that company. This letter of request for an interview is practically identical with the letter of application described in this chapter and, of course, it would be accompanied by a full (multipage) résumé.

> **Caution:** If your campus placement office has a policy that prohibits the use of such letters prior to campus interviews, follow that policy. The penalty for violating such policies is severe.

2A. Waiting between the Job-Application Letter and the Interview. After you've sent in your job-application letter and résumé, what do you do? Wait for the telephone to ring? Here's another chance to use communication to further your cause and show your prospective employer your willingness to write. Why not send a follow-up letter—say, after two weeks—requesting information on the status of your application? You don't want to make a big deal out of the letter, but it's nice to know where you stand with each application letter you send out, and a simple letter of request is not uncommon. If two weeks seems too short a time, send your routine follow-up letter after three or four weeks.

2B. Waiting between the Application Form and the Interview. After you send in your form applying for an interview, what do you do? Wait? If you're notified you're to have an interview, what do you do until the interview happens? Wait? Especially if you haven't already sent the interviewer a letter, why not do so now? If you're waiting to hear the response to the interview-request form you sent in, send the letter described for 1B. If you've already been notified you're to have an interview, why not send a letter introducing yourself and saying you're looking forward to the interview? Either letter may help advance your cause; both will say something positive to the interviewer about your willingness to write.

3A and 3B. After the Job Interview or Campus Interview. Once you've had your interview, it's entirely appropriate to send a "goodwill" letter to the person who interviewed you. The letter says that you enjoyed the interview, found it informative, and are more convinced than before that you'd like to work for that company. You don't "do business" in this letter; you're just building goodwill. And once again you're saying, between the lines, "Writing comes easily to me, and I use it freely."

4A and 4B. Waiting After Your Job Interview or Campus Interview. Once again, if two (or three or four) weeks pass and you've not heard from the person who interviewed you, it's entirely appropriate to send a simple letter requesting updated information on the status of your application. You don't want to seem to be overreacting—just making a routine request to keep your life in order.

5. The Office or Plant Visit. When you are invited for an office or plant visit, it may well be done by a telephone call. A number of details may be covered: what flight to take, how to pay, who will meet you, and so on. How much of that information will you jot down exactly correctly during the call? Why not follow that call with a short letter, "just to confirm the details of the visit"? Not only do you take a wise precaution against confusion surrounding an important event in your life, but you once again demonstrate how well you can make writing work for you.

6. Waiting After Your Office or Plant Visit. Two weeks, three weeks, or four weeks after your office or plant visit, if you haven't heard any information, it's a good idea (for your own personal sanity and self-esteem) to send that simple letter requesting an update on the status of your application. Maybe you can offer to provide any additional information that might be required to make the decision making easier. You gain some peace of mind and remind the reader that you're a fluent writer.

7. Job Offer or Rejection. When you are offered a job, it too may be done by a telephone call. Details may be covered—how much you will be paid, when you will begin work, and so on. How much of that will you jot down exactly correctly during the call? Why not follow that call with a short letter, "just to confirm the details of the offer," and to say again how pleased you are that you will be coming to work there? Not only does this letter give you a paper trail of the offer and help straighten out any misunderstanding there may have been, but you once again demonstrate how well you can make writing work for you—especially when it's important.

If you are rejected for the job, you may want to write a short follow-up letter to build goodwill or to request information. To build goodwill, simply thank the firm for the consideration they gave you, and express the wish that you will have the opportunity to do business with them in the future. To request information, briefly thank them and then ask, in the light of the fact that you hope to apply again another day, if they might give you some indication of how you could make your credentials more attractive for your next application. You may not get too many answers to such a request, but your letter is still a reasonable request, and even a few feedback letters can provide invaluable information for the future of this and subsequent job searches.

Isn't This Too Much Writing? It may well be that no one will do this much writing as a part of their job search. These examples of opportunities to write demonstrate options that exist for written communication as a part of your job search. The point is to show your prospective employer not just the quality of your writing but the willingness with which you do it as well. *Only you can make the right judgment about how much writing is enough or too much.* You need to do enough writing to show that you are a good writer and to show that you are not just an able but also a willing communicator. Quite apart from these two points, each of the notes and letters described has its own reason for being; which one or ones might be best for you to use is your decision.

2.   *Writing Effective Job Applications*

What does a good job-application letter look like? Figures 6.1, 6.2, and 6.3 present three typical examples.

How do these letters differ from one another? The letter in Figure 6.1 uses another person's name as the contact phrase. The letter in Figure 6.2 comes from an applicant who has both work experience and academic training that exactly fit the job. Combining the qualifications from the first letter with the strong contact phrase of the second would produce the strongest possible letter. The letter in Figure 6.3 is a good letter but comparatively the weakest of the three. The writer has only academic training for the job and little relevant work experience. Because most students are probably in the same position as the writer of the third letter, this chapter will show you how such a person can make the most of the qualifications he or she does have.

How are these three letters alike? Each uses the same basic principles of business correspondence, adapted especially for job-application letters.

2.1   Principles

The job-application letter is basically a persuasive letter of request. The following elements are important to it:

- Solicited versus unsolicited correspondence
- Reader benefits
- "You" attitude
- Goodwill
- Caution about unnecessary negative messages
- Positive emphasis
- Action close

Figure 6.1   A "Namedrop" Letter
This technique will work only if Mr. Yount knows Dr. Joiner and respects his recommendation.

```
                                        1402 Dale Street
                                        Dayton, OH 45405
                                        October 3, 1992


        Mr. N. J. Yount
        Recruiting Director
        North American Producing Division
        Atlantic Richfield Company
        Post Office Box 2819
        Atlanta, Georgia 30327


        Dear Mr. Yount:

            At the recommendation of Dr. Jerry Joiner, I would like to
        apply for the job of Professional Accountant in Atlantic Rich-
        field's Accounting Development Program. I will receive my
        B.B.A. in Finance in December 1992, and will complete my concen-
        tration in Accounting the following May.

            My course work at A&M fulfills the requirements listed in
        your announcement. I currently have a 3.34 grade point (on a
        4.0 scale), with a 3.60 in my major. My training here included
        courses in Managerial Finance, Investment Analysis, Cost Ac-
        counting, Auditing, Income Tax, and Investment Accounting,
        among others. As your notice requested, I have enclosed an unof-
        ficial transcript with this letter.

            In addition to my degree, I have work experience in areas re-
        lated to this job. Currently I am working as the teaching assis-
        tant for Dr. Joiner's Advanced Accounting classes. By helping
        his students, I am developing a better understanding of the con-
        cepts of consolidated statement preparation, international ac-
        counting policies, and financial statement analysis.

            The enclosed résumé will tell you more about my background.
        I look forward to the opportunity to talk with you about the
        contributions I can make to Atlantic Richfield's continued
        success.

                                        Sincerely,

                                        Jay Thomas

                                        Jay Thomas


        Enclosure
```

Figure 6.2 Letter from Applicant with Education and Work Experience
A strong letter, but it could have been even better with a person's name in place of the "Dear Sir."

> Sara Hayes
> 328 Krueger Avenue
> San Rafael, CA 94901
> February 1, 1992

Roth Young Food Specialists
5344 Alpha Road
St. Louis, MO 63103

Dear Sir:

   I wish to apply for the food-technologist position for the development of dairy products announced in the January 1992 issue of Food Technology. In May I will graduate from State Tech with a double B.S. in food science and chemistry. In my food-science curriculum I have emphasized courses in dairy science, and from these classes (listed in my enclosed résumé) I have gained a thorough understanding of the principles and practices of dairy-product manufacturing.

   Working as a technical assistant in a food-chemistry laboratory has taught me the scientific method of dairy-foods research and development. For my undergraduate project, I developed a technique for isolating lipase, the chemical compound that makes milk go sour. This laboratory also gave me experience in the standard methods for chemical analysis of dairy products.

   I have also worked as a technician in the food research and development laboratory of a large grocery chain. My responsibility was to determine the source of the problem in inadequate processes or defective products, to propose a solution to the problem, and to determine the economic feasibility and practicality of that solution.

   I would be happy to come to St. Louis to meet with you at any time. I look forward to the opportunity to discuss the contributions I can make to Roth Young's continued growth.

> Sincerely,
>
> *Sara Hayes*
>
> Sara Hayes

Enclosure

Figure 6.3   A Weaker Letter Presenting Only Weak Qualifications
This applicant has only academic training, with little directly relevant work experience.

883 Cherry Street 211
New Orleans, LA 70117
February 2, 1992

Box DN-145
Wall Street Journal
1233 Regal Row
St. Louis, MO 63123

Dear Sir:

I would like to apply for a job as one of your staff consultants as advertised in the <u>Wall Street Journal</u> on Wednesday, January 29, 1992. This May I will graduate from State University with a Bachelor of Business Administration degree and a specialization in Finance.

The curriculum at State includes an equitable mix of theory and practice combined to produce a good understanding of the fundamentals of business administration. In addition to my specialization in finance, my curriculum has included courses in basic managerial decision making, business simulation, marketing, business analysis, and accounting.

I have gained experience working at 1st Bank here in New Orleans twenty hours a week for the last year. My duties have included general bookkeeping, clerical work, and (my current assignment) working in the proof department. The enclosed résumé provides more information about my work experience and my education.

I would be happy to come to St. Louis any time for an interview. I look forward to discussing with you ways I can contribute to your firm.

Sincerely yours,

Thomas Baxter

Thomas Baxter

Enclosure

Figure 6.4 presents another sample letter, with notations showing its use of the elements. Briefly, here is how the letter uses each one:

1. *Solicited versus Unsolicited Correspondence.* The phrase "as described in the bulletin posted in the Geophysics office at Texas A&M University" makes it clear to Mr. Armstrong that this letter is specifically aimed at that particular job; the letter is responding to a job notice sent out by the company. If Mr. Armstrong is like most readers of business letters, the fact that the letter is in that sense solicited gives him a much more favorable predisposition toward it.

2. *Reader Benefits.* The explicit reader benefits in this letter are in the phrase "the ways in which I can contribute." There are also implicit reader benefits throughout the letter: the reader (Mr. Armstrong) is getting exactly the kind of job applicant he wants. When you deliver exactly what the reader wants, that is reader benefits.

3. *"You" Attitude.* The phrases "meet with you," "discussing with you," and "contribute to Texaco" (at the end of the letter) focus the letter on the reader. Everything about the letter suggests it was written just for Mr. Armstrong.

4. *Goodwill.* Offering to go into more detail in the enclosed résumé alerts the reader to your having adapted just for this reader what goes in the letter and what is better left to the résumé. Some writers put everything, trivia and all, into their letters, and others only generalize. It is a courtesy to your reader to put in just the amount of detail he or she needs and then remind the reader about the résumé.

5. *Caution About Needless Negative Messages.* It would be a needless negative message to claim "my classes give me the background for this job" without backing it up with specifics. To list all the classes here would also be a needless negative message.

6. *Positive Emphasis.* The fact that each of the summer jobs mentioned was with a different company, none of them with Texaco, doesn't faze this writer at all. She turns it to her advantage with a good positive attitude.

7. *Resale/Action Close.* It's clear what the writer expects of the reader; their next communication will be at an on-campus interview, for which the writer has apparently already signed up.

The careful use of these elements separates the letter written by a skilled writer from the letter written by an amateur. The biggest difference is in effect: one makes the most of your opportunity, and the other leaves your future to chance.

Figure 6.4   A Strong Application Letter
The letter's use of the principles for job-application letters is marked.

Kim Solman
1902 Southwest Parkway
Abilene, TX 79601
February 5, 1992

Richard B. Armstrong
Senior Geophysicist
Texaco Incorporated
P. O. Box 430
Bellaire, TX 77401

Dear Mr. Armstrong:

Solicited versus unsolicited —   I would like to apply for the position of geophysicist with Texaco as described in the bulletin posted in the Geophysics office at Texas A&M University. I will graduate from A&M this May with a Bachelor of Science degree in Geophysics, and I have worked the past three summers for major oil companies in various aspects of geophysical exploration.

Positive emphasis —   This summer work experience has guided my interest in exploration. The first summer I worked with Amoco Production Company interpreting gravity surveys with their gravity and magnetics group. The second summer I worked with Atlantic Richfield's Exploration Resources Group in Dallas preparing synthetic seismic logs using computers. This last summer I worked for Mobil in Dallas doing seismic interpretation, which I find to be most challenging.

Caution about unnecessary negative messages —   My classes at A&M give me the necessary background for this kind of work. These classes include Seismic Exploration, Geophysical Data Processing, Structural Geology, and Stratigraphy and Sedimentation. The enclosed résumé gives more details about

Goodwill — my education and work experience.

Resale and action close —   I hope to meet with you when you arrive on campus for interviews March 3. I look forward to discussing with you the ways in which I can contribute

Reader benefits and "you" attitude — to Texaco's exploration effort.

Sincerely,

Kim Solman

Kim Solman

2.2 Patterns

How are the sample letters in Figures 6.1 to 6.4 similar in their application of principles? Notice that they share the same four-paragraph structure (Fig. 6.5).

First Paragraph. The first paragraph contains the contact phrase that clarifies the letter's solicited nature and clearly states the purpose of the letter. The first paragraph also identifies the writer's single biggest claim to the job.

It is important to make it clear to the reader early in the first paragraph that this letter is solicited correspondence. Whether your contact phrase is "At the suggestion of . . . ," "As advertised in . . . ," or one of many others, place it in that first sentence. Also, get right to the point of the letter in the first sentence: apply for the job. "I would like to apply for the position of . . . as advertised in . . ." is a good way to say it. Your reader will appreciate your good business letter sense and straightforwardness. Often the person who opens your letter only routes it somewhere else; getting right to the point at the beginning helps your letter find the right desk fast.

The rest of the opening paragraph should be a brief statement of your biggest claim to the job. "I have both college education and practical experience in . . ." is probably the best phrase. Many new college graduates, however, can only say, "I will receive my B.S. degree from XYZ University on May 19___." Whatever your case, remember to put your best foot forward in the first paragraph; take your best shot first!

Second Paragraph. The second paragraph expands on your biggest claim to the job: academic training, job experience, or a combination of the two (whichever is best for the job). This may include a brief summary of your relevant background and a reference to more detailed information on the enclosed résumé.

If your college education is your biggest claim to the job, go into more detail about that education in the second paragraph. "As a management major, I took xx hours of management courses and xx hours of [whatever your second field of study is: computer science, accounting, or something else]. These xx hours of management included . . ." List a few of the most relevant courses, being sure to use descriptive titles: not "Management 4630," but "Theories of Organizational Behavior." If you haven't already done so, this is a good place to refer to your résumé: "Please refer to the enclosed résumé for more information on the courses I have taken that are relevant to this job." You can also refer to any extra information that might connect your education with this job, such as, "In my computer classes we used a Phrase Master 370/3601, which I know is the same hardware your company uses, so I should have no trouble adapting to . . ."

Figure 6.5   The Conceptual Outline of a Job-Application Letter
The letters shown in this chapter share this underlying structure.

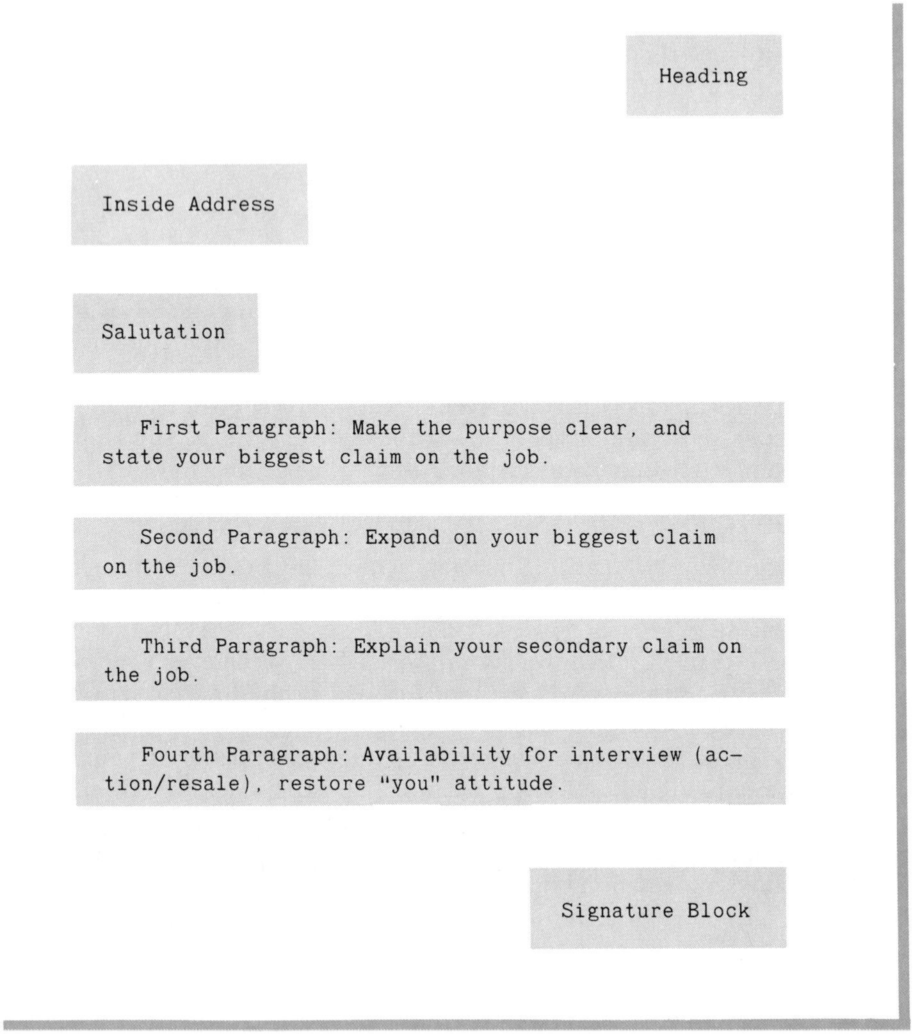

Heading

Inside Address

Salutation

First Paragraph: Make the purpose clear, and state your biggest claim on the job.

Second Paragraph: Expand on your biggest claim on the job.

Third Paragraph: Explain your secondary claim on the job.

Fourth Paragraph: Availability for interview (action/resale), restore "you" attitude.

Signature Block

Third Paragraph. The third paragraph explains your secondary claim on the job—typically either work experience or education. If you have education but no experience, finding a positive way to phrase this weaker claim is especially important.

Your reader may construe this paragraph—potentially the weakest—as a negative message; therefore, place it in the third paragraph, so it can have stronger buffer paragraphs both before and after it. If you have relevant work experience, even on a nonpaying, volunteer basis, mention it positively here. If the only work experience you have is not relevant to the job you're applying for, bring it up here to demonstrate that you are capable of showing up on time, getting along with people, and taking responsibility: "Every summer since my junior year in high school I have held a full-time job for three months. Ranging from building maintenance to insurance sales, these jobs have given me the ability to . . . and the confidence to . . ." That may be weak, but it is better than applying for the job on the strength of your education alone.

Here is another good touch: "Through part-time jobs during the school year and summer jobs between terms I have earned xx percent of my college expenses." That is the kind of thing too many job applicants seem to apologize for when they should be proud of it. Most employers respect that kind of stamina.

Fourth Paragraph. In the fourth paragraph, mention your availability for an interview and make clear just what you expect the next step in the process to be. If you have inside information about what this company is looking for from its applicants and you haven't used that information yet, this is the place. Finally, you need a closing that restores the letter's "you" attitude and prepares you and the reader for the next stage of the process.

If the next step is an interview, you may want to bring the subject up by discussing your availability: "I am available for an interview at any time from March 15 on," or "I will be in Philadelphia for Christmas (from December 19 to January 12) and will call to arrange an appointment then."

At the very end of the letter, turn the emphasis directly back on the reader with "you" attitude. "You" attitude is very difficult to achieve in a job-application letter because it naturally focuses on the writer more than on the reader. More than anywhere else, in the last lines you need to get the reader's needs and wants foremost. Here is one way to do that while also accomplishing the action closing and resale: "I look forward to the opportunity to discuss with you or a representative of your firm the ways in which I can contribute to your company's continued [success, growth, etc.]."

As with all the other typical forms presented in this book, the best letter for you may be somewhat different. Yours may have two, three, four, or five paragraphs, and in rare cases (less rare if you've been out of school a few years), your letter may be more than one page long. Write the letter that strongly presents *you* as a competent, educated individual, and place that presentation clearly in the context of what you can do for that company. The sample letters presented here suggest the main elements of the most effective job-application letters, but don't rule out other possibilities. **Use the sample letters to define the center of the set of all possible effective job applications, not to establish the boundaries on that set.**

2.3   Questions

The directions given here so far should give you a good basic letter. Of course, you have many questions; the following are the ones students ask most frequently:

Question: How do I say I'm qualified for the job? I really feel I can do a good job for these people, but where and how do I say so?

Answer: That's a tough question. The answers will vary from person to person, from job to job, and from profession to profession. What you want to watch out for is seeming to come on too strong. The way out of this is as follows. You would certainly not want to say something like, "I feel sure I can be the best lab technician Accu-Labs has," without giving the evidence to back that claim up (either before or after it), something like "My combination of two years as a military lab technician and four years' education in biology has given me . . ." Because you must give the evidence anyway, why not just state the evidence in a clear and suggestive enough way to make it inevitable that the reader will come to agree with your claim on his or her own, and then leave the claim unstated? To make that realization inevitable in your reader's mind takes writing skill, but the reward for that skill can be high.

Question: Do I really have to say, "I would like to apply for . . ."?

Answer: Yes. Everyone in business and industry has bad reactions to letters that beat around the bush. A solicited letter of request should be carefully planned and straightforward.

Question: Why such a formal ending as, "I look forward to discussing . . ."? Won't it be recognized as a stock ending?

Answer: Yes, it may well be recognized. But it is a comfortable, functional kind of formality, and that helps. And that particular ending works on your reader's psychology: many people who employ new college graduates have a negative attitude toward them. Some employers consider new graduates too "me" centered and not enough company centered, always asking, "What will the company do for me?" rather than "What can I do for the company?" The phrase "the ways I can contribute to your company's . . ." makes it clear that you are not "me" centered.

2.4   Tactics and Strategies

Carefully consider every word in your letter not only from your point of view but also from that of your reader. Again, remember that anything that *can* be misunderstood *will be*. Use the words that will present exactly the right message, whether your reader skims the letter or carefully considers each word and phrase.

Three principles of business communication deserve special consideration in this regard:

- Convey "you" (versus "me") attitude.
- Emphasize the positive rather than the negative.
- Avoid unnecessary negative messages.

Although all of these have been discussed already, they are important enough to be reemphasized, with specific application to the job-application letter.

Convey "You" Attitude. Try to build into every line of your letter your willingness to serve your employer. Show that you are much more interested in that than in having your employer benefit you. Check each sentence for accidental intrusions of "me" attitude that can easily be removed; be alert for chances to insert more "you" attitude.

Here are examples of "me" attitude turned into "you" attitude:

"Me" Attitude	*"You" Attitude*
"I know that the EDG Department is giving me the best possible training in the drafting profession."	"The Engineering Design Graphics Department has a national reputation for producing excellent drafting designers."
"This should be suitable to qualify me for your current job opening."	"This work experience should make me a strong candidate for the job you have advertised."
"I would like to tell you of my goals as a member of the construction industry."	"I can contribute to your company by . . ."
"The job you have described seems to fit my desires perfectly."	Either omit this kind of statement entirely, or, if "desires" and "qualifications" mean the same thing for you then you can say, "I have both education and training for a position in . . . ."
"I think my credits are more than sufficient to fill your job opening, and I look forward to employment by your company."	"I hope you find that the credentials I have listed here make me a strong candidate for the job you have advertised and that I can have the opportunity to discuss with you the ways in which I can contribute to . . ."

Emphasize the Positive. Word your phrases positively, not negatively. You want to describe your cup as half full, not half empty. Do not say, "I have not had any full-time jobs," but rather "through my part-time jobs I have earned 75 percent of my college expenses." Here are more examples of negative emphasis turned positive.

Negative Emphasis	*Positive Emphasis*
"While reviewing several job notices, yours caught my attention."	"I would like to apply for . . ."
"During my five years of industrial experience I have gone from job to job, . . ."	"I have five years of industrial experience at a variety of jobs, ranging from . . . to . . ."
"My major is General Industrial Technology with no particular area of specialization."	"My major is General Industrial Technology."
"There are a few courses in my curriculum which deal with software systems."	"I have taken *X* hours of courses which deal with software systems, including _____, _____, and _____."
"My job experience is confined to summer jobs."	"Through my summer jobs I have been able to pay ___ percent of my college expenses."

Avoid Unnecessary Negative Messages. Many instances of negative emphasis can also be viewed as unnecessary negative messages. They are things you say that will strike the reader the wrong way, and on further examination you may find you do not need to say them at all. If you have no full-time work experience, leaving the whole subject of full-time experience out of your letter may be better than anything you could say that would begin with, "Although I have no full-time work experience . . ." If the subject will not advance your cause, why bring it up at all?

Obviously, if you are asked directly about some such subject, you will need to respond honestly. But as long as you have the option of not bringing up such a subject, it is probably to your advantage to exercise that option.

The following list gives seven examples of typical unnecessary negative messages in job-application letters written by students. They do not require revision; they should be left out entirely.

Unnecessary Negative Messages

1. I would like to get established within a company soon after I graduate.
2. I have been looking into several job prospects but so far I haven't got a job.
3. I wish to apply for this position to gain agronomic experience on an international level.
4. It is true that as a whole my major areas of study deal with large industries, but I will gladly work in a smaller company.
5. I am interested in getting started in the engineering field and would like to start with your company.
6. I am sorry this application is so late.
7. Although I missed your recruiters when they were here . . .

Your job-application letters will never be exactly like the ones described. In some ways, all job-application letters are the same, but in other important ways, they should all be different. Each one should reflect the unique person who sends it *and* the unique person and job it is addressed to. This brings us to the next part of writing effective job-application letters: analyzing your audience and adapting your letters to them.

3. Individualizing Your Application

The more you know about your audience and the more you adapt your letter to them, the more effective your letter can be. Your letter is already individual in the ways that mean the most to *you* because the key features of your life, ability, and personality are described in it. But to be really effective, your letter also needs to be individual in the ways that are most significant to your *reader*. That means you need to think about what kind of person your reader is—what he or she knows or doesn't know, needs or doesn't need, is or is not interested in. Carried to its logical extreme, this means that each letter you write for each job is different from every other letter you write. We will look first at ways to analyze your audience and then at ways to adapt your letter to that audience (short of writing a totally different letter for each job).

3.1 Visualizing Your Audience

To write a fully effective job-application letter, you need to visualize your audience. Try to visualize how your reader looks, dresses, talks, acts, drives, reads; get as clear a picture as you can. At the least, make the best guesses you can about your reader; at the most, get definite information. The more you can learn about your reader and store in your mind, the more your subconscious will do your audience adaptation for you.

If you are writing to a specific, named person (rather than "Box 152" or "Personnel Manager"), try to find out as much as you can about that person. If you are about to enter a professional field, you probably already know quite a bit about the kind of person who does the hiring in your field. Even a stereotype can be useful here, if it helps you to visualize your audience.

The people who run professional advertising agencies do the most sophisticated audience analysis, and you can take advantage of it. Find a journal aimed at the kind of person you are writing to (such as a professional or trade journal in that person's field) and look at the people in the ads. Of course the people are professional models, but they are also the images your prospective employers have of themselves. If you have no other information to go on, write your letter as though it were to one of the people you saw pictured.

Successfully visualizing your audience gives your subconscious the chance to help you adapt your job-application letter (or any other document) to each particular audience. You also need to use the other analysis and adaptation techniques described in this book. There will also be specific characteristics about each job you apply for that will require your letter to be specialized.

You will probably still have questions that give you problems: Should I bring this point up at all? Should I make more of this item—or should I omit it entirely? The answers to such questions can come only through knowing more about and visualizing your audience; do what you believe will have the best effect on your audience.

3.2 Avoiding "Dear Occupant" Writing

If you fail to do audience analysis and adaptation for your job application letters, you are probably doing "Dear Occupant" writing, the same kind of writing that offends most people who get mass-mailing advertisements. Writing to "Dear Sir" or "Personnel Manager" with no idea of who that person is or what that person needs is like dropping a pebble into a deep, dark well—you may never even know for sure that it even hit bottom. And you may never know if your "Dear Occupant" letters reached anyone, because they were never specifically aimed at anyone.

By now you have probably heard stories of people who graduated ahead of you and sent out 100 or 200 copies of the same job-application letter. Those stories usually feature such details as having the letters done on good-quality paper, with their inside addresses and salutations individualized by the word processor. The price is high, and those stories almost never mention whether that person actually got a good job or whether any job at all resulted from the letter.

This shotgun approach to writing job-application letters usually backfires. As countless mass mail advertisers have found, it is practically impossible to be successful at simultaneously mass mailing and individualizing. Most people resent such letters—thus, the descriptive phrase "junk mail."

The precise name for such a letter, one with only a word or two changed here and there before it is sent out to many different people, is *generic*. No employer worth his or her salt will be deceived by generic job-application letters. But the alternative, to write every letter fresh from scratch, may be impossible for you. There is an attractive middle ground between these extremes, however: Many students find that they can write two or three basic letters (the number depends on the range of jobs they are applying for) and then write the letters they actually send out based on those basic (or *modular*) letters. The letters sent out are composed of parts of the two or three basic letters, but they are individualized in more detail than just changing the receiver's name and address. The individualizing makes it clear to the reader that the letter is not mass produced. It may be one line in one paragraph, a change in the order of paragraphs, or the creation of a brand new paragraph for that one letter.

An example will make this clear: modular letters written by an accounting student who has a strong second emphasis in computers. She writes three basic letters: One (Fig. 6.6) emphasizing her education and experience in accounting; another (Fig. 6.7) emphasizing her education and experience in computers; and a third (Fig. 6.8) emphasizing a combination of both. Following these examples, we will see what one of the letters she actually sent out looked like. Some of the letters Ms. Jones sends out may be word for word like one of those above, but many of them will vary slightly. Figure 6.9 shows an example of a letter she sent out, based on the letter in Figure 6.8.

Most people find it hard to write these basic letters starting from scratch because they're addressed to no one in particular. The way to do it is to write letters to your best job possibilities first, and then extract and modify your basic letters out of those specific examples.

Using basic letters still means you will have to look at each letter separately, making changes from letter to letter on a word processor. But by working from basic letters you will certainly be able to *compose* each letter much faster than otherwise, and these partially tailor-made letters are much more effective than mass-produced letters. You can also save the basic letters you wrote this year and use the best of them (the ones that get good results) again, with appropriate modifications should you want to change jobs.

Figure 6.6 A Basic Letter
Notice how anonymous and weak this letter seems. It lacks the specific details that come with aiming it at one particular job, but it makes writing the specific letters go much faster.

Dear _____:

I would like to apply for the position of Accountant which you have advertised in _____. I will graduate this May from State Tech with a Bachelor of Arts degree in Accounting. I also have summer work experience in accounting.

My course work in Accounting includes courses in _____, _____, _____, and _____. I have also taken _____ hours of courses in Computer Technology.

Each of the last three summers I have worked for Jones and Williams Agency, a local public accounting firm here in Los Angeles. The nature of my duties ranged from office assistant the first summer to assistant bookkeeper this last summer. Through these experiences I have learned more about how accountants actually work than college itself could ever teach me.

To provide you with more information about my work experience and education, I have enclosed a résumé with this letter. I can be in _____ for an interview any time at your convenience. I look forward to meeting you and discussing with the ways in which I can contribute to _____.

Sincerely,

Susan Jones

Susan Jones

Enclosure

Figure 6.7   A Basic Letter

This letter is anonymous and weak, like the one in Figure 6.6, but it brings out some interesting ideas about computers and accounting. It also shows the difficulty of applying for a job other than the one your degree leads people to expect.

Dear _____ :

    I would like to apply for the position of Computer Technologist you have advertised in the _____ . My summer jobs have given me experience with many aspects of computer technology, and I have taken twenty-one hours of computer technology courses in college.

    For the past three summers I have worked for Jones and Williams Agency, a public accounting firm here in Los Angeles. During those summers a major part of my duties has been to assist the firm's rapidly expanding computer operations. Thus I have had on the job experience with statistical analysis, interactive statistical processing, systems programming, and interactive computer graphics.

    I will graduate this May from State Tech with a Bachelor of Arts degree, majoring in Accounting. This accounting background makes me especially aware of the business marketplace's needs and problems with computers, and that awareness should help me contribute to the field of Computer Technology. I can see the computers from the point of view of the accountant as well as the point of view of the computer specialist.

    The enclosed résumé shows more about my education and work experience. I can be in _____ for an interview any time at your convenience. I look forward to. . . .

                Sincerely,

                Susan Jones

                Susan Jones

Enclosure

Figure 6.8 A Basic Combination Letter

If she can find a job that asks for this combination of skills, this student stands a good chance. Here the combination of skills is unique and so strong it starts to make up for the fact that the letter still is not addressed to any particular person. But this would almost by itself define a particular job.

Dear Mr. _____:

    I would like to apply for the position of Accountant which you have advertised in _____. I will be graduating from State University this May with a Bachelor of Arts degree in Ac-counting. In addition to the accounting curriculum, I have taken 21 hours of courses in Computer Technology. I also have summer work experience which combined accounting with comput-ers, giving me the exact credentials your job notice called for.

    My course work in Accounting includes courses in _____, _____, _____, and _____. Among my courses in Computer Technology were courses in COBOL, Pascal, ALGOL, BASIC, and Interactive Graphics. My overall grade point is 3.2 on a 4.0 basis.

    Each of the last three summers I worked for Jones and Wil-liams Agency, a public accounting firm here in Los Angeles. My duties ranged from office assistant to assistant bookkeeper. During those summers a major part of my duties was to assist the firm's rapid expansion of its use of computers. I gained on the job experience with statistical analysis, interactive sta-tistical processing, systems programming, and interactive com-puter graphics.

    The combination of accounting and computers in my background allows me to see computers from the point of view of accoun-tants and accounting from the point of view of computer special-ists. This perspective and my combination of abilities will, I hope, make me a strong candidate for the position you have ad-vertised.

    The enclosed résumé shows more about my education and work experiences. I can be in _____ for an interview any time at your convenience. I look forward to. . . .

                               Sincerely,

                               Susan Jones

                               Susan Jones

Enclosure

Figure 6.9   A Real Letter Based on the One in Figure 6.8.
Once Ms. Jones realized that her combination of accounting and computer skills was what really made her special as a job applicant, she went looking for a job that called for just that combination.

Dear Mr. Hermening:

    I would like to apply for the position of accountant you have advertised in the Houston Post. I will be graduating from State Tech this May with a Bachelor of Arts degree in accounting. I have also taken twenty-one hours of computer-technology courses. My summer work experience has combined accounting with computers, so I have exactly the credentials you asked for.

    My course work in Accounting includes the three standard accounting courses--cost accounting, managerial accounting, and accounting theory--among others. Among my courses in Computer Technology were . . .

    [Paragraphs 3 and 4 of Figure 6.8 would go here.]

    The enclosed résumé shows more about my education and work experience. I can be in Houston any time for an interview at your convenience. I look forward to discussing with you the ways in which I can contribute to your agency's continued success.

                    Sincerely,

                    Susan Jones

                    Susan Jones

Enclosure

3.3   Creating Yourself, Creating Your Reader

Why are individualized letters easier to write and more effective than mass-produced ones? The answer to this question contains the essence of the psychology of human beings as writers and readers. Most people who receive mass-produced writing of any kind resent the depersonalization they feel it brings. People want to be treated as individuals, to have their unique needs and desires taken seriously. But mass-produced, "Dear Occupant" writing creates a picture of a reader who is nameless, faceless, and characterless. For the reader, it is like looking in a mirror and seeing an empty silhouette staring back. Everyone you ever write to will have individual needs and desires. If you as a writer ignore the uniqueness of your reader, thus producing "Dear Occupant" writing, you risk having your writing ignored at best, actively resented at worst.

The reader you create in your writing needs to match your actual reader's own best self-image, because that is what your readers will respond to most favorably. You may also find that your writing comes easiest and pleases you most when the image that it creates of you as a writer matches your own best image of yourself. If you are composing "Dear Occupant" writing, the kind of self-image you are creating may be so psychologically unpleasant to you that you actually find it hard to write. Skilled writers are those who not only have become conscious of their own styles but also have realized the value of style. Professional writers are those who learn to vary their styles easily and successfully, to create different implied writers and implied readers as the situation demands. Just as you need to consider the reader your writing creates, so you also need to think about the kind of person it implies that you, the writer, are.

4.   *Creating Effective Résumés*

At some point in the job application process, nearly everyone has to prepare some kind of ordered visual presentation of abilities, education, and work history. This may be called a *data sheet,* a *vita sheet,* a *résumé,* or a *bio;* you may use it with your initial letter, at the campus interview, or at the on-site interview. You may also hear these kinds of documents classified as historical, biographical, or functional. Whenever you use such a document, and whatever you call it, this visual presentation is a vital part of your job application. It is something you will need to keep—and to update—the rest of your working life. If you wait to make one until the week before your first interview, chances are it won't be done nearly as well as it could be.

Three ways of presenting yourself on paper will be discussed here: the one-page résumé, the multipage résumé, and the qualifications sheet. The most challenging of the three to produce is the multipage résumé, and because all of the principles involved in it are also to a lesser extent involved in the shorter forms, most of the attention here will be paid to the multipage résumé.

Each of the three forms presented below employs the same five elements:

- Placement on the page
- Typeface for emphasis
- Compartmentalization
- Psychological structure
- Audience analysis and adaptation

4.1   The One-Page Résumé

Figure 6.10 shows a one-page résumé in a popular format. It illustrates how to use each of the five elements listed above:

Placement on the Page. The most important material (name and professional goal, for instance) is at the top, in the most prominent position.

Compartmentalization. The document is divided into categories to suit the person it describes and the job he or she is applying for. Each compartment has contents adjusted for exactly those needs.

Psychological Structure. The document is designed to appeal to the psychology of the reader. The underlined items in the body (*Engineering Support Technician*, etc.) are set to catch the reader's eye as it scans the page. The most important and responsible job is listed first, not last. High school graduation is not mentioned; if he were applying for a job in his home town, it might well be.

Audience Analysis and Adaptation. The job Mr. Lynn is applying for is designing and building electrical equipment as part of a group. Thus, it is no coincidence that his professional objective is worded that way. He has interpreted this to be a very down-to-earth kind of company, and that's how he presents himself in this one-page résumé.

4.2   The Multipage Résumé

With the multipage format you can add categories, delete categories, move categories, and move items within categories, giving you more

Figure 6.10   A One-Page Résumé
This is one of many popular possible forms.

MATTHEW LYNN

1431 Rudder Road   Single
Knoxville, TN 37920   Age 23
(615) 577-5902   Willing to Relocate

PROFESSIONAL   To participate in the design and building of
OBJECTIVE   sophisticated electronic equipment.

EDUCATION   Bachelor of Science in Electrical
                      Engineering, The University of Tennessee,
                      May 1991. Emphasis in Electronics and
                      Instrumentation. Grade point 3.0 (on 4.0
                      scale).

Important   Electronics, 9 hours
Coursework   Instrumentation, 9 hours
                      Computer Programming, 6 hours
                      Physics, 8 hours
                      Calculus, 6 hours
                      Technical Writing, 6 hours

WORK   Engineering Support Technician, Petkus
EXPERIENCE   Corp., Tullahoma, Tennessee.
1991-1992   Participated in development of production
                      line tests and construction of sequential
                      electronic test systems. Performed all
                      Quality Control procedures during
                      production.

Summers,   Electrician's Assistant, Welch Electrical
1990-1991   Contracting, Tullahoma, Tennessee.

1989-1990   Framing Carpenter, Collier Construction
                      Company, Tullahoma, Tennessee.

PERSONAL   Matthew Lynn is a native of Tennessee. He
BACKGROUND   enjoys travel, hunting, fishing, and
                      camping. His other hobbies include gardening
                      and home computers.

REFERENCES   Available upon request.

                                                    March 2, 1992

variables than can be fully discussed here. Figure 6.11 shows a good multipage résumé; it uses each of the elements discussed earlier as follows:

Placement on the Page. This form uses headings centered within a wide left margin (as well as horizontal and vertical spacing) to achieve emphasis and divide categories. In the multipage form the left margin headings are easier for a reader's eyes to follow than a series of centered headings would be. The most important items are placed in the most prominent positions: tops of pages, bottoms of pages, and left margins. The least important (or potentially negative) items are buried in the centers of pages. Thus, the placement of items on the page appeals to the reader's psychology. To a greater degree than the shorter forms, the multipage résumé organizes its information psychologically (to satisfy the reader) rather than logically (which may make sense only to the writer).

Typeface for Emphasis. One of the keys to success with a multipage résumé is to use the resources of your word processor to your advantage. Use words typed with all capital letters, initial capital letters, and underlining—each in a planned way. Combined with vertical and horizontal spacing, the varieties of typeface allow a skillful writer to guide the reader's attention across the page in the way most favorable to the writer.

Compartmentalization. The categories and subcategories used in the sample résumé may not be exactly those that will best reflect your own abilities and background. Also, the categories and subcategories—or the ordering of them—that work best for you on one application may not work best on another. Sometimes you need to enlarge a category, shorten it, move it to a more or less prominent position, introduce a category, or delete a category, all based on changes in your own qualifications or in the nature of the job you're applying for. Compartmentalization of the résumé allows you to choose how to present yourself to your reader. The kind of person you make yourself seem to be is (within the bounds of truth) completely optional.

Psychological Structure. The most important principle is to make the whole presentation a psychological sculpture, reflecting your reader's needs and interests rather than your own preconceptions. For example, it may never have occurred to you that from the reader's point of view it makes more sense to list your most recent work experience first, because it is usually closer to the job you're seeking, both in kind and in level of responsibility. Besides using reverse chronological

Figure 6.11   A Multipage Résumé
Such résumés can be adapted and modified almost endlessly.

```
TERRY WRIGHT

Box 253, Rudder Road                          Single
Knoxville, TN 37919                           Age 21
(615) 577-4711                                Will Relocate


PROFESSIONAL        To become Production Control Manager,
OBJECTIVE           supervising the completion of
                    industrial projects.

EDUCATION           Bachelor of Science in Engineering
                    Technology, The University of
                    Tennessee, May, 1992.

Courses in          Manufacturing Processes
Major               Industrial Safety
                    Machine Production Technology
                    Mechanics of Materials
                    Statics and Dynamics
                    Engineering Design Graphics (I & II)
                    Plane Surveying (I & II)
                    Statics and Strength of Materials
                    Foremanship and Supervision
                    Electricity and Electronics
                    Communication for Industrial
                    Personnel
                    Computer Programming for Engineers

Supporting          Purchasing
Courses             Industrial Marketing
                    Business Law
                    Technical Writing
                    Statistics

WORK                Air-conditioning repair technician,
EXPERIENCE          Johnny's Appliance and Parts,
(part-time,         Knoxville, Tennessee. In addition to
summers)            repairing air conditioning and all
                    types of heaters, duties included
                    inventory, billing, and book work.
```

TERRY WRIGHT
(Résumé, p. 2)

1991   Designer and Drafter, Ramtag (a
                    division of Bemco), Knoxville,
                    Tennessee. Duties included designing
                    and drafting custom bakery equipment
                    (tandem bun slicers, conveyors).

1989   Maintenance Worker, Taggart Research
                    Division, Knoxville, Tennessee. Duties
                    included overhauling equipment and
                    assisting in building construction.

1988   Carpenter's Helper, Pyramid
                    Construction, Knoxville, Tennessee.

SCHOLARSHIPS   Through various part-time and summer
and   jobs, paid for 80% of college
AWARDS   education.
                    Distinguished Student Award,
                    1987-1988, 1989-1990, for achieving a
                    grade point of 3.25 or better.

                    Carl and Florence Wilson Scholarship,
                    1987-1988, for placing in the state
                    Ready Writing Contest.

MEMBERSHIPS   American Society of Civil Engineers,
                    student member
                    Engineering Technology Society
                    Intramural Sports: tennis, basketball,
                    football
                    Fellowship of Christian Athletes
                    Knoxville Boys Club

TERRY WRIGHT
(Résumé, p. 3)

REFERENCES   Mr. Joe E. Braberson, owner
Johnny's Appliance and Parts
5423 N. Broadway
Knoxville, TN 37925
(615) 523-8736

Dr. Tom Gunderson, Assistant
Professor
Department of Engineering Technology
The University of Tennessee
Knoxville, TN 37912
(615) 974-8015

Dr. James Hall, Associate Professor
Department of Engineering Technology
The University of Tennessee
Knoxville, TN 37912
(615) 974-8023

Mr. William Hefflin
Production Engineer
Ramtag, Inc.
6897 Clinton Highway
Knoxville, TN 37880
(615) 563-9880

Mr. Michael Hannert
Foreman
Taggart Research Division
6800 Oak Ridge Highway
Knoxville, TN 37985
(615) 534-1885

order in work experience and perhaps in other areas, you may want to separate your work experience into *full time* and *summers,* or you may want to sort your experience by the kinds of duties you performed—clerical, supervisory, accounting, and so on. You need to decide what it is in your background that will most appeal to your prospective employer, and structure your résumé to bring that material into the forefront.

Audience Analysis and Adaptation. To give your résumé a structure that has a psychological appeal to your reader, you need to analyze the kind of person your reader is. Based on what you know and can learn about the profession, the company, the job, and the people who will do the hiring, what do you think your strongest points are? How can you emphasize those points? What else can you add that will help you get the job? Are there areas you are forced to bring up, but that you want to deemphasize? How much detail do you need in each of the various categories? You need to think through these questions to make your résumé as effective as possible, and you can answer such questions accurately only if you analyze your audience.

Constructing an effective multipage résumé never ends: whatever form you are happy with today, whatever form best reflects your unique qualifications today, you will probably want to change next week or next month as your own image of yourself and the kind of professional work you do changes.

4.3 The Qualifications Sheet

A new and different kind of one-page presentation of your abilities is called a *qualifications sheet.* Some people call it a functional résumé, as opposed to the historical or biographical form described above. The traditional résumé tends to look backward, telling what you *have* done. The qualifications sheet looks forward, telling what you *can* do. This lets you include the kinds of subjects you want to steer an interview toward, subjects that might not even be covered in the traditional résumé. A qualifications sheet is especially useful if your interviewer has already seen your multipage résumé. As an elaboration of the professional goals section of your résumé, the qualifications sheet lets you go into detail about exactly what you choose to have the interview focus on.

The same principles of construction apply to the qualifications sheet as the résumé. Figures 6.12 and 6.13 show how typical good qualifications sheets look. Because these particular qualifications sheets are to be used in interviews, where the applicant's address is already available, they do not carry addresses.

Figure 6.12   A Qualifications Sheet
Notice the emphasis on skills.

Qualifications Sheet

JOHNNIE MACK SAMPSON

Bachelor of Science, Agronomy

Land Use Planning   Livestock Production Planning
Farm Equipment Design   Farm Equipment Construction
Farm Equipment Repair   Crop Production Planning

PROFESSIONAL OBJECTIVES

Short Term: To manage a large, production—oriented farm or
ranch, and to develop a high degree of professionalism for
production and marketing of agricultural products.

Long Term: To own and operate a farm of about 3,000 acres
around the Crowley, Texas, area.

PROFESSIONAL SKILLS

Farming
Experienced and qualified to operate most types of farm machinery including:
      Tractors——I.H., 1066, 1486, 1566, 4166, 4366
      Land preparation——moldboards and subsoilers
      Cultivators——rolling and sweep
      Cotton stripper——John Deere 484
      Planter——I.H., Cyclone 500
      Combine——John Deere 7700

Shop Work
Metal Duplication——lathe, milling machine, drill press
      Metal Forming——shear machine, Devorak iron bender
Metal Cutting——hand held torch, pattern torch
      Welding——standard rod welder, wire welder

PERSONAL DATA
JOHNNIE MACK SAMPSON is 22 years old and a native Texan.
He was born and raised on a six—thousand—acre irrigated
farm near Crowley. He is single and active in community activities such as Methodist Men and coaching a little
league softball team.

Figure 6.13   A Qualifications Sheet
Notice what the "biography" adds to your picture.

Qualification Sheet

LINDA A. WOODRUFF

M.S. in Wildlife Biology

PROFESSIONAL OBJECTIVE

To work as a biologist conducting variable wildlife research in areas where the impact of humans will alter habitat and adversely affect wildlife.

<u>Professional Skills</u>

Statistical Analysis   Plant Taxonomy
Computer Applications   Vegetation Sampling
EIS Preparation   Aerial Photo Interpretation
Technical Writing   Telemetry

EDUCATION

Master of Science, <u>Wildlife Biology</u>, The University of Tennessee at Knoxville, 1992. Strong background in botany,
statistics, and computer science. Bachelor of Science,
<u>Magna Cum Laude</u>, Wildlife Biology, The University of Ten-
nessee, Knoxville, 1990.

<u>Important Courses</u>

Wildlife Biology–18 hrs.   Statistics–15 hrs.
Botany–12 hrs.   Computer Science–9 hrs.

EXPERIENCE

Graduate Research, University of Tennessee, 1991–1992
Graduate Research Assistant, University of Tennessee,
1990–1991
Supervisor, Youth Community Services, Knoxville, Tennessee, 1989
Research Assistant, Great Smoky Mtns. Wildlife Forest,
1988–1989

BIOGRAPHY

LINDA A. WOODRUFF is 25 years old, a native of Knoxville.
She is single and willing to relocate. She enjoys backpack-
ing, canoeing, cross–country skiing, sewing, and photography.

4.4   Questions About Résumés

Question: My grade point average isn't so hot. What should I do?

Answer: If you feel you must mention something about it on paper, try including only your grade point average in your major. Or place yourself in the appropriate percentile of your graduating class—"top 25 percent of class." Remember, as a negative message it may be better communicated orally than on paper.

Question: What about references?

Answer: Every professional handles references differently. Some include reference letters with the résumé, some put letters on file with placement services, and some prefer the employer to write or call the references. Whatever the case, you need to arrange with three to seven people at least to be prepared to write letters in support of your application if the need arises. Choose these people carefully; try to diversify your references so that each person will be able to speak to a different aspect of your individuality—weighted toward your professional abilities, of course. Don't hesitate to ask more people to be prepared to write letters for you than you may actually use. It is flattering to them, and the more people who know you are looking for a job, the greater the chances are one of them will be able to help you as the contact for a possible job.

Question: My school's placement service has a standard form we all fill out when we sign up for interviews. Do I still need a résumé?

Answer: You need a résumé even more. Standardized forms, such as those used by placement services, tend to make everyone look the same. You want to come across as an individual. See if you can attach your multipage résumé to that standardized form; if not, attach your short one. If even that isn't allowed, inquire about changing the placement director's policy.

Question: I've heard we don't need to put things like our age, gender, and so on in these résumés and qualifications sheets, but the examples in this book often include that information. Should I include this information on my résumé?

Answer: Legally you cannot be forced to answer (and should not even be asked) questions about your age, gender, marital status, overall health, and so on. But that does not mean you are forbidden to volunteer such information when you believe it will help you get a job. The students whose résumés these are modeled on included exactly the personal information they believed would help them in the job market, and no more than that. If you choose to provide no such information, you can still use the upper right-hand corner of your résumé productively by including an availability date or a statement about your willingness to relocate (where applicable).

Question: I've attached the résumé to my application, but I feel I need something to hand directly to the interviewers at the beginning of the interview. What do I do?

Answer: Use a qualifications sheet.

5. Succeeding in Interviews

The best way to train yourself to handle interviews effectively is through mock (practice) interviews. Check with your placement service and the student chapter of your professional society for such interviews. The second-best way is to view films of interviews and to critique them. Also, some advice to help make your interviews more successful is presented in the following sections as answers to these questions:

What should be the *goal* of your interview?
What *preparation* is best?
What is a *script* for an interview?
What *questions* usually come up?
What *feedback* can be obtained?

There are two broad classes of interviews—those on campus or in some other neutral setting (such as a professional meeting) and those at the prospective employer's place of business. The former is usually the shorter, and it is the one discussed here. But everything said also applies to the longer, on-site interview—just more so.

5.1 Determining the Goal

Interviewers typically have three goals:

1. To evaluate the candidate.
2. To stimulate the candidate's interest in the firm.
3. To maintain the candidate's goodwill.

You can increase your chances of success in interviews by making your goals compatible with the interviewer's goals. The candidate who goes in with one and only one goal—to get the job or go down in flames trying—risks too much emotionally to be able to survive more than two or three such encounters. That very single-mindedness also works against the candidate's success because it goes so strongly against the grain of the interviewer's goals. Go in trying to make a good accounting of yourself; tell yourself, "If I don't get the job, I still want to feel as if I took my best shot, that they saw me at my very best." Aim for these goals:

1. To be evaluated at your best.
2. To stimulate the interviewer's interest in you and your ability to contribute to the interviewer's firm.
3. To build and maintain the interviewer's goodwill.

5.2   Doing the Right Preparation

As with goals, preparation that mirrors that of the interviewer is best. When you walk into the interview room, the interviewer will probably have a folder of information on you—professional goals, school background, letters of reference, application forms—and a good idea of just where you might fit into the firm's structure and how you might best contribute to the firm's success. You need at least as much information about each firm you interview with.

From resources in your library, your placement service, and other published sources, try to find answers to questions like these:

What does the firm do?
What is its past and future?
Who owns it?
Who runs it?
Where are its various offices, branches, and manufacturing facilities?
What is its financial structure and status?
Who is the interviewer, and where does he or she fit into the corporate structure?
Where does that firm usually start entry-level employees (financially, geographically, and so on)?

The more homework you do for your interview, the greater are your chances for attaining your goals.

The best preparation is that which tells you the most about what you need to know regarding the firm, the interviewer, and your possible contribution. Doing that kind of preparation should also increase your confidence in your ability to handle the interview successfully. And that by itself should be enough to repay the effort you put into doing the homework.

5.3   Taking Advantage of the Script

Suppose that after you get your first paper back in class, you decide you want to talk with the instructor and explain that while you don't quarrel with the grade, you want her to know that the paper really isn't representative of your ability as a writer. You had tests in other classes to prepare for, which kept you from doing as well as you hope to in the future. That's your *script.* Or your teacher may call you in to tell you that if you don't start being more careful with written assignments, you'll fail the course. That's *her* script.

Failures in communication occur when two different scripts collide. Most interviewers have scripts for their interviews; in many cases the script was written down, and the interviewer has it memorized. Figure 6.14 shows a typical script for a thirty-minute interview.

To be a successful job candidate, you should be sensitive to when the interviewer is following a script and try to adjust your own script accordingly. You need to have your own script—things you want to bring up, points you want to make. It will be different from the interviewer's script, but you don't want the two to clash. They must be compatible in order for the interview to be successful, for both you and the interviewer.

Figure 6.14   An Interviewer's Typical Script
This is for a thirty-minute interview.

The Interviewer's Plan

Objectives of the Interviewer: 1. Evaluate the candidate
2. Stimulate the candidate's interest in the firm
3. Maintain the goodwill of the candidate

Structure of a Good Interview (30 min.)

Time	*Interviewer's Activity*	*Interviewer's Purpose*
	1. Reading résumé	1. Notice grades, academic achievement, extracurricular activities
4–5 min.	2. General comments • résumé • weather • "Why did you decide to study at ____?	2. Establish rapport. Put candidate at ease. Ask questions candidate can answer easily so he or she gains confidence.

3. Information gathering
   (a) Open questions requiring explanatory answers
       - "How do you spend your leisure time?"
       - "Tell me about your college experience."
       - "What courses did you like?"
   (b) Closed questions requiring short factual answers
       - How does a candidate act in certain situations?
       - Focus on past behavior

16 min.
   (c) Listening techniques
       - echo (repeat phrases)
       - eye contact
       - silence
   (d) Behavior observations
       - interested?
       - enthusiastic?
       - poised?
       - assertive?

3. Decide whether to ask candidate back for a second interview
   (a) Determine candidate's motivations and habits
   (b) Determine if candidate's apparent attributes are reflected in reality
   (c) Show acceptance
   (d) Pierce through candidate's nervousness to substance of what the candidate is like

5 min.   4. Answer questions

4. Sell the organization

1 min.   5. Summary and close

5. Tell the candidate when and how he or she will hear from the organization

4 min.   6. Write up evaluation

6. Write while impression is fresh

(From Joan W. Rossi, "Making Your Students Interview-Ready," *ABCA Bulletin*, September 1980, pp. 2–5. Reprinted by permission of Joan W. Rossi.)

5.4 Anticipating Questions

Of course there are too many possible questions to list here. Let's first focus on ten typical questions, and then look at five especially tough ones. There is no one right way for everyone to answer these questions that might come up, so you should rehearse your own answers to them.

Ten Questions Interviewers Usually Ask

1. What are your short- and long-range professional goals and objectives?
2. Why are those your goals, and how are you trying to achieve them?
3. What are your professional goals for where you want to be five or ten years from now?
4. Why did you choose this career?
5. Why should we hire *you*? What specifically can *you* contribute to our company?
6. What qualities should a successful (engineer, accountant, etc.) possess?
7. Why are you seeking a position with *this* company in particular?
8. Do you have geographical preferences for your work? Why?
9. What do you like best about your field? Why?
10. What major problems have you encountered in your life (or education, or work experience), and how have you dealt with them?

Five Especially Tough Questions

1. Suppose X happened. How would you deal with it? (X is some tough, work-related problem, typically a personnel problem.)
2. How well do you work under pressure? Give examples.
3. Which is more important to you: the amount of your salary or the type of job?
4. Why aren't you married?*
5. When do you plan to have children, and how will that affect your job?*

Practice dealing with these and other foreseeable questions. They are often the ones that interviewers ask of each person interviewed and are thus inevitable bases for making comparisons among candidates.

---

* If you think these last two questions are of questionable legality, you're right. The problem is that they still seem to get asked. Including them here is not meant to be a contradiction of the principle that people should be hired and promoted based on their abilities—with equity, not prejudice—but rather an acknowledgment of the reality that people *do* ask those questions, and wise job applicants will decide *in advance* how to deal with them.

Remember that although you don't *have* to answer any question, failing to do so may jeopardize your chances at a job. Sometimes you can deal most effectively with a challenging personal question by laughing it off or responding lightheartedly. Remember, too, that many times an interviewer will ask such questions just to see how you deal with tough situations. That is, the content of your answer doesn't make as much difference as the way you deal with the whole situation—being on the spot or being asked a possibly impertinent or insulting question. Thus, *any* answer may be right, as long as it does not show you to be a person who cannot deal with tough situations.

5.5   Getting Feedback

It's almost impossible to get feedback from interviews that fail. If you can find a class where mock interviews are done, you have probably found the best way to obtain feedback from your own interview behavior. Studies that have been done of interview failures point to these ten common areas in which candidates fail.

Common Reasons Why Interviews Fail

1. Poor personal appearance.
2. Overbearing, overaggressive, conceited, or know-it-all behavior.
3. Inability to express self clearly.
4. Lack of planning for career, no purpose or goals.
5. Lack of interest and enthusiasm.
6. Lack of confidence and poise.
7. Failure to participate in activities.
8. Poor scholastic record.
9. Overemphasis on money.
10. Unwilling to start at bottom.

You can do an effective job of creating your own feedback. Right after each interview, while it's still fresh in your mind, jot down notes:

• What questions were asked that you hadn't anticipated or that you didn't handle as well as you might have?
• What did you fail to say that could have helped you?
• Did you do enough homework, and was it of the right kind?

Use your answers to help you prepare for your next interview. The next list contains a preparation checklist.

An Interview Checklist

1. Have you focused squarely on making the best possible presentation of yourself?
2. Have you taken sufficient care with your personal appearance?
3. Have you done enough homework so that you can approach the interview with a strong, positive attitude?

4. Have you rehearsed answers for the predictable questions? for possible tough questions?
5. Do you have your own script for the interview, and are you prepared to be sensitive to (and adapt to) the interviewer's script?

Finally, although you almost never get feedback from an interviewer, be sure to give it. Send a courtesy letter to the interviewer, thanking him or her for the time and trouble. Remember not to do business overtly in the letter; it is designed primarily to build goodwill.

6. *Finding Jobs Other Ways*

More jobs are acquired through personal contacts than through letters (although even personal-contact jobs usually involve letters at some point). How can you place yourself in a better position for such jobs? First you need to have the personal contacts:

• Join the student affiliate of your major's professional association.
• Involve yourself in internship, co-op, or volunteer programs connected with your chosen profession.
• Become active in your school's government; make yourself visible.
• Get to know your teachers, and let them know you.
• Keep those professional relationships in good repair; don't be in the position of looking someone up after a two- or three-year total lapse in communication to ask for a letter of reference.

Once you have formed the personal contacts, let those people know you're looking for a job. Some students act as though looking for a job is something to be ashamed of; it's not. You didn't get fired—you graduated; people who graduate look for jobs. A nice way to let people know you're job hunting is to ask them to be prepared to be one of your references. You don't necessarily have to use that person's name on your résumé's list of references. What you're fishing for is the chance that your potential reference will know of an opening just right for you and suggest you use his or her name in applying for it. Another possibility for letting someone know you're beginning to look for a job is to ask that person for advice: How can you best enhance your credentials as a job seeker, and what are that person's suggestions for successful job seeking in today's market?

EXERCISES

➤ 1. *The Job Application.* This assignment has very little of the classroom in it. It is designed to simulate the real job-application process as closely as possible.

(a) Find a job notice advertising an opening for the kind of job you can legitimately expect to apply for after you graduate. Make a copy of the notice. Now find a piece of advertising aimed at the kind of person the job notice asks you to write to. Make a copy of the advertisement. Think about what kind of person the advertisement is designed to appeal to. Can you create a psychological portrait of that person? Now write a letter applying for the job; include the job notice and the copy of the advertisement with your letter.

(b) Write a one-page résumé to go with the job-application letter.

(c) Write a multipage résumé to go with the job-application letter.

(d) Write a qualifications sheet to go with the job-application letter.

(e) Assume that four weeks have passed since you sent in the letter and résumé and you have not heard anything about the job. Write a letter inquiring about the status of your application.

(f) Assume the identity of the person you wrote the letter to. Write a letter turning down the job applicant. Pay careful attention to human goals.

2. There are hundreds of books and articles that contain advice on how to write job applications and résumés. Find five to ten such pieces, and write a brief report summarizing the similarities and differences in their approaches. Your teacher or a librarian can help you with ideas for places to look.

3. If your university has a job placement center, pay it a visit and learn the way the staff there recommends that students go about job searches. Ask specific questions, such as how many students with your major get more than one offer, how many interviews the typical student in your major has, what the biggest things to look out for in the process are, and so forth. Write a brief (300-word) report describing what you have learned.

4. Get the names of two or three students in your major field who have graduated in the last year or two. Contact each one and conduct a brief interview about his or her experiences in the job market and what lessons may be learned from those experiences. Write a brief report (150 words) detailing what you have learned.

5. Critique the following job-application letter, based on the job notice provided above it and on the information provided in this chapter.

A ➤ indicates a case study exercise.

Dear Sirs,

    Your ad in the <u>Collegeville Times</u> caught my eye as I was
browsing the Sunday paper. I am a senior history major at
State University, with a minor in business. I have a couple
of Incompletes to make up from last term, but I hope to graduate this May.

    Although my major isn't business, I have some experience
in retail sales. Each summer for the last three years I sold
general merchandise for the Woolworth's store downtown. Now,
as I approach graduation, I am looking to begin a career in
retail sales, and I feel certain computers are a growth area.

    My other interests include skiing and running. In addi-
tion, I am treasurer of State University's student government.

    I look forward to hearing from you soon.

Sincerely,

Lee Wilson

Lee Wilson

PART THREE

Elements of Professional Writing

To a layperson's eye, the heading of "reports" might well cover most of the writing that professionals in business, industry, government, science, agriculture, and technology do. Within that grouping appear documents that are nearly as different as day is from night—from multimillion-dollar proposals to routine quarterly progress reports, from audit reports to annual stockholders' reports. (See the Part Five introduction on page 442 for a full list of kinds of documents.) All of these documents tend to have certain physical elements in common, however, and those elements are the subject of Part Three.

- *Abstracts* are only occasionally a part of papers that undergraduate college students write; *executive summaries* are rarer still. But few professional reports have neither, and many have both. Chapter 7 explains what abstracts and executive summaries are, how they are different, when to use them, and how to write them.

- Most report-style documents in professional life also have separate *introductions* and *conclusions*. Many of these documents have multiple introductory elements—a letter of transmittal, an executive summary, an abstract, and an introduction. Chapter 8 presents one particularly effective way to write an introduction and fifteen different ways to write conclusions.

- *Headings* and *subheadings* comprise perhaps the single most important visual organizer in professional reports. Not only do headings help cue readers into more efficient reading, but headings also encourage writers to organize their writing in a logical, orderly fashion. Chapter 9 explains the benefits of headings and subheadings in detail, how to use them, and how writers sometimes abuse them.

- Surely the most striking difference in appearance between the writing students do and the writing professionals do is the presence of *visuals* in professionals' writing. School writing is often 100 percent verbal— page after page of nothing but words. Professional writing is often 50 percent verbal and 50 percent visual. Something designed to teach people how to perform a task may well be 80 to 90 percent visual—and the visuals will properly come *before* the words. Visual elements of professional writing may be tabular, graphic, structural, or representational. Chapter 10 explains when and how to use each variety of visual— what each type does best and what it does worst.

Are these important physical elements the essence of what makes the writing of professionals different from the writing of students? No, not really. Each of these four elements has an important function, that's true, but it's still the content, style, and organization of what you write that really marks you as a student or a professional. Still, readers see your content, style, and organization first and foremost through the lenses of these four elements. If the lenses aren't well made, the resulting images won't be clear.

7. Abstracts and Executive Summaries

In professional life, nearly every report-style document you write will be introduced by an executive summary or an abstract (or both). The abstract or executive summary is often the first thing your reader sees; sometimes it's the *only* thing. The decision to read or not to read your report may well be made on the basis of this material.

Abstracts are rarely more than a page in length and often only a paragraph. Many people believe that an executive summary should not be longer than a page or two to be effective. Despite the brevity, your executive summary or abstract strongly affects key decisions readers make about your report—in particular, how much and what kind of attention to give it. Thus, the ability to write an effective abstract or executive summary is especially important.

1. Abstracts

Abstracts and executive summaries are brief overviews of the contents of a report. Executive summaries will be discussed later in this chapter. We first take up the simpler form, the abstract.

1.1 What Are Abstracts?

There are two main types of abstracts: *informative* and *descriptive*. Confusingly enough, the informative abstract is often also called a "summary"; that will not be done here. The informative abstract tells the reader what the report says. The descriptive abstract names the topics the report covers but not necessarily what the report says about them. The informative abstract's point of view is *internal*, as though the abstract itself is talking to you. An informative abstract of Chapter 12, "Writing Definitions and Descriptions," might begin, "There are fourteen different definition-and-description techniques," and then go on to list and possibly to explain them. The point of view of the descriptive abstract, on the other hand, is *external*, as though someone has read the report and is telling you about it. The descriptive abstract of Chapter 12 might begin, "This chapter lists and explains fourteen different definition-and-description techniques," but the descriptive abstract might well *not* go on to explain what those techniques are. Of the two forms, many people find the informative abstract more useful. Perhaps the major use of the descriptive abstract today is in indexing (discussed later in this chapter).

To make it easier for you to see the differences between the two forms, the list on page 186 compares the basic features of informative and descriptive abstracts, followed by an example of each.

1.2 Why Are Abstracts Important?

Abstracts have important functions for both writers and readers. Writing an abstract of a report you have just completed helps you to clarify in your mind what the important parts of your report really are. What do you want your reader to see while deciding whether to read the whole report? Like a topic outline of the report (or its sequence of headings and subheadings), the abstract helps you, the writer, test the organization of your report.

Abstracts are especially important for readers. First, they help the reader grasp your report's organization. Second, because reading entire reports takes more time than many professionals are willing to spend, anything you as a writer can do to help your reader make a good decision about whether to read your whole report, read only a part of it, skim it, or send it to someone else to read will usually increase your reader's goodwill toward both you and your report. Often the abstract is a major factor in that decision.

Informative Abstract	**Descriptive Abstract**
Tells readers what the report says	Names topics the report covers
Uses internal point of view: "Fifty pine seedlings were germinated under two different sets of controlled conditions . . ."	Uses external point of view: "This report presents results of two experiments on fifty pine seedlings . . ."
Useful for all readers	Useful mainly for indexers and for index users

Example:	*Example:*
Fifty pine seedlings were germinated under two different sets of controlled conditions to determine the effects of acid precipitation on germination rates. Results, though limited by the size of the sample, suggest that other factors (such as the robustness of the seedling stock) may play relatively bigger roles than that played by acid precipitation.	This report presents results of two experiments on fifty pine seedlings designed to determine the effects of acid precipitation on germination rates.

On a deeper level, a good abstract will help your reader read your report faster and understand it better. When people are given brief, accurate forecasts of what they are about to read, their reading is more efficient. If your abstract is well written, gives an accurate sense of your report's content, and addresses your reader on a level the reader understands, it will help him or her read the report more quickly and comprehend it better and more easily.

1.3   When Should Abstracts Be Used?

You should use an abstract at the beginning of any report-style document you write. Place it on a separate page, double or triple spaced, and clearly labeled. (If it's on the cover page, it will be single spaced.) An abstract can be as short as one or two lines or as long as a page.

Many people also use abstracts to keep up with current developments in their field. Many journals and other organizations solicit and circulate abstracts of completed work (or work in progress) to inform their readers or members of work they need to know about. This use is one of the reasons that your abstract must make sense as a separate document. It is more than a miniature of your report; it is a miniature report, self-contained and able to stand on its own. Envision your abstract published with 100 other abstracts, and ask yourself whether it still would present your report accurately and effectively.

Abstracts are often used as parts of proposals; they are often also used by themselves as miniproposals, as a way of describing a project you would like to undertake (such as a report you would like to write). In that situation it's important to know whether you are expected to submit a descriptive or an informative abstract. The descriptive abstract can be *speculative*. In it you can, for example, promise to cover a great number of topics even though you may not know exactly what you will say about each one at the time you write it. But the informative abstract requires you to do more than just summarize the points you intend to cover; you must also state the conclusions you intend to reach about those points. Thus, the informative abstract requires you to know what points you are going to talk about *and* what you are going to say about them.

1.4 How Are Abstracts Used in Research?

One of the most important sources of information today is the computerized indexing service. Private companies (and the federal government) collect abstracts of published reports, speeches, and all kinds of research, and organize (or *index*) them in computerized data bases so as to make them accessible in a number of ways, such as by subject, combinations of subjects, and author. For a fee, potential users of that information can access that data base through computer and telephone links anywhere in the world and inquire what is available on a certain subject or subjects, or by certain authors, and so forth. The computer will then send any abstracts it holds in the requested category, and the user will screen them and decide which reports are worth acquiring. Or users can be placed on regular mailing lists to receive paper copies of abstracts on a certain subset of the larger topic. The users then screen those abstracts and decide which reports (if any) they need to order copies of.

For example, the Office of Scientific and Technical Information, the national information center for the Department of Energy, collects roughly 7,000 reports a month. One of its main functions is to maintain the Energy Data Base (EDB), a vast collection of bibliographic citations, abstracts, and indexes to world literature on energy. The EDB contains more than 700,000 entries, and it adds about 150,000 entries each year. For most users, getting a listing of all the information newly indexed in any particular month would mean getting much too much information. So the Office of Scientific and

Technical Information subcategorizes the reports it receives into many different groupings (such as fossil energy, fusion energy, solar energy, etc.) and subgroupings, and then mails out regular summaries of the work done in each area to people who request lists for those areas. An authorized user can also receive more extensive summaries, but for most applications the specific, subcategorized listings are quite sufficient.

This kind of compilation of sets of information into various data bases goes on in every field. It has created a totally new research technique, the Boolean search, described in Chapter 16, one that takes advantage of computers to scan quantities of information beyond anything any one person could hope to digest on his or her own without the computer. Because of the popularity, speed, and comprehensiveness of this research technique, abstracts have become even more important in research than they were before computers revolutionized the growth of scientific and technical information. The abstract becomes the primary means by which readers select which reports to read and which ones not even to look at.

1.5   What Are the Qualities of a Good Abstract?

From the discussion in this chapter so far, we can see that a good abstract needs to satisfy at least three criteria; it must

- give an accurate sense of the report's content,
- address the reader on a level he or she can understand; and
- make sense as a separate document.

The fourth criterion a good abstract must satisfy is that it must

- be well written.

A well-written abstract has the same qualities of style that characterize all effective professional writing; clarity, economy, and straightforwardness (Chapter 3 explains these qualities in more detail). *Clarity* here means choosing the exact words you need to get a precise meaning across to your specific audience. It also means avoiding the telegraphic style that some writers fall into. That is, avoid writing, "Report summarizes results of experiments conducted fall, 1991, in game species' native habitat"; instead write "This report summarizes the results of experiments conducted in the fall of 1991 in the game species' native habitat."

Economy in abstracts requires that you be careful about the details you include and exclude and that you be especially careful about using economical sentence structures. Most readers have an intuitive feeling (reinforced by experience) that abstracts should be no longer than one page (usually about 200 words). That means you must choose your words and expressions very carefully. (You may want to review the section on "Wordy Expressions" in the Appendix when you get to the point of tightening up the wording in your abstract.)

   Straightforwardness means that both the sequence of elements in each sentence and the sequence of sentences must be in the order that is easiest for the reader to read, which is not necessarily the order that is easiest for the writer to write. You cannot merely pull one sentence from each major section of your report and string them together into an abstract. You must look at that sequence of sentences as a separate work and ask yourself if that sequence, by itself, will make sense to a reader. Figure 7.1 shows a poorly written descriptive abstract. Figure 7.2 shows the same abstract revised for clarity, economy, and straightforwardness.

Figure 7.1   A Poorly Written First Draft of an Abstract
Notice the telegraphic style, the weak verbs, and the way the last sentence comes in almost as an afterthought.

"Using Research Libraries" (Chapter 16 of *Effective Professional and Technical Writing* by Michael L. Keene.)

Chapter is description of library research process. Preliminary planning, card catalog, periodical indexes, government documents, and technical reports are included. Both research and writing of reports are covered.

Figure 7.2   A Well-Written Abstract
Notice the way the style, ordering of elements, and choice of verbs have been improved over Figure 7.1.

ABSTRACT

"Using Research Libraries" (Chapter 16 of *Effective Professional and Technical Writing*, by Michael L. Keene.)

This chapter describes the process of library research, including record keeping and report writing. It describes preliminary planning, using the card catalog, searching periodical indexes, accessing government documents, and finding government documents.

1.6   How Are Abstracts Written?

The easiest way to write an abstract is to work from an outline of the report. If the report makes extensive use of headings and subheadings, they can be the beginning of your abstract. Of course, if you wrote the report yourself, you may well have an outline already available to work from. If you are writing an abstract of someone else's report, the best way to proceed is to reread the report carefully, underlining key sentences and making notes to yourself in the margins. Of course it's easier to write a descriptive abstract than an informative abstract, because the descriptive abstract requires explaining much less about the report's meaning.

Figure 7.3 presents a short article, with key lines underlined in preparation for writing an abstract of it; Figure 7.4 is an outline made from the underlinings; and Figures 7.5 and 7.6 show informative and descriptive abstracts written from the outline.

Although abstracts take relatively little time to write and very little space to print, their importance continues to increase as society grows more and more saturated with scientific and technical information and more reliant on computers.

2.   Executive Summaries

If you're writing a problem-solving report—someone is expected to make a decision based on what you write—consider putting an executive summary at the beginning of the document. (Just as some people use "summary" to mean "informative abstract," others use "summary" to mean "executive summary.") Often there are people who want to see only the executive summary. Perhaps the typical use of executive summaries, however, occurs when those in top management read the executive summary and then—if what it says appeals to them—hand the full report to someone on the staff with instructions such as, "Read this report carefully, and see if it bears out what the executive summary says." That is, *management* reads the executive summary and is interested in the *policy* it offers, but first there needs to be a *technical* review by *professional staff.*

Because your executive summary will be the first part of the report your boss reads, and may in fact be the *only* part of the report your boss reads, you need to know about executive summaries: what they are, when you should use them, and how you should write them.

2.1   What Are Executive Summaries?

An *executive summary* is a brief description of a report's contents that focuses on their significance for the report's readers. The executive

Figure 7.3 An Article to be Outlined and Abstracted
Original article.

CRITERIA FOR EVALUATING BINOCULARS

Optical Criteria

To estimate optical quality, examine binoculars for each of the following features. Every pair of binoculars will have particular strengths and weaknesses. Focus on an object with fine parallel lines. Do not look through a window. The criteria do not include interpreting the specifications—things like magnification and objective lens size.

Brightness of Image

A bright image indicates how well the optics gather and transmit light. The lenses and prisms should be coated on all glass-to-air surfaces—100 percent coated.

Resolution of the Image

Look carefully through the binoculars for several kinds of defects in their ability to form a clearly resolved image. Defects in resolution create serious problems in long-term enjoyment of the binoculars in the field. The more severe the defects, the less reliable the optical quality of the binoculars.

Edge-of-field defects. There will always be some lack of resolution around the edges of the field of view, even in sharply focused glasses. But margins should be as sharp as possible, like the center of the field. Edge-of-field defects are a ready guide to relative quality in the optical system of binoculars.

Pincushion distortion. Parallel lines crossing the image may appear to curve slightly toward the center.

Curvature of the image. A flat surface, such as a wall across the street, may appear concave, or slightly bowl-shaped.

Spherical aberration. Caused by imprecise lens grinding, spherical aberration.

Range of Resolution

Excellent optics will seem to go in and out of focus very slowly. This means the optical system has a wide range of resolution. Since binoculars with a wide range of resolution do not have to be perfectly focused to be useful, they are more useful and less aggravating in the field.

Alignment

The images formed by the two barrels of the binoculars should merge imperceptibly into one unified image. Watch especially for a shadow down the center of the image and for vertical misalignment.

Eye Relief

A full field of view should be afforded when the ocular lenses are placed to your eyes or eyeglasses. If dark shadows blot part of the field and adjustments in the distance between lenses and eyes are required, eye relief is at fault. Eyeglass wearers should rigorously test the retractable cups on binoculars for eye relief. Quality in eye relief for eyeglass wearers varies considerably. Look for a full field of view.

Construction and Casing

Axis Hinge

A good gauge to the construction of the casing, the hinge should be strong and work smoothly; it should be neither too loose nor too stiff. Compact binoculars often employ double hinges of various design. Double-hinged compact binoculars may be tiring to hold for long periods of time, and it may be difficult to maintain proper alignment.

Focusing Mechanism

A center-focus mechanism is more convenient than the old-fashioned individual focusing mechanisms for each barrel. A well-designed focusing mechanism should be conveniently located, easily reached without having to look for it, and smoothly adjustable.

Balance and Comfort

In well-designed binoculars, weight, size, and balance combine to make an instrument that fits comfortably in your hands and is enjoyable to feel.

Source: "Criteria for Evaluating Binoculars" from "The Glass of Fashion" by Charles A. Bergman from *Audubon* magazine, November 1981. Reprinted by permission of the author.

Figure 7.4   A Scratch Outline of the Article in Figure 7.3

CRITERIA FOR EVALUATING BINOCULARS

I. Optical Criteria
   A. Brightness of the image--indicates how well the optics gather and transmit light.
   B. Resolution of the image--possible defects:
      1. Edge-of-field defects
      2. Pincushion distortion
      3. Curvature of the image
      4. Spherical aberration
   C. Wide range of resolution--goes in and out of focus very slowly
   D. Alignment--images formed by each barrel merge imperceptibly into one
   E. Eye relief--full field of view
II. Construction and Casing
   A. Axis hinge--strong and smooth-working
   B. Focusing mechanism--center focus, well designed
   C. Balance and comfort--weight, size, and balance combined

summary is under no obligation to mention all the parts of a report or a research project, to cover elements in the same order as does the report, or to cover elements in the same proportions. The obligation is to answer the questions decision makers will ask about its subject: What's the problem? What are its adverse effects? What should be done? What are our benefits from doing it? Because of that obligation, the writer of an executive summary must know more about the report than just its contents, especially the report's audience and the report's purpose. Only then can an effective executive summary be written.

Here are two examples of short executive summaries. The first is from a student paper, "A Recommendation for Improvement of Current Waste Disposal Patterns" (Fig. 7.7). The second is modeled on one from a proposal submitted by one department of a hospital to the National Cancer Information Service (Fig. 7.8).

Figure 7.5   An Informative Abstract of the Article in Figure 7.3

Although the following abstract was written from the outline in Figure 7.4, you may notice that the abstract does not include all of the outline's details, which would have made the abstract much too long.

> "Criteria for Evaluating Binoculars," from "The Glass of Fashion" by Charles A. Bergman, *Audubon* magazine, November 1981, p. 77.
>
> Binoculars can be evaluated on the basis of optical criteria, construction, and casing. Optical criteria include the brightness of the image, the clarity of resolution of the image (absence of edge-of-field defects and other distortions of the image), the range of resolution of the image (it should go in and out of focus very slowly), the alignment of the two images the barrels provide into one image, and the fullness of the binoculars' field of view. Criteria for construction and casing require that the axis hinge be strong and smooth working, the focusing mechanism be a center focus, and the weight, size, and balance combine to provide good balance and comfort.

Figure 7.6   A Descriptive Abstract of the Article in Figure 7.3

Compared to the informative abstract, the descriptive abstract is shorter, easier to write, and contains much less information.

> "Criteria for Evaluating Binoculars," from "The Glass of Fashion" by Charles A. Bergman, *Audubon* magazine, November 1981, p. 77.
>
> This article gives two broad sets of criteria for evaluating binoculars: optical criteria and construction and casing criteria. The article gives five different optical criteria and three different criteria for construction and casing.

Figure 7.7   An Executive Summary from a Student Paper
The 20-page report this executive summary introduced was submitted in a senior-level professional writing class. The audience the report was aimed at comprised city council members who had asked for a preliminary recommendation report concerning whether they should create a committee to investigate formally the creation of an incinerator that would supplement or replace the city's current system of waste disposal.

```
          Improvement of Current Waste Disposal Patterns:
                        Executive Summary

     In the last five years, the population of Collegeville has
     grown by 12 percent. Over the same time period, the amount
     of solid waste disposed of in Collegeville landfills has
     risen 20 percent. Even at the current rate of disposal,
     all available facilities will be 80 percent full in ten
     years. If the rate of disposal continues to accelerate as
     it has for the last five years, facilities will be full in
     ten years.
          Each of the last three landfill sites purchased by Col-
     legeville has been farther out of town than the previous
     one. The Possum Valley Landfill, the most recently pur-
     chased one, is 30 miles outside of town, making a 60-mile
     round trip for every truck that goes there. And 80 trucks
     go there every day. New locations that are nearer are un-
     available, and, as surrounding towns and counties exercise
     more and more control over their own environments, it is
     increasingly doubtful whether any new landfill sites will
     be found, at any feasible distance. Thus, Collegeville is
     facing, within ten years, a solid waste disposal crisis of
     major proportions.
          A number of options exist for dealing with Col-
     legeville's impending solid waste disposal problems: re-
     cycling, source reduction, and incineration. The combina-
     tion of recycling and source reduction can be implemented
     relatively easily and inexpensively. Both methods, how-
     ever, require phasing in because they depend on changing
     householders' attitudes and merchants' practices. Even
     working in combination, recycling and source reduction may
     produce only a 20-25 percent change. The only long-term so-
     lution to solid waste disposal, then, is incineration.
     With current technology, an incinerator that takes two
     years to plan and two years to build will be able to han-
     dle Collegeville's solid waste disposal for approximately
     30 years (i.e., the life of the incinerator).
```

Incineration has a number of advantages over an endless (and possibly futile) search for landfills, especially landfills that require longer and longer hauling. The incinerator's major cost is up front and can easily be handled with a bond issue; continuing costs of operation are almost totally covered by sales of by-products and recycled materials. Landfills present more environmental problems every year, while the incinerator is tightly regulated from day one and hence environmentally preferable. Today's "fluidized bed" incinerators burn waste at such high temperatures that all that's left is a small amount of ash, most of which gets sold for use in highway construction. With the choice of an incinerator, then, Collegeville begins to put an end to its landfill problems. By combining incineration, source reduction, and recycling, Collegeville can handle its solid waste disposal problems in a financially practical and environmentally safe way for years to come.

2.2   When Should Executive Summaries Be Used?

Executive summaries are used in writing for decision makers—writing in which you are proposing a solution to a problem and asking readers to endorse that solution. In terms of this book's chapters, executive summaries would be expected for proposals, problem-solving reports, and recommendation reports. In terms of student writing, a term paper entitled "Design of a Car to Maximize Ethanol Fuel Economy" might well use an executive summary; a term paper entitled "Artificial Intelligence Applications in Engineering" wouldn't need one.

2.3   How Are Executive Summaries Written?

In preparing an executive summary to be placed at the beginning of a relatively short (say, 20-page) problem-solving report, think in terms of a four-paragraph format:

Figure 7.8   An Executive Summary from a Professional Proposal
This executive summary was written to accompany a 300-page proposal origi-
nating with University Hospital and addressed to the National Cancer Institute (NCI), which had announced that several grants of up to one million dollars each were available to establish or expand local and regional cancer information hot lines.

The Mid-State Cancer Information Service:
Executive Summary

The five-county area surrounding Collegeville and served by University Hospital currently includes no NCI-funded cancer information
services. Yet the role of education and communication in promoting
cancer prevention and cancer survival has long been recognized. As
people know more about subjects like life-style changes, cancer
screening, up-to-date treatment options, and support services, their
chances for disease-free outcomes are significantly improved. Five
million Americans who have been diagnosed with cancer are
alive today.

The over 600,000 people in the five-county Collegeville region
have higher than average risk factors, incidence rates, and mortality
rates for certain cancers (including lung cancer and colon cancer).
These counties contain significant populations that are very rural,
are 65 years of age or older, are African-American, are poor, have
completed fewer years of schooling than national averages, and are
unemployed. Each of these factors increases cancer risk.

In its one year of existence, University Hospital's Cancer Hotline
has answered over 2,000 calls and sponsored over 100 cancer information and education activities in Collegeville. This proposal seeks
funding to expand those activities to serve the entire five-county
area. The aim is to provide education that will lead to early diagnosis and prompt treatment of cancer, and thus to help move the five-
county area toward NCI's Year 2000 goal of a 50 percent reduction
in cancer.

The University Hospital Cancer Hotline currently serves one
county of 250,000 people. Expansion of the service into the five-
county area will enable us to serve approximately 600,000 people
from one central site as the Mid-State Cancer Information Service.
With additional funding from this grant, the experienced and capable staff can make a significant contribution to realizing NCI's Year
2000 goal.

- **What is the problem?** Here you need to know how much your readers already know. Do you need to go back over the territory in detail or just hit the high points? At a minimum, you should hit the high points—give the most striking numbers, tell the one story that exemplifies the situation. In this section, you're not just setting out background; you're also defining the situation in such a way as to make clear that action needs to be taken.

- **Why does it need to be solved?** Here you explain the adverse consequences of maintaining the status quo. But you're doing more than accomplishing what a speechwriter might call an "urge to action"; you're also describing the negative consequences in such a way that when you propose your solution (in the next section), it's obvious to any reader how that solution prevents these consequences.

- **How should it be solved?** The solution you propose is not *your* solution; it is the *right* solution or the *company's* solution. It needs very clearly to fit the problem pattern you set out in your problem definition. If the problem had three parts, the solution needs (at least) three corresponding parts, and so on. The way the solution is described also needs to make clear how it heads off the negative consequences you described earlier. If there are other, competing solutions, you probably won't have time to dispose of them here, but you can certainly state that on examination (as detailed in the report proper) this solution is the one that needs to be implemented.

- **What are our benefits for enacting this solution?** You may well begin these benefits with the most immediate (often also the smallest) and end with the most distant in time (often also the biggest). But you only want to focus on the major benefits here. Although you have in fact been selling your solution all along, this section is where you make your point. If you have written the three previous sections properly, this section will seem to follow easily and logically from what they say. If it doesn't, your executive summary may not be as effective as you want it to be.

Because any executive summary will inevitably get a very wide circulation, you need to write yours clearly, carefully, and thoughtfully. This one- or two-page piece can make or break your whole project, so give it the time, effort, and consideration it deserves. Make sure other people have seen it and given you equally careful feedback on it. If you can get your executive summary informally reviewed by someone at the level of management it is written for, so much the better.

EXERCISES

1. Write a descriptive abstract and an informative abstract of Chapter 4 of this text. Assume students like you are the audience for each.

2. Write an executive summary of Chapter 4 of this textbook. Assume students like you are the audience.

3. Write an informative abstract of an article in *Scientific American* specified by your teacher. Your teacher will also specify the audience.

4. Using the resources of your school library, write a short report for your teacher on how a person in your field can find abstracts of research articles in your field without having to look up the articles themselves.

5. If your class requires a major report, write an executive summary and a descriptive abstract of it, based on your current knowledge of your subject. It is understood that this summary and abstract will be tentative and speculative and that you do not know exactly what your report will say until you finish your research.

➤ 6. Write a 50-word descriptive abstract of the section of the *Value of the Energy Data Base* report presented below. Assume the audience is a taxpayer for whom you are preparing a report on how DOE-spent money is justified (or not justified). Then write a 150-word descriptive abstract on the same subject for the same audience.

➤ 7. Write an executive summary of the portion of the *Value of the Energy Data Base* report. Assume your audience is a group of taxpayers who want an explanation of why their tax dollars are being spent on that kind of information activity. Try for a length of no more than 150 words.

The example in Figure 7.9 comes from a report entitled *Value of Energy Data Base*. The report was written by King Research, Inc. (Rockville, Maryland), under subcontract to Maxima Corporation, for the Technical Information Center (now the Office of Scientific and Technical Information), the national information center for the Department of Energy. The purpose of the report was to assess the dollar value of the Energy Data Base (described earlier in this chapter; see Section 1.4). The report's primary audience was the individuals in the federal government who make funding decisions about the continuation of the Energy Data Base; its secondary audience was the scientists and technicians who use the data base; its tertiary audience was people who are interested in the growth of the information industry. The portion of the report shown in Figure 7.9 contains a figure and a table not printed here; this version is shorter by about one-third. If you want to see the whole (81-page) report, your librarian should be able to find it for you; its document number is DOE/OR/11232-1 (DE82014250).

A ➤ indicates a case study exercise.

Figure 7.9   One Portion of *Value of the Energy Data Base*

The U.S. Department of Energy currently expends about $5.8 billion annually on research and development in the program areas of defense, nuclear science, basic research, and others. The return on this investment to the nation is basically achieved through the accomplishment of the specific goals and objectives of the R&D and through the use of the knowledge gained from the R&D. There is abundant evidence that energy information plays an important role in current research and development activities. A survey of the 60,000 scientists and engineers funded by the Department of Energy shows that annually they read* about 7.1 million journal articles and 6.6 million technical reports. The total of 13.7 million readings includes 2.5 million by researchers in the defense program area, 2.2 million in the nuclear area, 3.0 million in basic research program areas, and 6.0 million in other program areas.

Clearly, use of existing information saves researchers considerable time and effort. In a survey of DOE-funded scientists and engineers, many indicated that a recent reading of a technical report or article led to a savings of time and/or equipment. For example, from reading a report on steam electric plant construction costs and production expenses which make it unnecessary to repeat the report's calculations, a nuclear scientist reported savings of about $1,000.

The total annual savings attributable to reading by DOE-funded scientists and engineers is estimated to be about $13 billion. This is one estimate of the consequential value of primary energy information found in articles and technical reports. An estimate of the apparent value, or the cost of the 13.7 million readings, is $500 million. One could look at the consequential value from the standpoint of a return on investment. The Department of Energy annually expends about $5.8 billion on research and development. Of that amount, about $500 million is expended in information processing and use, and the remaining $5.3 billion is spent on other research-related activities, which generate information. Thus,

Generation of Information	+	Information Processing and Use	→	Future Saving to DOE Scientists
$5.3 billion	+	$500 million	→	$13 billion

---

* "Reading" means examining an article or technical report beyond its title.

This suggests that an investment of $5.3 billion in the generation of information and about $500 million in processing and using information yields a partial return of about $13 billion in terms of savings to scientists and engineers in their time and in equipment. Overall, this partial return on investment is about 2.2 to 1. One way of expressing this relationship is that the DOE paid $5.8 billion for the research, information processing, and use, which in turn has been found to be worth at least $13 billion.

The DOE Technical Information Services program managed by the Technical Information Center (TIC) is an Energy Program that has the responsibility of managing the information products from the multibillion-dollar R&D program and maximizing their use by Department staff and contractors. For example, TIC helps increase use of technical reports by making copies available in paper copy and in microfilm. It is estimated that reports distributed by TIC are read [an aggregate of] 6.8 million times by DOE and non-DOE researchers. If $1,280 is the average savings value derived from each reading, then this service has a potential of yielding savings of over $8 billion to the overall energy community. TIC also enhances the use of information through systems that provide effective access to energy-related technical reports, journal articles, and other materials produced worldwide. The major resource containing access information to these materials is the Energy Data Base (EDB). In 1981 it was estimated that the cumulative amount of the international R&D investment represented in the EDB was over $139 billion. The savings value derived by DOE researchers is enhanced by the ability to carefully select relevant items for reading from this data-base resource. A number of TIC products and services are related to the EDB and aid this selection process. These include the published index, *Energy Research Abstracts* (*ERA*), and the on-line bibliographic retrieval system, RECON. In fact, it was found that for DOE researchers, there are about 70,000 searches performed annually using RECON and an estimated 244,000 searches performed annually using *ERA* and other TIC-produced indexes. About 2.6 million of the total 13.7 million readings of energy articles and technical reports are directly attributable to use of EDB secondary information in some form. DOE scientists and engineers also conduct searches using other on-line systems and printed indexes, making a total of 4.6 million readings derived from bibliographic searches.

In determining the value of the Energy Data Base, only those searches and readings that are directly attributable to the EDB are included in the analysis. There are a total of 1.0 million readings from 70,000 RECON on-line searches, 600,000 readings from 40,000 on-line searches of the NTIS data base, and 1.6 million readings from 244,000 searches of *ERA* and other TIC indexes. The values directly attributable to the EDB are given in Table 1.1. [Table not shown here: value to searchers = $20 million; value to readers = $117 million; value to DOE organizations/funders = $3.6 billion.] The numbers given do not include the value associated with reading of technical reports received on standard distribution from TIC. These 3.3 million readings are associated with an additional $122 million in reading value, for a total of $239 million, and $4.2 billion in value to DOE, for a total of $7.8 billion. Of the total funds expended for energy information, the values directly attributable to TIC products and services are 43 percent of the total for searchers, 48 percent for readers, and 60 percent for the funding organization (DOE).

For those funding EDB searchers, one could evaluate searching by treating the $20 million amount paid by searchers as cost and the consequence in readings as benefits. Apparent value of the reading directly attributable to EDB searching is $117 million. The benefit-to-cost ratio of EDB searching is thus 5.9 to 1, and the net value is $97 million. Similarly, evaluation of reading would compare the EDB reading costs of $117 million against the $3.6 billion savings, which results in a benefit-to-cost ratio of about 31 to 1, or a net value of $3.5 billion. For all reading of energy materials, total costs of $500 million can be compared against the $13 billion savings, which results in a benefit-to-cost ratio of 26 to 1 and a net value of $12.5 billion.

The way in which we calculate the apparent and consequential values of EDB products and services is to determine the direct effects of the searches that would be lost by substituting other products and services and assuming a fixed total budget. There are three types of secondary products and services that use the EDB information: RECON, ERA, and other printed indexes; other on-line services, such as Lockheed Dialog and others, that have the NTIS data base (which includes many EDB items); and BRS, which provides access to the EDB. By dropping all these products and services and substituting others, there would be 354,000 fewer searches and therefore 2.5 million fewer readings. This results in the reduction of searcher apparent value by $15 million, consequential value to readers by $90 million, and the value of savings in time and equipment by $3 billion.

The latter figure, $3 billion savings in labor and equipment from the EDB products and services, can be roughly translated into productivity. If it is assumed that, with the EDB, research and development costs $5.8 billion for a given level of output, without the EDB the same output would require an investment of $8.8 billion. This is an increase in productivity of about 52 percent. This says that to accomplish the same R&D output without TIC information services, the R&D budget would have had to have been $3 billion higher.

Value assessments were also conducted separately for five major research areas funded by DOE: defense; nuclear science; and basic energy science, fusion, and health and environmental research, components of the basic research program. Separate reports are provided elsewhere for each of these areas with summary results given in this report. Differences were found in the level of reading by researchers in the different program areas, in the methods used to identify materials, and in the value associated with reading. Average annual researcher's readings of technical reports and articles ranged from 203 in the defense area to 276 in the fusion era. Printed index searching was used more heavily in the areas of nuclear, health and environmental research, and on-line searching in the basic energy sciences. Fusion researchers made greater use of standard distribution copies of reports from TIC. Value associated with report reading was estimated to range from $930 in the health and environmental research area to $1,840 in the fusion era. Looking at individual research budgets within the program area, we see that increases in productivity as a result of EDB products and services were calculated as ranging from 24 percent for health and environmental research to 94 percent for nuclear research programs.

Source: Ten paragraphs from *Value of the Energy Data Base,* King Research, Inc. (Rockville, MD) for the Office of Scientific and Technical Information, Dept. of Energy, DOE/OR/11232-1 (DE82014250). Reprinted by permission.

8. *Introductions, Transitions, and Conclusions*

Any piece of writing longer than one page should make special provisions for its readers simply because of its length. This requirement is especially true in professional writing, where writers put a large amount of content into a small number of pages and readers tend to be busy, tired, and frequently distracted. If you want a reader to go beyond page one, you must make some special efforts in your writing. Some of these you have seen in previous chapters (paying attention to the *real* audience and employing "you" attitude). This chapter begins with the most important part of any piece of writing: the introduction. If your introduction shows special concern for your readers, the whole report stands a much better chance of being read. The second part of this chapter shows you how to write one kind of transition that is especially important in professional writing. The third part of this chapter explains conclusions, another special provision that long reports make for their readers.

Here you will learn what makes a good introduction, transition, or conclusion, how to write a good introduction, and how to adapt your introduction to your specific audience.

1. *Introductions*

Haven't there been times when the first few paragraphs or pages of something put you off so much that you decided not to read the rest—or, at most, grudgingly and hastily to skim it? The same is true of readers in business and industry. They often have a deskful of other business, ringing telephones, and knocks on the door to attend to, a hundred distractions inviting them away from your work. If your introduction does not lead your working reader into the report and implicitly promise to make the content easy to understand, you must expect your words to remain unread.

If your introduction is too long, vague, hard to understand, or too far above (or below) the reader's level of knowledge or interest, you are saying to the reader, "I don't really care very much about you." That kind of negative message, once sent, is hard to retract.

1.1   What Qualities Comprise Good Introductions?

Most successful writers seek to lead their readers from a general understanding of a subject to one particular point. For example, a report may seek to lead me from a general understanding of how forklifts are used in lumber yards to a specific recommendation to buy one particular model of forklift. Or the report may lead me from the need to remodel industrial buildings to take advantage of opportunities to save energy (a general understanding) to what should be done to one particular building (a specific understanding).

Many writers produce reports that dive right into their subjects, beginning with a recitation of facts, figures, and formulas. Those are, after all, the things that are right at hand when a professional begins to write. A good introduction, on the other hand, one that leads the reader from general knowledge into particulars, is generally not ready at hand; it has to be *made*. A good introduction does not consist of simply telling what you did or simply relaying the facts. A good introduction moves the reader from general awareness to specific points and by doing so unites the reader's purpose with the writer's.

Any good introduction does at least three specific things for the reader:

1. It *creates a context* shared by reader and writer.
2. It *clearly establishes the purpose* of the report.
3. It *forecasts the organization* of the report.

Create a Context. Creating a context for your report means meeting your reader at a level of knowledge and interest in the subject you both share. This need to establish some common ground between reader and writer at the paper's opening expresses a basic principle of communication: for two people to communicate, their lives must touch somewhere, somehow. If their two worlds do not touch, they do not communicate. You are aware of this whenever you try to establish a conversation with a stranger. The first impulse is usually to establish some common ground. "What's your home town [major, hobbies]?" are typical college conversation starters. Any answer that establishes some common ground usually leads to attempts to build on that ground. Suppose the two of you have the same home town. The next questions are, "What high school did you go to?" or "Did you know . . . ?" or "Who did you have for   . . ?"

These follow-up questions that attempt to build on the one point of shared ground lead to this theory's application to writing effective introductions: You can maximize your chances to communicate with your audience by establishing some common ground at the beginning of your report, and the more extensive that common ground is, the better are your chances to communicate effectively.

By making the most of whatever you have in common with the reader, you improve the chances for effective communication. By meeting the reader at his or her level of knowledge and interest in your subject, you become more able to lead the reader to the level you want in the report. To do this effectively, you have to gauge not only your reader but the professional community your reader is part of (scientists, technicians, government bureaucrats, engineers, and so on) and that community's values and attitudes toward your subject. (For more on professional communities as readers, see Chapter 17.)

The first quality of a good introduction, then, is that it creates a context that unites writer and reader at the level on which their different knowledges of the subject most nearly coincide. The other two qualities of a good introduction, establishing purpose and forecasting organization, also help ensure that writer and reader are coming into the subject with at least approximately the same goals.

Establish the Purpose. Sometimes your statement of purpose is easily accomplished, as in, "The purpose of this report is to explore the feasibility of expanding operations into the Houston area." But behind such an apparently simple statement, important issues are being decided.

The phrase, "explore the feasibility of" means that the writer is, at least initially, leaving open the question of whether operations should be expanded into the Houston area. This *open stance* may or may not be what the reader expects. If the reader has strong expectations of a report that takes expansion into Houston for granted and presents a detailed plan for that expansion, this report's tentative opening will make that reader unhappy. The reader's frustration may override anything else the report subsequently does, even if the report actually does present such a detailed plan. When the introduction doesn't establish a purpose that pulls reader and writer together into a common task, the reader may not read the report at all.

The opening statement of purpose gives the first explicit form to both (1) the nature of the reader's purpose and (2) its relationship to the writer's purpose. So although your statement of purpose may comprise only one line, you need to give its exact wording careful consideration. Is it talking about your purpose for writing the report, the reader's purpose in reading it, or the way those two purposes interact to establish the paper's purpose?

Forecast the Organization. Your introduction should give the reader a clear idea of how the report is organized. The reasons are rooted in human psychology. Listeners and hearers understand incoming information more quickly and more efficiently if they first receive a brief idea of what they are about to hear and how it is organized. People who do lots of reading use a trick based on this principle to increase their reading speed: they read any abstract or summary printed with the text, glance at the table of contents to see the order of ideas, and leaf through the pages quickly, noting the sequence of headings and presence of figures or illustrations. Only then will an experienced reader actually start reading. By including in your introduction a forecast of the organization of the whole piece (or in longer reports, a summary) you can help your readers in the same way. For your reader, the shortest part of your introduction, the statement of organization, may well be the most important part.

1.2 How Are Good Introductions Written?

The easiest way to learn how to write an introduction that includes context, purpose, and organization (a *cpo* introduction) may well be to observe how those three elements appear in each of the examples, taken from student reports, in Figure 8.1. Each student was asked to define something in his or her own field of study for a layperson/ executive reader. Notice how each introduction, in its use of context, purpose, and organization, establishes a tone and approach the student feels is appropriate to that writing task. Notice, too, how those tones and approaches are different in each introduction.

These student-written examples show the kind of orientation that writing for a layperson often requires. Of course, professional reports are often written for people with expert knowledge, and the introductions of those reports reflect that fact. The two introductions from professional reports in Figure 8.2 show not only how the tone varies with the changing audience but also how the forecast of organization can become a brief summary of the report.

Once you see how the forecast of organization can become a brief summary of the report, you can understand another adaptation to audience that shapes how reports begin: the use of an executive summary. This adaptation, especially useful in writing for decision makers, was discussed in Chapter 7, "Abstracts and Executive Summaries."

Figure 8.1   Three Examples of *CPO* Introductions

DEFINITION OF THE NICOSIA CONSUMER
BEHAVIOR MODEL

Buyer behavior models are usually divided into the two basic areas of *microscopic* and *comprehensive* models. In each
area, several behaviorists have developed their own ideas
as to how the consumer acts before, during, and after the
time of a purchase. Through the use of several descriptive
methods, I would like to describe for you one of the comprehensive models developed by Francesco M. Nicosia.

The Nicosia model of consumer behavior is a comprehensive look at what the consumer experiences during and
after a decision process . . .

DEFINITION OF AN INSECT

Many people often refer to any small, multilegged creeping
animal as a bug. Many of these "bugs" are not true insects but rather animals closely related to insects, such
as spiders. Within this paper, I will present some of the
distinguishing characteristics of insects so that you can
be better informed and more able to identify them properly
in the future.

Generally speaking, an insect is a small animal . . .

THE DEFINITION OF AUDITING

American businesses have become so large and complex as to
inspire awe and confusion in the average man or woman. As
businesses grow larger and more complex, the public wants
more and more assurance that businesses are acting honestly and openly in all their dealings. The one method of
assurance to which we are looking with the greatest interest is the audit. My purpose in this paper is to use several definition techniques to help you understand what
auditing is.

Figure 8.2   Two Examples of *CPO* Introductions in Professional Reports

MANPOWER REQUIREMENTS AND SUPPLY FOR MAGNETIC FUSION ENERGY, 1981–2000

This study has been conducted to help the U.S. Department of Energy (DOE) complete a task assigned to it by the "Magnetic Fusion Energy Engineering Act of 1980" (P.L. 96-386). This law requires the Secretary of Energy to "assess the adequacy of the projected United States supply of manpower in the engineering and scientific disciplines required to achieve the purposes of this Act, taking cognizance of the other demands likely to be placed on such manpower supply."

The law also requires that the Secretary make recommendations regarding the need for increased support for education in engineering and scientific disciplines.

This study focuses on estimating current employment and future personnel requirements for magnetic fusion energy until the year 2000. It also examines sources of labor for fusion energy and competing demands for the disciplines most important to the future growth of fusion energy.

DATA QUALITY IN THE 1978 & 1979 ANNUAL PATIENT CENSUS: AN ANALYSIS OF DATA EDITING PROCEDURES

Each year the Department of Medicine and Surgery and the Reports and Statistics Service, Office of the Controller, conducts a census of beneficiaries who are either in VA health care facilities or in non-VA health-care facilities under VA auspices. The census provides medical and administrative data on a cross-section of these beneficiaries for annual reports, program review, and planning purposes.

This report analyzes the procedures for editing data collected in the 1978 and 1979 censuses, with particular emphasis on the quality of data—determining whether data items are "good" or "bad" as judged by "high" or "low" error rates. The report begins with a description of the data collection and editing procedure, discusses error rates and their analysis, and ends with future research topics and a summary section.

Contexts to Use with *CPO* Introductions. As the sample introductions shown here demonstrate, there are many possible contexts: the way things are usually done, the delivery of new information, the needs of some group of people, the response to a need, or the evaluation of a set of data, to name just a few. Whatever context you use, it should tell the reader the background or perspective from which the report should be viewed, and it should operate on about the same level as you judge your reader's knowledge and interest in the subject to be.

Some statements of context are used so often as to be recognizably recycled. This recognition factor does not mean that you cannot use such statements but rather that you should learn to use them carefully. A typical example is the high school graduation speech—or *crossroads* context: "Each of you has come to a crossroads in your life . . ." If you use such a familiar image, you must either anticipate a ripple of recognition on the part of your readers, or you must weave the familiar image into the fabric of your subject so thoroughly that the image's familiarity is not immediately apparent. The following list gives fifteen such images; you can use it to suggest openings for your own introductions.

Fifteen Useful, Standard Statements of Context

1. *The Offer of Something New:* "New developments in fiber optics have now made possible the smallest and least expensive . . ."

2. *Historical Narrative:* "For the first 50 years of our company's history, we were known primarily for . . . Now, however, with the increasing demand for new energy supplies, we find the marketplace demands that we change to . . ."

3. *Comparison of Old and New:* "Until the recent increases in transportation prices, efficient warehousing and distribution of our products was not a central concern of management. In recent years, however, the cost to us of moving products by truck has more than doubled. Therefore . . ."

4. *Reference to Authority:* "Will Rogers's famous saying, 'I never met a man I didn't like,' could be taken as the motto for the image we want our salespeople to project."

5. *Crisis:* "Because of growing public awareness that cigarettes do indeed cause serious health problems, our company's sales of cigarettes have ceased to grow. We now find ourselves at a crisis."

6. *Disappointment:* "When we first introduced our XZ-3000 model, we had great hopes that it would soon capture a large share of the market. Unfortunately . . ."

7. *Opportunity:* "The failure of our competitor's XZ-3000 model presents us with a rare opportunity to . . ."

8. *Challenge:* "These are the times that try men's souls. The summer soldier and the sunshine patriot . . ."

9. *Blowing the Whistle:* "It is time to stop kidding ourselves into believing that our XZ-3000 model will . . ."

10. *Adventure:* "Space—the final frontier. These are the continuing voyages of the starship *Enterprise,* whose five-year mission . . ."

11. *Response to an Order:* "We are all aware that paper and printing costs are continuing to increase at an alarming rate. Because of those increased costs, our department's copying bill this quarter is 50 percent higher than last quarter's. The administration has decided that those bills can no longer be afforded. Therefore, in response to the administration's order, we are terminating all copying privileges."

12. *Revolution:* "The advent of digital transmission is revolutionizing the telephone business, and it is the duty of each of us to adjust to that revolution as it comes. Therefore, the following short courses will be offered to all employees beginning . . ."

13. *Evolution:* "Since the founding of XYZ Corp. in 1880, the company has always changed to meet the changing needs of the times. This process of evolution and change has continued to today, and now . . ."

14. *Confession:* "As Head of the Marketing Department for XYZ Corp., I have to confess that the government's predictions of a mild recession in the last quarter of this year had, for a time, not convinced me. But now . . ."

15. *The Great Dream:* "I say to you today, my friends, that in spite of the difficulties and frustrations of the moment I still have a dream. It is a dream deeply rooted in the American dream. I have a dream that one day this nation will rise up and live out the true meaning of its creed: 'We hold these truths to be self-evident; that all men are created equal.'"*

---

* Martin Luther King's, "I Have a Dream" speech, delivered at the conclusion of the march on Washington, August 28, 1963.

Times to Use *CPO* Introductions. You may find that once you have learned how to write this kind of introduction, you are tempted to use it everywhere. You obviously do not need a *cpo* introduction on a one-page business letter, nor do you want to use the same pattern at the beginning of every report you write. In general, the need for a *cpo* introduction increases with the report's length and the reader's unfamiliarity with your subject matter. The longer the report or the less familiar the reader is with the subject, the greater is the psychological need for the sneak preview a *cpo* introduction provides.

Adapting Introductions to Real Readers. As the introductions shown so far in this chapter demonstrate, the kind and amount of detail you include in the introduction will vary depending on your audience, your purpose, your topic, the available space, and a number of other variables. One particular adaptation, *reader benefits,* shows how *cpo* introductions can vary according to audience. For many lay audiences, you will do well to include a clear statement of reader benefits as part of your introduction. By telling readers just what they can hope to gain from reading your report, you increase the chances that your report will be read and understood.

Including "reader benefits" means making an explicit statement of why your reader should read the report and of what your reader can gain from reading it. Chapter 4, "Forms and Principles," explains reader benefits in more detail. Figure 8.3 shows two of the introductions from Figure 8.1, rewritten to include reader benefits (shown in italics).

2. Transitions

As your reports grow longer, the need to make special provisions to help your reader grows greater. The same principles that make *cpo* introductions necessary and effective also can be applied to key places *inside* your report. Every few pages, you will want to begin a new major subsection of your report; when you begin a new subdivision of your topic, you may want to use a kind of transition that is especially important in professional writing: an *internal summary* and a *fresh forecast.*

Both recall where the report has been so far and remind the reader where it is going. At crucial places in the report, they ensure that the reader and the writer are together. An internal summary is merely a brief review of the main points that have been covered so far; a fresh forecast reminds the reader of what remains to be covered. The examples in Figure 8.4 come from the middle of long

Figure 8.3   Two Examples of Introductions Presenting Reader Benefits

DEFINITION OF THE NICOSIA CONSUMER
BEHAVIOR MODEL

Buyer behavior models are usually divided into the two basic areas of microscopic and comprehensive models. In each area, several behavior theorists have developed their own ideas as to
how the consumer acts before, during, and after the time of a
purchase. Through the use of several descriptive methods, I
would like to describe for you one of the comprehensive models
developed by Francesco M. Nicosia.

The Nicosia model of consumer behavior is a comprehensive
look at what the consumer experiences during and after a decision process. *As a consumer yourself, you will find that an understanding of the Nicosia model of consumer behavior gives you
insight into your own purchasing behavior and into the ways
that behavior can be taken advantage of by businesses.*

THE DEFINITION OF AUDITING

American businesses have become so large and complex as to inspire awe and confusion in the average man or women. More and
more, the public wants assurance that businesses are acting honestly and openly in all their dealings. The one method of assurance to which we are looking with greatest interest is the
audit. My purpose in this paper is to use several definition
techniques to help you understand what auditing is. *The more
customers and investors such as yourself understand about the
audit process, the easier business's public relations jobs will
become and the more confidence the public will have in business's financial dealings.*

reports; notice the two different ways the examples present the relationship among the internal summary, the fresh forecast, and the
heading.

Internal summaries and fresh forecasts are easy to write. The only
difficult point is to remember to use them and to decide whether they
should be used before, after, or around your new subheading. That
decision, however, is much less important than remembering to care
enough about your reader to help him or her over the hurdles involved in reading your report.

Figure 8.4   Two Examples of Internal Summaries and Fresh Forecasts

.
.
.

```
As the comparison above has shown, the economics of some situa-
tions favors the use of solar over fossil fuels. Another factor
must be considered, however, and that is the availability of so-
lar energy systems.

                    MAJOR SOLAR DISTRIBUTORS

Information on availability of solar energy systems was re-
ceived from Arkla Industries in Indiana, Helio Association in
Arizona, and Solar Energy Digest in California. According to
those sources . . .

                 ISOLATION OF THE CHOLESTEROL

Centrifugation has prepared the plasma for the next procedure,
which is isolation of the cholesterol from the plasma. The iso-
lation entails three steps: saponification, extraction, and
evaporation . . .
```

3.   Conclusions

Just as any document longer than a couple of pages needs an introduction, such a document also needs a conclusion. Most readers remember best what they read last. The conclusion gives you one last
chance to do what you want to do for your reader, to move your
reader in the direction you choose. Twelve different techniques for
conclusions are given in the following list.

Twelve Techniques for Writing Conclusions

1. *State the results of the investigation or study.* Of course, stating the
   results is what most readers expect a conclusion to do. Usually
   you will want to combine this technique with one of the others
   listed here.

2. *Recommend what action should be undertaken.* Although it is a fallacy
   that any recommendation report must *necessarily* end with the
   recommendation (see Chapter 15), this is another typical pattern
   for a conclusion. This pattern is also frequently combined with
   one or more of the others listed here (such as 5, 6, 7, and 8).

3. *Repeat the major points, for emphasis.* The longer and more complicated your report is, the more important it is to conclude by summarizing its major points.

4. *Briefly summarize the entire document.* Sometimes a report doesn't break down into major points very conveniently. In this case, write a summary to be used as a conclusion (Chapter 7, "Abstracts and Executive Summaries," shows you how to write a summary).

5. *Extend the implications of the current work.* Often your conclusion will go beyond the scope of your report proper, extending the report's temporal or causal sequence further than the range covered by the report's body (as in 6, 7, 8, and 9).

6. *State what the further problems or concerns might be and indicate how to deal with them.* One way of moving your report into the future is to foresee possible problems or events that could mean that the report's subject needs reexamination.

7. *Tell how to evaluate whether the process or action your report describes or recommends is proceeding properly.* Although presenting evaluative techniques in detail may be beyond the scope of your report, a brief description of them may be a real help to your readers.

8. *Describe alternative steps or troubleshooting procedures.* Another way to make your report more useful is to present as a conclusion a brief explanation of whatever alternative readings of the situation there might be or of any particular actions that need to be taken, should the future course of events not go as the report suggests.

9. *Suggest other entirely different approaches.* Depending on how tentative or firm your report is, you may want to make your presentation of other approaches to the problem more detailed.

10. *Restate cautions and safety warnings.* Although it is never a good idea to have the necessary cautions and safety warnings *only* at the end of a document, in circumstances of extreme risk (or liability) it may well be a good idea to restate the appropriate cautions or safety warnings as part of your conclusion.

11. *Reemphasize the importance of the topic.* This ending and technique 12 are less frequently used than 1–10 in scientific and technical writing and more frequently used in popularized science writing. Especially when writing for laypeople, reemphasizing the importance of the topic at the end of the document may be a good idea because laypeople especially may not really be aware of it.

12. *Add a sense of an ending.* Most readers, especially lay readers, sub-consciously expect the things they read to have some kind of an ending. Besides the more functional kinds of endings described in techniques 1–11, you can add the *sense* of an ending by

- ending with an appropriate quotation.
- echoing a key phrase from the introduction.
- adding an emotional appeal.

EXERCISES

1. Select five important subjects in your field of study. For each one, write an introduction to a paper that will define each one of them for a lay/executive audience. Then rewrite each introduction for an expert audience.

2. Choose ten articles at random from your field's professional or trade journals and examine their introductions. Do they conform to the principles described in this chapter? Write a brief report describing the way introductions are done in your field, based on these samples. (Be sure to turn in copies of the introductions with your report.)

3. Compare the types of introductions discussed in this chapter with five introductions from popularized science magazines (*Psychology Today,* for example). How do the popularized science introductions differ from the ones in this chapter? What accounts for the differences? (Be sure to turn in copies of the introductions with your answers to these questions.)

➤ 4. Carefully consider the following professionally written introduction. What are its strong points? its weak points? Be prepared to discuss in class how to make it better. (The report has three chapters.)

Educational and Income Characteristics of Veterans
(March 1989)

This report focuses on three basic social indicators with respect to the population of male war veterans 20 years old and over. The analysis describes the attained education of veterans in 1989 and their personal and family incomes in 1988. These data are discussed in comparison with nonveterans within five-year age intervals.

Chapter 1, Median Family Income in 1988, shows the average income of families headed by veterans in comparison with that of families headed by nonveterans of similar age. Data are broken down by age of head and type of family, income status of the wife of the family head, and veteran status.

A ➤ indicates a case study exercise

Chapter 2 discusses the personal income of individual men by age and by veteran status, with trend data going back to 1978.

Educational attainment of veterans is the subject of Chapter 3. The data demonstrate the relationship between years of school and median income of male veterans by their age. Data are also presented for nonveterans.

5. You have volunteered to assist your local zoo by helping produce a manual to guide the work of the many volunteers who help out at the zoo. The volunteers conduct tours for school groups, help stage special events (the spring zoofest, the oldest elephant's birthday, etc.), staff the chldren's petting zoo, help stage fund-raising events in the community, staff the information booths, and perform a variety of other functions. This manual will be especially useful for new volunteers who do not know the customary procedures and practices around the zoo, but it will also go to the "old hands," who once in a while need to be reminded that they can't rewrite policy on the spur of the moment to suit their own wishes. Your task is to write the introduction to this manual (100–150 words); invent any other information you may need (within the bounds of reason and common sense).

6. Write a short introduction for the report in exercise 7 at the end of Chapter 7.

7. Write a short introduction for the report, "House of Representatives Research Budget Requests," described in problem 3 of the Exercises at the end of Chapter 10.

8. Write a short executive summary and a short introduction for the report, "State University Contributions to the Capitol City Economy," described in problem 4 of the Exercises at the end of Chapter 10.

9. Headings and Subheadings

One of the most immediately apparent characteristics of professional writing is the use of headings and subheadings. Their presence does more than make this type of writing visibly different from other kinds of writing; it also signals important structural and psychological differences. Professional writing is designed to be easy for readers to use, and that has several implications, as shown in earlier chapters:

- The style must be readable.
- The real audience must be considered.
- Delicate features, such as negative messages, must be handled carefully.

The key implication, for this chapter, is this one:

- The different parts of the document must be easy to find.

Most readers will not read *all* of a report—only the sections they need to read. Other kinds of writing, including that which is done in most English classes, stress an overall unity that writing in business and industry often sacrifices in favor of giving the reader ready access to particular sections. Using headings is the most visible method of furnishing that access. This chapter will show you why and how to use headings and discuss the common problems associated with them.

1. Why to Use Headings

Headings have four basic functions—two for the reader and two for the writer:

1. *To help readers find specific parts of the report.* In professional settings, readers want to be able to find *at a glance* the specific part of your report that most concerns them. Headings make that possible. Have you ever tried to find a particular part of a book or report, maybe while reviewing for a test, and spent much more time finding that section than you did reading it? Most likely that document did not make full and frequent use of headings and subheadings. One of the ways in which writers show their concern for readers is by labeling specific parts of the whole with headings and subheadings, so that readers can find those parts easily.

2. *To present your outline and organization to readers at a glance.* If your reader can understand the way you have structured your report and see the principle behind its organization, the time and energy needed to read the report will be reduced. Most working readers thumb through any multipage document before they decide whether or how much to read. If you handle headings properly, a reader thumbing through your report will see the headings and subheadings as a clear and logical outline of your presentation. That outline will be valuable for the reader's speed in reading and understanding the report.

3. *To ensure the writer's logical organization.* When you use headings, you will find they make you think consciously and carefully about the structure of what you write. Does the structure make sense to you, and will it make sense to your reader? Does the organization that you (the writer) find comfortable to work with prove equally comfortable to your reader? Dealing with such questions is an important stage in the development of any writer, a stage that can come with the careful use of headings and subheadings.

4. *To announce new topics.* The ability to make smooth and effective transitions usually comes to writers only after much practice and hard work. With headings and subheadings, such transitions can be less important. Instead of having to lead the reader slowly and subtly from one topic to the next, headings and subheadings can simply announce the change in topics. (See Chapter 8, Section 2,

for information on the most common transitional devices.) Practice and hard work are still necessary, but headings and subheadings can help.

2. How to Use Headings

There are many different schemes for using headings and subheadings. This section explains four different levels of headings, offers several common variations, and then discusses other ways to make your material's organization visible to your reader.

2.1 Four Levels of Headings

Many reports use four levels of headings, sometimes called *major headings, minor headings, subheadings,* and *paragraph headings.* You may find it less confusing to remember them as A, B, C, and D levels (where A is the major heading). Typically, the headings are indicated as shown in Figure 9.1.

Figure 9.1   Four Levels of Headings
Headings alert readers to a report's structural elements.

A-level heading ——————— MAJOR HEADING

Xxxxxxxxxxxxxxxxxxxxxxxxxxxxxxxxxxxxxxxxxxxxxxxxxxxxxxxxxxxxxxxxxxxx
xxxxxxxxxxxxxxxxxxxxxxxxxxxxxxxxxxxxxxxxxxxxxxxxxxxxxxxxxxxxxxx .

B-level heading——————— Minor Heading

Xxxxxxxxxxxxxxxxxxxxxxxxxxxxxxxxxxxxxxxxxxxxxxxxxxxxxxxxxxxxxxxxxxxx
xxxxxxxxxxxxxxxxxxxxxxxxxxxxxxxxxxxxxxxxxxxxxxxxxxxxxxxxxxxxxxx .

Subheading —— C-level heading

Xxxxxxxxxxxxxxxxxxxxxxxxxxxxxxxxxxxxxxxxxxxxxxxxxxxxxxxxxxxxxxxxxx
xxxxxxxxxxxxxxxxxxxxxxxxxxxxxxxxxxxxxxxxxxxxxxxxxxxxxxxxxxxxxxx .

⌐ D-level heading

Paragraph Headings. Xxxxxxxxxxxxxxxxxxxxxxxxxxxxxxxxxxxxxxxxxxxxxxx
xxxxxxxxxxxxxxxxxxxxxxxxxxxxxxxxxxxxxxxxxxxxxxxxxxxxxxxxxxxxxxx .

The purpose of headings is to use the typography and the placement on the page (vertical and horizontal spacing) to alert the reader to structural elements in your report. You can reinforce the visibility of your headings—and make your writing more attractive visually—if you use the spacing above the headings to reinforce their levels of importance: skip five lines above an A head, four above a B head, and so on. Many writers emphasize the structural importance of A heads even more by beginning a new page whenever one occurs. But the most visible characteristics of the different levels of headings are their typography (all capital letters, initial capitals, underlined, not underlined) and their placement on the page.

2.2   Variations on the Four Levels

There are a number of ways to vary the four levels of headings. You can have five levels, delete a level, use the decimal system, use "talking" headings, or use any of a number of other kinds of visual organizers.

Use a Fifth Level. The four levels of headings presented here can be shortened to three or expanded to five. If you leave out the C or D level, you have three levels. If you use a heading that is centered, all caps, and underlined, you have five levels, and the new one becomes the A-level head.

<u>FIFTH LEVEL</u>

Or you can insert another level between the B and C levels, with initial caps, underlined, at the left margin. But keeping track of heading levels should not be allowed to draw too much of the reader's attention; if you use too many levels, the complexity of the scheme may become self-defeating.

Use the Decimal System. In many fields, writers are encouraged to number the sections of their reports, either letting the numbers stand alone in the place of headings, or putting numbers in front of the headings. This common usage is illustrated in Figure 9.2. The combination of numbers and headings may be the most popular variation of this scheme and is used throughout this book.

2.3   Talking Headings

In your first draft of a report, you may well use general headings (*topic* headings) such as "Background," "Procedure," and "Results."

Figure 9.2   Use of Decimal-Numbered Headings in Reports

(a) The Decimal System, One of Several Possible Varieties

1. MAJOR HEADING

1.1   SUBHEADING
1.1.1 Minor Heading
     1.1.1.1 Paragraph Heading

(b) Another Version of the Decimal System

4.0 SYSTEM DEFINITION AND BASELINE REQUIREMENTS

4.1 Functional Requirements
4.2 Performance
    4.2.1 Reliability
    4.2.2 Availability
    4.2.3 Maintainability
4.3 Preliminary Assessment of Interface Requirements
4.4 Critical Elements
4.5 Constraints
4.6 Plans for System Tests

Source: Adapted from DOE sample Feasibility Report Table of Contents.

(The sample DOE feasibility report structure in Fig. 9.2 uses such topic headings.) Your headings will be much more useful to the reader if in the finished version of your report you make them fully descriptive and specific. Rather than just name the topic, indicate what the section says about it. The four headings in Section 1 of this chapter, "Why to Use Headings," are talking headings. Here are three more examples:

Background: Steady Growth from 1880–1950
Procedure: Survey of Industrial Growth Since 1950
Results: Economic Growth Slow or Totally Lacking

Specific headings such as these can do much more for your reader than the shorter, topic headings.

3.   *Cautions About Headings*

There are several conventions about using headings—*do's* and *don'ts* that you should be aware of.

3.1   Stacked Headings

Your headings will be more useful for your readers if you do not have one heading following right after another, stacked with not even one sentence in between them. Professional writing is designed more for clarity than for stylistic elegance. In most situations you should use at least one sentence between any two headings, even if the sentence is only a brief forecast of what is to come next (as with the sentence between the headings for Section 3 and Section 3.1, above). The exception occurs when one heading, such as the title of the report, is by itself on one page, followed on the next page by another heading, such as "Introduction" or "Executive Summary."

3.2   Pronoun Reference

It's not a good idea to use a pronoun in the sentence after a heading to refer to the heading. Consider this example:

> *Pronoun Reference.* This is often a problem with headings . . .

Rewrite the sentence to clarify the pronoun reference, even if it means repeating the words of the heading right after the heading.

> *Pronoun Reference.* Pronoun reference is often a problem with headings . . .

If this kind of repetition bothers you, recast the sentence in question to move the repeated words away from each other:

> *Pronoun Reference.* Writers often have a problem with pronoun reference . . .

3.3   Parallelism

To give your writing a polished look, make the headings on each respective level grammatically parallel to the other headings on that level (unless a higher-level heading intervenes). That is, balance a noun in one heading with a noun in the next heading on the same level, balance a noun plus prepositional phrase with a noun plus prepositional phrase, and so on. Figure 9.3 illustrates this use of grammatical parallelism.

Does every heading of any particular level have to be grammatically parallel to every other heading at that level? No. The rule is that intervention by a higher-level heading can start a new parallelism.

Figure 9.3   Grammatical Parallelism
Notice the parallelism can be broken each time a higher-level heading
intervenes.

VISUAL ORGANIZERS

HEADINGS	GRAPHICS
How to Use Headings	Different Kinds of Graphics
Typeface	Pictorial Graphics
Spacing	Numerical Graphics
Why to Use Headings	Different Uses for Graphics
Reader's Reasons	
Writer's Reasons	

That is, all the B heads under the first A head must be parallel to
one another, but they don't have to be parallel to all other B heads.
In Figure 9.3, this is why the two D heads under "How to Use Headings" must be grammatically parallel to one another but not to the
two D heads under "Why to Use Headings."

3.4   Frequency

At first, you may have problems deciding just when to use a new
heading or subheading. How frequently does your subject need to
be divided? There is no one universal answer to this question; the
frequency with which you use headings will depend largely on the
nature of your topic, the length of your report, and the nature of
your reader. Two general guidelines can help you decide:

- *Use a heading of some level at least every two or three pages,* if only to
  help the reader remember where in the report's structure he or she
  currently is.

- *Try to avoid having more than about seven headings on any one level
  without another, higher-level heading intervening.* That is, if your report
  has fifteen major sections, break it up into two or three larger segments. This is because humans understand hierarchical structures

Figure 9.4 A One-level List
It's difficult for anyone to comprehend such a structure without close study.

> Definition of Accounting
> Size and Importance of Accounting Profession
> Governance of Accounting Profession
> Pre-1800 History of Accounting
> History 1800–1900
> Problems with Financial Growth
> Rising Professionalism
> Accounting at the Turn of the Century
> The Federal Government Increases Its Role
> The Crash of 1929
> Conflicting Government Policies
> Role of Professional Organizations
> The Era of Overruns
> Current Controversies
> The Future of Accounting

better than long, one-level structures. Figures 9.4 and 9.5 show two versions of a report's table of contents to illustrate this principle.

4. *Other Visual Organizers*

In the classroom, you can experiment with more techniques to organize the report visually for your readers. As with headings and subheadings, the point is to use the resources at your disposal to make your message reach the right readers in the right way. *But before you use any of the following techniques, consider whether and how your writing is going to be reproduced* (conventionally word processed, photocopied, or photo-offset printed). As with other kinds of visuals, some of the techniques described here are possible only with certain types of manuscript reproduction. Most important, though, you should consider whether your specific reader may react negatively to less than common methods of visually organizing the text. If you are writing for the Department of Energy, you probably cannot use any of these; if you are writing for a marketing agency, maybe you can.

As you can see in the sampler in Figure 9.6, the use of too many visual organizers in one paragraph is a little disconcerting, but the careful use of one or two of those techniques is something too many

Figure 9.5   The Hierarchical Table of Contents
This version makes sense at a glance.

The Accounting Profession Today
    Definition
    Size and Importance
    Governance

The Accounting Profession's Past
    History Pre-1800
    History 1800–1900
    Problems with Growth
    The Rise of Professionalism
    The Turn of the Century

The Accounting Profession in This Century
    The Role of Government
    The Crash of 1929
    The Conflicts among Government Policies
    The Role of Professional Organizations
    The Era of Overruns
    The Controversies of Today

The Accounting Profession Tomorrow

Figure 9.6   Examples of Visual Organizers

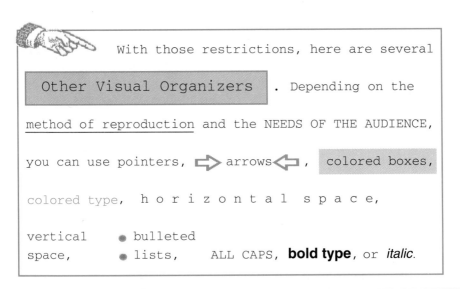

practicing writers ignore. For the most part, though, you will have to rely on typographic variations that can be done on an ordinary word processor, with perhaps a laser printer, assisted by vertical and horizontal spacing.

EXERCISES

1. Describe and analyze the use of headings in a popular science magazine of your choice (*Omni, Scientific American, Discovery, Smithsonian,* etc.).

2. With careful attention to the need for parallelism, make a detailed outline (at least three levels of headings) of a report in which you define in detail a thing or a concept in your field for an executive/layperson audience. The entries in your outline should function as the report's headings and subheadings.

3. Survey the top three scholarly, professional, or trade journals in your field, and write a brief description and analysis of each journal's use of headings and subheadings. How do they differ, and which is most useful to you as a student?

4. Perform the survey described in problem 3 and write up its results into a brief report on the merits of using visual organizers in reports. Include the value of the full range of visual organizers—visuals, displayed lists, underlining, italics, etc.—in your report.

➤ 5. Assume you are a student hired as a summer intern in a local bank. Having taken a professional writing course in school the previous year, you were surprised to see that the bank's reports were written without visual organizers of any kind. Your internship supervisor suggested you write a brief report on the value of headings and other devices as tools for better writing. The audience of the report is your internship supervisor (a middle-management person), but you're also aware the report may eventually go much higher up into the bank's executive structure.

➤ 6. Photocopy the report on pages 67–68, cut the paragraphs apart, and insert appropriate headings and subheadings.

➤ 7. Photocopy the report on pages 200–203, cut the paragraphs apart, and insert appropriate headings and subheadings.

A ➤ indicates a case study exercise.

10. *Visuals*

Most people in the United States learn to write as an offshoot of drawing. Somewhere around the third or fourth grade, you were asked to draw a picture of something you saw last summer that you really liked, to color the picture in, to stand up in front of the class and tell the story that goes with the picture, and—finally—to write that story down. Even before that, the way you first learned to sign your name was as a form of drawing. If you watch a second grader write her name, you see someone *drawing* one careful letter after another. Writing thus evolves from drawing.

By the time you reached high school, drawing and writing were probably two completely separate activities for you. English class papers were solid walls of words, and drawing may have been left for only those who were "artistic." Chances are your freshman composition class in college involved no visual presentations as part of its student papers.

Thus, it may seem odd to you that now you are going to be expected to use the visual presentation of information as an important part of every project. Yet that is exactly what being an effective professional communicator will require of you. Today's professional documents may well have a fifty-fifty verbal-visual mix as the norm, and documents that are more visual than verbal are becoming more common every day.

Visuals can make the documents you write clearer, more attractive, more functional, and more frequently read. No other feature you can incorporate into your writing with so little effort carries such great rewards as proper visual presentation. Some writers use visuals only when the subject cries out for them; some use just a few visuals in exactly the right places; and some use visuals whenever they can. This chapter is designed to show you when to use visuals, what kinds of visuals to use, how to use them, and where to place them in your documents.

The communication situation (discussed in Chapter 1)—your purpose, subject, audience, and role as a writer—plays an important role in your decisions about when, what, how, and where visuals should be used. Other, very particular factors also need to be considered: how your readers will use this document, your access to professional help with the visuals, the kind of paper and the document reproduction process, to name just a few. Sometimes the use of multiple colors is too expensive; other times the use of color at all is too expensive. Some audiences frown on anything beyond the most restrained use of visuals; others respond better to visuals than to words. Some writers have access to sophisticated desktop publishing equipment that makes creating visuals easy; other writers don't. Some subjects need visuals more than others; some absolutely require visuals. *Successfully determining all these variables is the key to the effective use of visuals in professional writing.* Unfortunately, most of these variables are ignored in student writing. Thus, the first step in coming to deal with those variables is to consider *when* visuals should be used.

1. When Should Visuals Be Used?

There are three primary uses of visuals:

1. To support textual material
2. To convey information themselves
3. To direct the reader's actions

Within this section, each of those primary uses is explained in detail, accompanied by examples.

1.1 To Support Text

The use of visuals to support textual materials is perhaps the one that is most immediately apparent to students. A passage in a document may be accompanied by a visual either to arouse the reader's interest, to emphasize a point, or to clarify a point.

To Arouse the Reader's Interest. Figure 10.1 shows an example of a visual used to catch a reader's eye. In this case, a newsletter produced by a professional society is including an article titled, "How Is Advanced Technology Affecting Our Future?" The title of the article spans all three columns of the newsletter, and the figure, a *conceptual visual,* is placed directly beneath the title (at the top of the middle column). Its function is only—but importantly—to catch the eyes of people skimming the newsletter and entice them into reading the article.

Figure 10.1   A Visual to Catch a Reader's Eye
This *conceptual visual* accompanies an article entitled, "How Is Advanced Technology Affecting Our Future?"

Source: STC *Intercom* 37, no. 4 (November 1991): 6.

Figure 10.2   A Visual to Emphasize a Particular Point
"Gamma Radiation of the Pool Surface Directly above the Reactor (16.5 ft. above Core)."

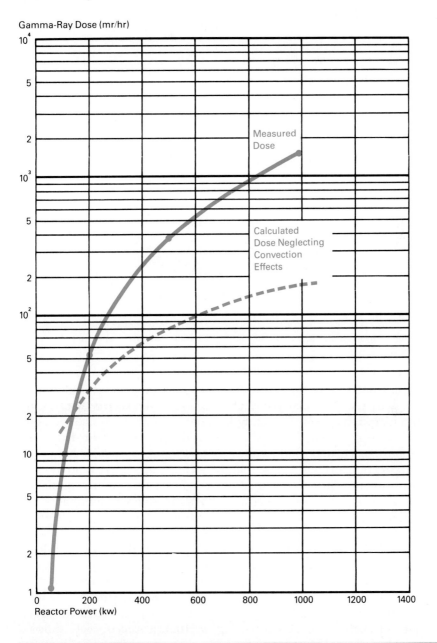

Source: Oak Ridge National Laboratory.

Figure 10.3   A Visual to Clarify a Point

Notice that the lines show interpolated relationships among the scattered data.

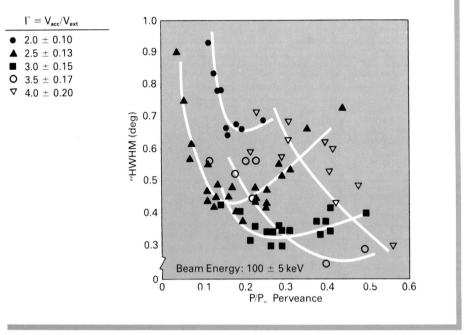

Source: Oak Ridge National Laboratory—Fusion Energy Division Annual Progress Report, Period Ending Dec. 31, 1977, p. 135.

To Emphasize a Point.   Figure 10.2 shows an example of a visual used to emphasize a particular point. One section of the report that this figure is from has been discussing how the amount of gamma radiation as measured experimentally is considerably different from the amount calculated mathematically. The figure, a *multiple line graph,* appears in the report where that point comes up. There is no thought that anyone will try to puzzle out the exact dosage at a particular power level based on this graph; it's there solely to emphasize the difference between dosage as calculated and dosage as measured.

To Clarify a Point.   The writing that professionals do—technical documents especially—at times needs to make a difficult point. Sometimes such a point will involve a set of complex relationships within data. At such times, a visual can be indispensable in clarifying exactly what the author's point is. Figures 10.3 and 10.4 show two different examples of such a use of visuals. Figure 10.3, a *scatter graph,* shows a trend (indicated by the relationship among the different white lines) reflected in five different sets of data (indicated by the five different

Figure 10.4   A Visual to Clarify a Point
The sophisticated graph to show ranges can be generated only by computer.

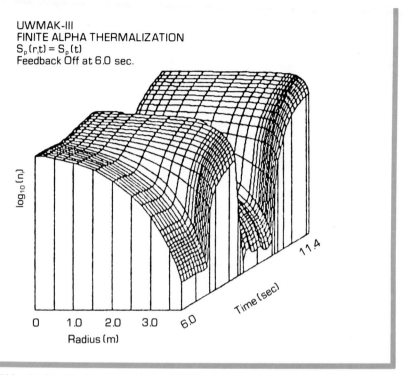

UWMAK-III
FINITE ALPHA THERMALIZATION
$S_p(r,t) = S_p(t)$
Feedback Off at 6.0 sec.

$\log_{10}(n_i)$

Time (sec)

11.4

0   1.0   2.0   3.0   6.0

Radius (m)

Source: Oak Ridge National Laboratory—Fusion Energy Division Annual Progress Report, Period Ending Dec. 31, 1977, p. 115.

geometric shapes). Figure 10.4, an *isometric graph*, shows ranges of temperatures over time at different radii.

1.2   To Convey Information

Visuals can be used to provide information that the text itself cannot. Such visuals may be employed to say something in fewer words than the author feels the text itself could afford, to go into greater detail than the text itself ever could, or to convey a mood or a tone that would, perhaps, be too difficult to put into words.

To Use Fewer Words. Sometimes an author needs to explain a point briefly but clearly. Perhaps the point could be made in words, but it would take too many to do so, given the point's importance. Figure 10.5, a *line drawing*, shows two details of an arrangement of apparatus for a laboratory experiment. Here the author could have used words,

Figure 10.5   A Visual to Convey the Point Using Fewer Words
How many words would it take to explain these arrangements?

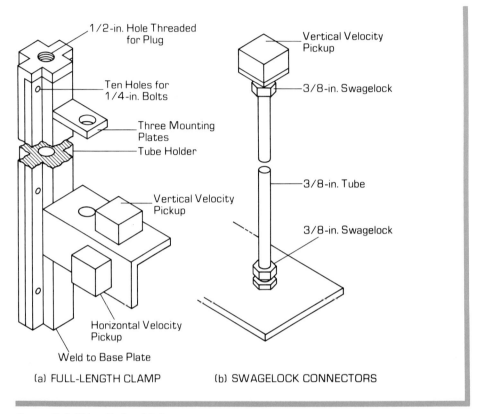

1/2-in. Hole Threaded
for Plug

Ten Holes for
1/4-in. Bolts

Three Mounting
Plates

Tube Holder

Vertical Velocity
Pickup

Horizontal Velocity
Pickup

Weld to Base Plate

(a) FULL-LENGTH CLAMP

Vertical Velocity
Pickup

3/8-in. Swagelock

3/8-in. Tube

3/8-in. Swagelock

(b) SWAGELOCK CONNECTORS

Source: Oak Ridge National Laboratory.

but it would have taken too many, and, frankly, the point being made isn't worth that much space in the document. The visual is clearer and more economical.

To Go into Greater Detail. At times a document will need to present more detail than words ever could—for example, with a whole set of figures presented in a complicated budget or with a complicated set of spatial relationships presented in a map or an electrical schematic. Figure 10.6, a *schematic diagram,* shows in the form of a wiring diagram a kind and level of detail that words perhaps never could approximate.

To Convey a Mood or Tone. Professional writing covers all kinds of human subjects and thus all kinds of human feelings. A site report for a proposed condominium development may need to show its readers how a certain piece of land "feels," or a description of a new product

Figure 10.6   A Visual for Greater Detail
This kind of detail can't be put into words.

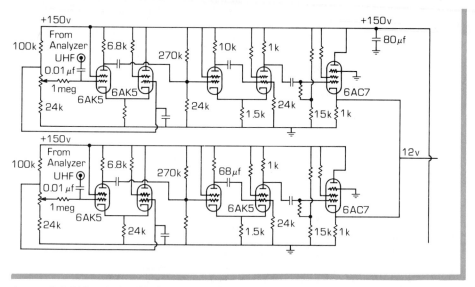

Source: Oak Ridge National Laboratory.

may need to evoke a feeling the manufacturer wants associated with the product in the buyer's mind. This function is one of those that visuals accomplish easily and well and words (for most of us) accomplish with difficulty or not at all. Drawings and photographs can be used to invite readers to share feelings they themselves may never directly experience. Figure 10.7, a *photograph* (turned into a *halftone* for use in this text), shows a person reacting to a mishap at the office.

1.3   To Direct Action

The third major use of visuals is to direct action—to show someone what to do (or what not to do) and/or how to do it. Such visuals may follow text (the words walk the reader through the actions, which the visuals then illustrate), precede text (on the assumption that humans characteristically look at the pictures first, try the actions next, and read the words only as a last resort), or replace text totally (especially in circumstances where the document needs to be used by speakers of several different languages).

To Follow Text. Sometimes visuals designed to direct action are only supplemental to the text. The assumption in this case is that a reader could perform the specified action just by reading the words and that the visuals are, in that sense, only supportive. In fact, such an

Figure 10.7   A Visual to Convey a Mood or Tone
This photograph speaks to readers of a person's reaction to an office mishap.

assumption flies in the face of most readers' experience of most documents: most readers look at the visuals first and the words second, regardless of their placement. Such an assumption also ignores the fact that human beings register pictorial stimuli more strongly than verbal stimuli, again regardless of which comes first. So even if your visuals follow your text, it is unwise to assume that the reader will grant the words primacy over the visuals—or even read the words first. Figure 10.8, a set of *line drawings,* is part of the instructions for playing a tape in a radio/cassette player. Note that the words precede the drawings.

To Precede Text. In documents designed to direct action, the visuals often precede the text. This placement conforms to what many communication professionals believe is the way human beings inevitably employ such documents. Figure 10.9, a set of *line drawings,* is part of

Figure 10.8 Visuals That Follow Text

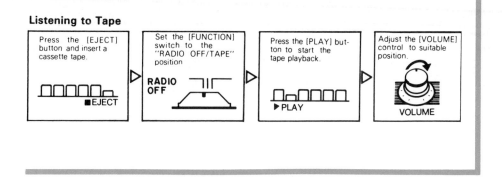

Figure 10.9 Visuals That Precede Text

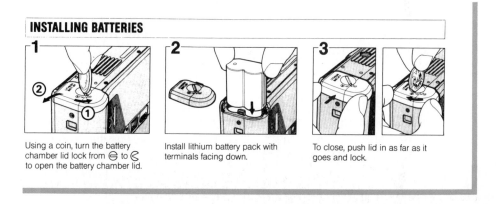

a set of operating instructions for a camera. Notice the relative emphasis the document's designers place on visual versus verbal communication.

To Replace Text. Sometimes, especially in documents that need to be understood quickly or by speakers of several different languages, the visuals in documents designed to direct action actually stand apart from—perhaps even replace—the text. The same set of camera instructions that was the source for Figure 10.9 also offers the visual in Figure 10.10 to show how to insert the film. The words accompanying the visual don't try to tell the users how to insert the film properly, only that proper insertion is important. The entire message of how to accomplish proper film loading is carried by the visual.

Figure 10.10 Visuals That Operate Independently of Text

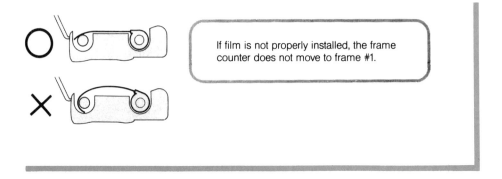

2. *What General Guidelines Must All Visuals Meet?*

Regardless of what use or type of visual a text contains, there are three general guidelines that should be observed for all visuals: *accessibility, appropriateness,* and *accuracy.*

2.1 Accessibility

Visuals add little to your writing if they are inserted poorly or in the wrong place. To make your visuals as accessible as possible, you need to make sure they meet these criteria:

> Reference in the text
> Framing with white space
> Identification of the visual
> Simplicity

- *Reference in the Text.* Every visual needs to be referred to in your text before the visual itself appears. There are any number of different ways to do this, including, "as Figure 1 shows," "(See Figure 1)," and "As shown in Figure 1." Which way you refer to the visual is relatively unimportant, but each visual *must* be referenced in the text. This reference is the only way a reader can know exactly at what point in the text to refer to the visual.

- *Framing with White Space.* A typical problem with visuals produced by laypeople is the tendency for the visual to be squeezed onto a page. Every visual needs to be framed with sufficient white space— at least an inch of white space all around as a bare minimum. For students making their own documents, a good alternative to worrying over this spacing is to put each visual by itself on a separate page immediately following the page in the text that refers to it. That way there's always enough white space around the visual, and sometimes more than enough. But too much white space is not nearly the problem too little is.

- *Identification of the Visual.* Every visual needs to be identified in three ways: with a number, a title, and (usually) a short explanation. The two most common kinds of explanations for visuals are either to summarize the relevant points in the visual or to focus the reader's attention on the most important points.

- *Simplicity.* Except for the most technical of applications, visuals need to be simple enough for browsing readers to grasp. It's always tempting to present extra information—additional levels of detail—in each visual, but unless the *point* of the visual is to present more detail than the text, you should resist the temptation. If a reader similar to your intended one browses through the document and looks dismayed by the sight of a visual, your visual is too complicated. Maximum simplicity should always be the goal of your visuals.

2.2   Appropriateness

Some types of visuals do particular kinds of things better than other types do. For example, to show the change in a set of figures over time, you might use a bar graph or a line graph. Which one will you choose? The strength of the bar graph over the line graph is that it's more visual, and hence easier for the reader to grasp at a glance. The strength of the line graph over the bar graph is its greater accuracy. Which does your particular document and its audience need more: at-a-glance comprehension or accuracy? Similarly, suppose you want to show the different percentages of your company's annual budget that its five different divisions comprise. Would you use a column graph or a pie graph? Both are easy to read, and both are about the same in terms of precision. But with a budget, you are talking about parts of a whole, and in most such situations a pie chart can show that aspect better than any other visual.

Another way to explain this aspect of appropriateness is to say that each visual in your document involves two decisions: the *decision to use* a visual and the choice of *what kind* of visual to use. Each of these choices is something to consider carefully. Section 3 explains the strengths and weaknesses of each type of visual in more detail; the point here is to emphasize the choosing.

2.3 Accuracy

The final set of general guidelines for visuals concerns their selection and their execution. The guidelines are grouped together here under the heading "accuracy" because they all speak to the amount of attention to detail you put into your document and, in that sense, send an important message to your readers about how carefully worked out and hence how accurate everything you say may be.

- Is the visual clear? Are its lines crisp?

- Is the important content of the visual summarized in the text, before the reader sees the visual? That is, you should avoid lazy introductions like this: "Table 1 presents information on 1991 earnings." If the information is important, don't force the reader to dig it out of the visual; put the key points into your text: "As Table 1 shows, 1991 earnings are significantly lower in all categories than 1990 or 1989 earnings." Do not use the visual to *make* your point; use it to *reinforce* your point.

- Is the source of the data or information used in the visual acknowledged, either in a source note that accompanies the visual's caption or in a reference note that is part of the document's larger referencing system?

3. What Categories and Types of Visuals Are There?

As the first ten figures in this chapter may have suggested to you, professional writers today can use many different types of visuals. As desktop publishing equipment becomes available to nearly all writers in professional settings and as the visual components of desktop publishing become more accessible to all users, the variety of visuals available to all writers will continue to grow. Thus, it becomes increasingly important to be able to sort the many types into just a few categories, both for writers to understand the range of choices available to them and for writers and readers to learn in general terms which kinds of visuals might work best (or worst) in which situations.

The discussion that follows divides all types of visuals into four categories: tables, graphs, structural visuals, and representational visuals.* Where there are specific guidelines for how to use a category

---

* In dividing visuals into these four types and in much of the subsequent discussion in this section, I am indebted to Carolyn Rude's *Technical Editing* (Belmont, Calif.: Wadsworth, 1991).

or type beyond those spelled out in the previous section, they are detailed under the appropriate category or type here.

3.1 Tables

Tables are the simplest of visuals. They are made up of columns (the vertical lines of numbers or items) and rows (the horizontal lines), each with a heading that clearly identifies its content. The *strength* of tabular presentations of information lies primarily in precision; no guesswork or interpolation is required by the reader. In addition, tables are extremely easy for writers to produce. Many commonly available word processing programs offer the capability to produce tables as part of their basic package, and any number of specialized software packages do even the most complicated spreadsheets reasonably easily. The *weakness* of tabular presentation is that tables need to be read just as much as text does. Most readers find tables hard to skim or to interpret at a glance; some other categories of visuals may be taken in easily and quickly. Some writers will attempt to make tables easier to use by highlighting the key data, either with color or with **boldface** type. Figure 10.11, a *table highlighted by color,* shows the

Figure 10.11 A Table Highlighted by Color

SUMMARY OF RADIATION DAMAGE IN STAINLESS STEEL STRUCTURAL COMPONENTS

Neutron wall loading = 1.47 MW/m²

Location	Atomic displacement rate[a] (dpa/yr)	Hydrogen gas production rate (appm/yr)[b]	Helium gas production rate (appm/yr)
First structural wall	20.9	913	276
Center of blanket	4.62	127	34.7
Rear of blanket	7.70×10^{-1}	7.43	1.91
Front of shield	2.16×10^{-1}	1.45	3.62×10^{-1}
Rear of shield	4.92×10^{-5}	1.11×10^{-3}	2.74×10^{-4}
Front of magnet coil	3.51×10^{-5}	5.13×10^{-4}	1.26×10^{-4}
Rear of magnet coil	6.26×10^{-9}	7.76×10^{-10}	1.67×10^{-10}

[a] Based on an effective displacement energy of 40 eV.
[b] Atom parts per million per year.

Source: Oak Ridge National Laboratory—Fusion Energy Division Annual Progress Report, Period Ending Dec. 31, 1977, p. 194.

use of this technique in order to highlight radiation damage to the front of a shield

Checkpoints for tables include the following:

- Make sure that the heading at the upper left-hand intersection of the columns and rows (called the *stub heading*) names the category to which the row headings belong.

- Arrange the table so that the data being compared are in the columns, not the rows. That is, if your point is to compare stopping distances of different models of cars and you have those data also at different speeds, the row headings running down the left side are the names of the different car models; if your point is to compare stopping distances at different speeds and you have those data also for different car models, it's the speeds that are the row headings down the left side.

- Minimize the number of black lines in the table. Include enough to guide the reader's eyes but not so many as to clutter the page.

3.2 Graphs

Graphs are the next easiest category of visuals for writers to produce. The *strength* of graphs resides in the variety of types writers can choose among, in their ease of production, and in the ease with which readers can digest the information graphs convey. The *weakness* of graphs lies in their general inability to present complicated data. The more data you want to present and the more different relationships you want readers to see within those data, the less useful graphs become. Moreover, each type of graph has its own strengths and weaknesses. The following sections discuss the various types of graphs, in alphabetical order.

Bar Graphs. A bar graph uses horizontal bars to compare quantitative information, especially information that is linear (the distance from point A to points B, C, D, and E, for example, or the time from conception to birth for species A, B, C, and D) or that is seen at the same point in time (such as military spending in 1991 for the United States compared with spending in Germany, England, France, and Japan for the same year). If the data are being compared over a span of time, a column graph is preferable (see the next section). The *strength* of a bar graph is in its ease of production and ease of comprehension; its *weakness* is lack of precision; unless you print the exact numbers next to (or on) each column, no one will get an exact reading from a bar graph (or a column graph). But to help your readers see how numbers within a set compare at a glance, a bar graph can be very effective, as Figure 10.12 shows.

Figure 10.12   A Bar Graph

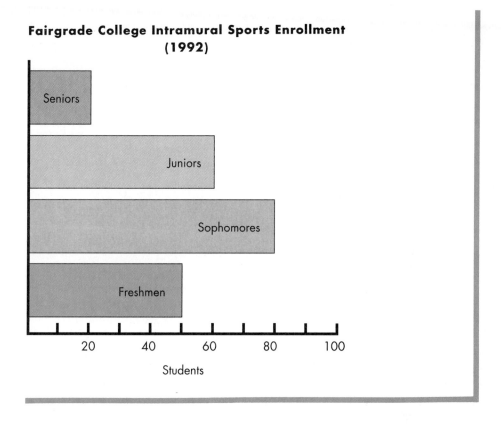

Fairgrade College Intramural Sports Enrollment (1992)

Seniors

Juniors

Sophomores

Freshmen

20   40   60   80   100

Students

Segmented Bar Graphs. This type of graph uses color, shading, or cross-hatching of one set of bars to allow you to compare several sets of numbers by, in effect, placing the bars for one set as extensions of the bars for the other set(s). Suppose you wanted to compare military spending in the United States, Germany, Japan, England, and France in 1990 with the same data for 1980 and 1970. A segmented bar graph using different colors (or different cross-hatching) allows you to put each country's numbers for 1980 as an extension of those for 1970 and those for 1990 as an extension of those for 1980. Thus, you have a comparison that works two ways (across countries and between years). Figure 10.13 shows an example of a segmented bar graph.

Column Graphs. A column graph works in much the same way as a bar graph and offers the same option to produce a *segmented column graph.* Column graphs are easy to produce and easy to read; like bar graphs,

Figure 10.13   **A Segmented Bar Graph**

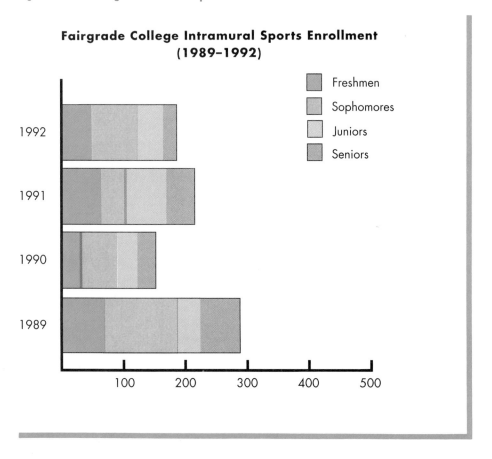

they have the weakness of not being precise. The major difference between the use of bar and column graphs—and it is an important one—is that column graphs are properly used to produce comparisons of sets of data over time in order to show trends. That is, if instead of showing military spending in five different countries, you wanted to show military spending in one country over five different years, you would need to use a column graph instead of a bar graph. While the bar graph is typically viewed all at once, so as to get a simultaneous comparison among its different numbers, a column graph is read more in a left-to-right manner, producing a sequential effect that reinforces the notion of change over time. The *segmented column graph* works like the segmented bar graph, except that the additional bars placed as extensions to the side become additional columns placed as extensions on the top. Figures 10.14 and 10.15

Figure 10.14   A Column Graph

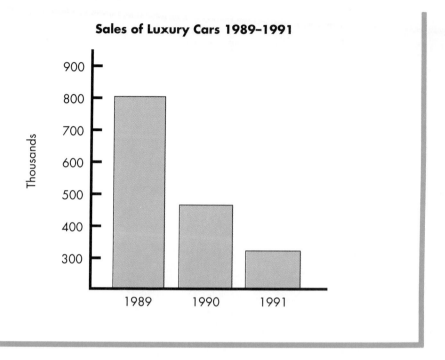

show, respectively, a sample column graph and segmented column graph.

Line Graphs. Line graphs show precise changes within sets of data— changes over time, in temperature, in altitude, or any other variable. Line graphs can be single (with only one line) or multiple (with several lines). Regardless, every line graph has at least three components: the lines comprising the *horizontal axis* and *vertical axis*, the *grid lines*, and the *data*. The *strengths* of line graphs are that they are easy to read (especially if, in the case of a multiple line graph, some distinction such as color is used to differentiate lines of data) and easy to produce with either a word processor or even a simple straight edge and colored pens. The *weakness* of line graphs is that they can be misleading (either through carelessness or deception on the writer's part).

Two problems that result in misleading line graphs happen frequently enough to warrant mention here. *Shifting scales* occurs when the axis of importance, usually the vertical one, uses a misleading or deceptive scale. For example, if a company's profits over a sequence of months (the horizontal axis) were to be stated in millions (the

Figure 10.15   A Segmented Column Graph

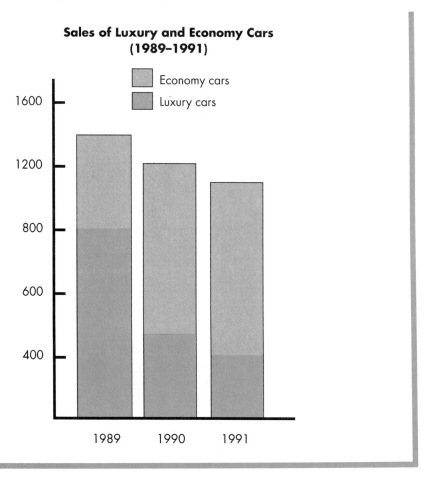

vertical axis), the variation within the vertical scale might be small. If the vertical scale increments were shifted to hundreds of thousands, the variation would appear much larger. Thus, the person who constructs a graph can, by shifting scales on the vertical axis, give very different impressions of what the graph shows as actually taking place.

The second problem, *quick generalization,* occurs when a writer uses a line to connect dots on a graph even though there are no data suggesting such intermediate values and no good reason, given the nature of the process, to expect that, if intermediate values were determined, they might not vary considerably from what the interpolated line suggests. Suppose we were graphing airplane miles traveled by a sales force for the first, second, third, and fourth quarters of the

Figure 10.16   A Multiple Line Graph

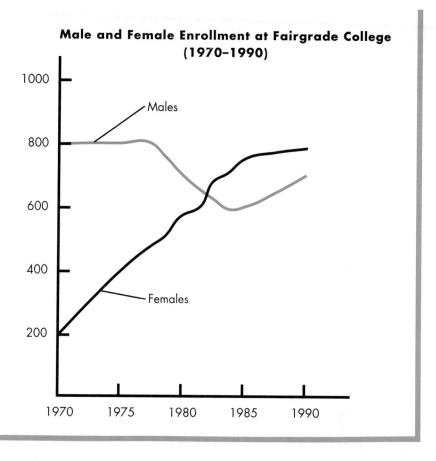

year, and the corresponding values were 2 million, 1.5 million, 1.2 million, and 1.7 million. We could graph those points, but how accurate would it be to connect the points with lines? The lines would suggest a steady decrease between Quarter 1 and Quarter 2, when in fact Quarter 2 could just as well have begun with an all-time peak that immediately plummetted to a record low and then evened out at a point that yielded a total of 1.5 million miles for the quarter. Thus, a careless (or intentionally misleading) graph maker may, by connecting points of data, suggest something the data themselves do not support.

Figure 10.16 shows a multiple line graph.

Checkpoints for line graphs include these:

• Every line graph needs to be set up to be read the same way, from the left to the right and from the bottom to the top. The zero point (or the position where the horizontal and vertical axes intersect) needs to be in the lower left-hand corner. If the zero point is not shown because the values plotted are so high, that fact needs to be called explicitly to the reader's attention, such as by leaving the horizontal and vertical axes unconnected at the lower left-hand corner and by printing some such phrase as "zero point not shown" at that point.

- It's customary for the horizontal axis to be the constant factor (such as Year 1, Year 2, Year 3) and the vertical axis to be the changing factor (such as annual sales); that is, the horizontal axis plots the *independent variable* and the vertical axis the *dependent variable*.

Pictographs. A pictograph uses simple pictorial images (or *icons*) to represent quantities in a manner similar to the bars in a bar graph or the columns in a column graph. If you want to show how many cars the United States imported from Japan for the ten years 1981 to 1991, you might have ten rows (or columns) of cars, with each car image representing 1 million imported autos. Even the simplest clip-art packages on a word processor make pictographs possible, and they offer the advantage of being visually appealing and easy to read. Their *disadvantage* is that they are not accurate (unless you print the exact totals beside each column or bar). Figure 10.17 shows a sample pictograph.

Pie (or Circle) Graph. Pie graphs are perhaps the best way to show the relationship of parts to a whole—for example, how the portion of the federal budget spent on education compares with each other portion. Pie graphs are simple for readers to understand and can be drawn or word processed quite easily. Their main weakness is that when there are too many wedges or when the sizes of the wedges are too closely similar, they become hard for readers to understand. When you make pie graphs, remember that, starting with the twelve o'clock position and proceeding clockwise, the biggest segment goes first, followed by the next biggest, and so on to the smallest. Figure 10.18 shows a sample pie graph.

3.3   Structural Visuals

Structural visuals, such as flowcharts, schematic diagrams, and organizational charts, emphasize organizations and organizational patterns. Their *strength* is their ability to help readers visualize complex patterns at a glance; their *weakness* is their tendency to become too complex.

Figure 10.17   A Pictograph
Pictographs combine numerical and visual impact.

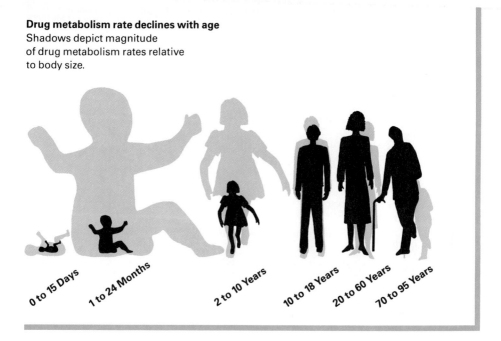

Drug metabolism rate declines with age
Shadows depict magnitude
of drug metabolism rates relative
to body size.

0 to 15 Days   1 to 24 Months   2 to 10 Years   10 to 18 Years   20 to 60 Years   70 to 95 Years

Flowcharts. Flowcharts can range from simple to complex. It's important to remember to make the chart flow from top to bottom and from left to right and to be sure to include arrows in the chart to show which way the flow goes. Figure 10.19 shows an example of a flowchart.

Schematic Diagrams. Schematics typically show the inner structure of something—the circuitry in a two-way radio or the pathways of the human nervous system, for example. Schematics are very difficult for laypeople to produce, but they offer an accurate and detailed view of their subjects. The *disadvantage* with schematics is that their highly technical appearance may drive some readers away; it's easy, that is, for a writer to overestimate readers' ability to digest information presented in a schematic format. Such schematics as architectural blueprints, engineering drawings, and wiring diagrams often can be understood only by experts. Thus, a common problem with schematic diagrams is that writers often forget to simplify the diagram appropriately in the light of the reader's knowledge and level of interest in the subject. Figure 10.20 shows an example of a schematic diagram.

Figure 10.18   A Pie Graph

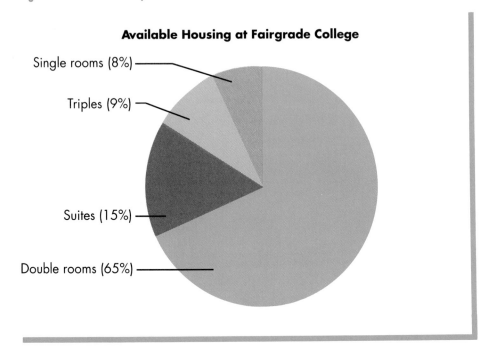

Available Housing at Fairgrade College

Single rooms (8%)

Triples (9%)

Suites (15%)

Double rooms (65%)

Organizational Charts. These charts show the structures of organizations at a glance. As they become more complex, they become less easily accessible quickly and are thus obvious candidates for the audience adaptation technique known as stairstepping (described in more detail in Section 4), in which the complex version of the chart is preceded by a simplified one that provides a quick overview to orient readers.

3.4   Representational Visuals

Representational visuals include photographs and line drawings. Depending on the method of duplication to be employed, photographs can be relatively easy for laypeople to include in professional documents. Line drawings, on the other hand, are very difficult for laypeople to do well and usually need to be drawn professionally. Figures 10.21 and 10.22 show examples of two popular kinds of line drawings: a *cutaway* drawing and an *exploded* drawing.

Photographs can be easy to include in your documents, but you need to plan on a trip to a printer first to have the photographs *screened* (converted to a pattern of tiny black dots) in order to turn them into *halftones,* which can then be printed in your report. With

Figure 10.19   A Flowchart

Such charts are popular for explaining computers.

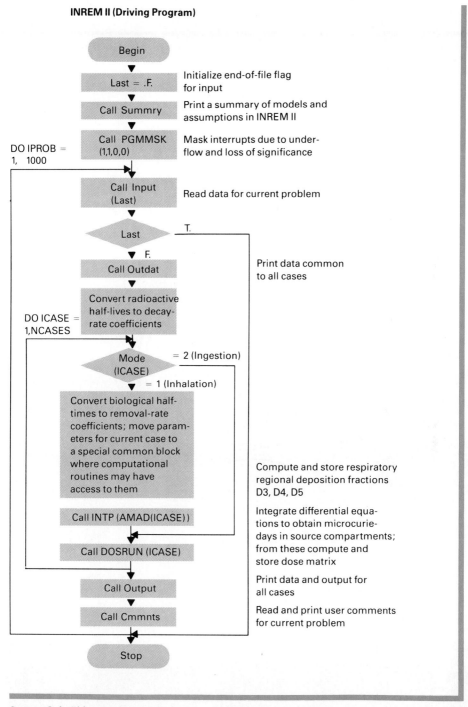

INREM II (Driving Program)

Begin

Last = .F.   — Initialize end-of-file flag for input

Call Summry   — Print a summary of models and assumptions in INREM II

Call PGMMSK (1,1,0,0)   — Mask interrupts due to under-flow and loss of significance

DO IPROB = 1, 1000

Call Input (Last)   — Read data for current problem

Last   T.

F.

Call Outdat   — Print data common to all cases

Convert radioactive half-lives to decay-rate coefficients

DO ICASE = 1,NCASES

Mode (ICASE)   = 2 (Ingestion)

= 1 (Inhalation)

Convert biological half-times to removal-rate coefficients; move parameters for current case to a special common block where computational routines may have access to them   — Compute and store respiratory regional deposition fractions D3, D4, D5

Call INTP (AMAD(ICASE))   — Integrate differential equations to obtain microcurie-days in source compartments; from these compute and store dose matrix

Call DOSRUN (ICASE)

Call Output   — Print data and output for all cases

Call Cmmnts   — Read and print user comments for current problem

Stop

Source: Oak Ridge National Laboratory—INREM II: A Computer Implementation of Recent Models, p. 44.

Figure 10.20 A Schematic

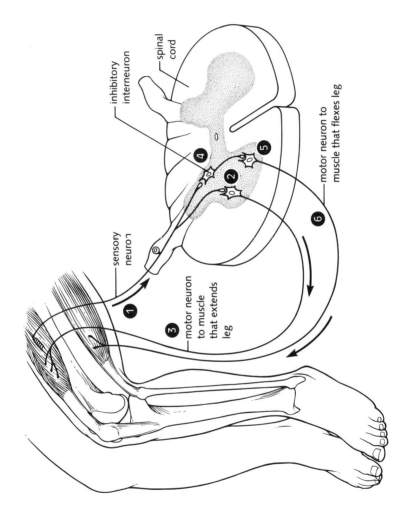

Source: Mikki Senkarik, *Biology: Discovering Life*, D. C. Heath and Co., 1991.

Figure 10.21 A Cutaway Drawing

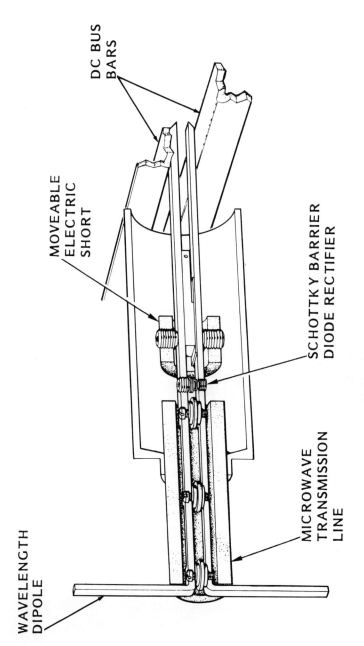

WAVELENGTH
DIPOLE

DC BUS
BARS

MOVEABLE
ELECTRIC
SHORT

SCHOTTKY BARRIER
DIODE RECTIFIER

MICROWAVE
TRANSMISSION
LINE

RECTENNA ELEMENT FOR EFFICIENTLY
CONVERTING MICROWAVE POWER TO DC POWER

Source: Raytheon Company.

Figure 10.22   An Exploded Drawing

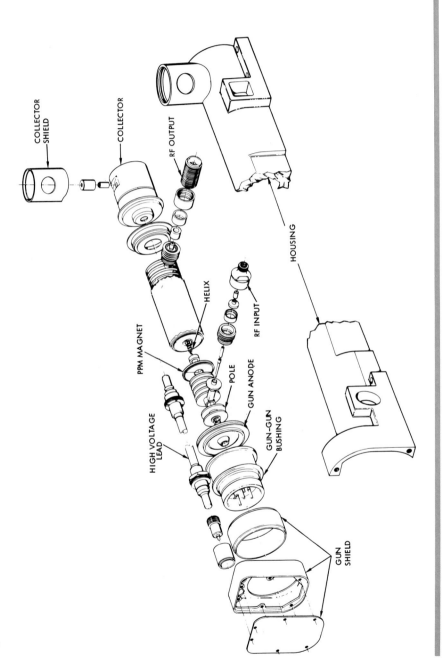

Source: Raytheon Company.

desktop publishing equipment, an increasingly popular alternative is to use a scanner to digitize the photographic image, converting it into a file in the computer that can then be altered (*cropped, reduced,* or *blown up*) and inserted into the document files any way you wish before being printed. The only limitation on this technique is the degree of resolution (number of dots per inch) of the scanner and printer and the size of memory of the computer. For some publishing purposes, today's scanners and printers are not of sufficiently high resolution to produce a crisp image; by the time you are reading this chapter, sufficiently high-resolution scanners may be readily available. Even within current technological limitations, the best desktop publishing equipment gives writers great flexibility to alter and enhance photographs using this technique. At worst, you can use this technique to show a printer exactly where a photograph needs to be inserted into the text and exactly how you want it altered. At best, you can avoid the trip to the printer entirely.

Checkpoints for photographs include making sure the image is appropriately sized, that it has been appropriately trimmed (cropped) to make whatever is supposed to be the most important feature in it in fact turn out visually the most important, that there is sufficient contrast between light and dark tones, and that the overall image is sufficiently crisp.

4. *How Are Visuals Adapted to Specific Audiences?*

As with the other key elements (content, organization, style) of documents, your use of visuals needs to be adapted to the audiences you write for. The key ways to accomplish this adaptation are through the placement of visuals, their frequency, and their kind and amount of detail (especially the annotations or callouts you choose to include).

4.1   Placement

Sections 1.1–1.3 of this chapter mentioned placement of visuals as a function of the three basic uses of visuals. Where you place visuals also says quite a bit about what kind of reader you're aiming at and what kind of purpose you're trying to fulfill with your writing. The worst option is to place all the visuals together at the end of the document. By doing that, you're saying, "I don't expect anyone to look at these," and that is probably what will happen. (An exception would be additional charts and diagrams, included to supplement those in the text proper. Such material can be placed in an appendix.)

Another option is to put the visuals before the text. This is normally done only in advertising materials or in a how-to-do-it manual.

That kind of placement tells the readers that the visuals are more important than the words and tell a story the words can't match.

The third option is to place the visuals directly after the text on separate pages, as in student papers or internal corporate documents that are routine (Class C or B documents, as explained in Chapter 11, Section 1.2). This placement is especially effective if the document is printed on both sides of each page, which often enables you to put the visual on the page facing its reference in the text. Placing the visuals on the page directly after the matching text is probably the norm in most documents in professional life. (This option is further discussed in Section 5.)

The final option is to place the visuals on the same pages with the text. This is called *integration of text and visuals* and is normally done only on prestige publications such as annual reports, prospectuses, and new product presentations (Class A documents). You should expect to have the help of publication professionals to produce this kind of document. However, as computer systems accomplish a fuller integration of visual and verbal elements in mainstream word processing programs and as the screens on word processors more and more show exactly what the printed page (visuals included) will look like, this full integration of text and visuals is becoming more accessible and more common. When done well, it sends readers a strong, positive message about how important the document is to you (and, by extension, should be to them).

One final word about placement of visuals. Sometimes the placement of visuals in your document may be out of your hands, depending on whether you are preparing something directly for publication (often called camera-ready copy), or you are submitting the manuscript to a publisher who will take it through the steps of production and printing.

Publication Style. If your pages will go directly into print—as they do when they go from your word-processor to your laser printer, perhaps to be plastic-comb bound on a machine in your own office—you have control over where your visual will appear within the document. In this case, apart from the various considerations discussed in earlier sections of this chapter, you would generally want to put each visual directly after its reference in the text. But what if the visual needs more room than is left on that page? Rather than squeeze the visual into the available space, you will do better to continue with your text until the end of that page and then place the visual at the top of the next page, resuming your text after it. You might also consider whether you should put every visual on its own separate page, with no text around any of them. (Of course, that means you need to "size" the visual to fill an appropriate proportion of the page.) If you're preparing material that will be joined with other pieces to go in some kind of collection (such as an article to go in a journal or in conference proceedings), such matters may well be specified by the collection's editor.

Manuscript Submission Style. If the document you are preparing will go through some process of editing and typesetting before it is printed, you may need to place your visuals somewhat differently. In such a case, you may be directed by published guidelines for the submission of manuscripts (such as to a journal or to conference proceedings), or by the editor you are working with, to place each visual on a page by itself immediately following its reference in the text; then you would insert a note to the editor indicating the ideal placement of the visual. Suppose the visual is figure 10; your note will look like this:

[figure 10 here]

During the publication production process, when the document is being readied for print, the visual will then be placed as close as possible to this position.

4.2   Frequency

The frequency of your visuals will vary with how visual your topic is. A report about the movements of river otters being reintroduced into the Great Smoky Mountains National Park and tracked with surgically implanted radio homing devices would be surprising if it didn't include a number of maps showing their travels. But apart from consideration of the nature of the topic, how many visuals you use tells your audience to what extent you expect them to be readers and to what extent viewers. That quality is determined somewhat by the audience themselves but also somewhat by you. If you intend your document to be looked at more than read, use more visuals.

4.3   Kind and Amount of Detail

The biggest way you can adapt visuals to readers is by adjusting the kind and amount of detail the visuals present. The tendency for most writers is to overdo the level of detail—after all, you have all these data, why not use them somewhere? To make visuals as effective as possible, however, you need to use exactly the right amount of detail for that reader and that visual. One way you can do this is by the number and kinds of annotations (some people name them *callouts*) you add to visuals. Generally, the more parts of anything you label, the more complex you make it for the reader and the more you seem to be assuming a technically trained or expert-level reader. A report written on a layperson level but containing highly detailed visuals with every part, however small, labeled may be presenting a contradictory message to the reader. Remember also that your annotations need to be oriented on the page the same way the rest of the document's text is. The only exception is for a visual that is turned 90 degrees counterclockwise before printing (known to editors as a *broadside* or *landscape* orientation) because it's too wide for the normal page. In that case, the annotations are similarly rotated. (If you do this, remember to set up the page so that readers turn the page clockwise to read it.)

Another audience adaptation technique is especially notable because it so clearly shows the principles of audience analysis and adaptation in operation. Whenever your subject matter is especially complex, and you suspect it may dismay or mystify your audience with its complexity, you should consider *stairstepping* visuals. Present a simplified version first, followed later by the fuller version (sometimes as much later as in an appendix).

Although there is no simple way to generalize about when a visual is too complicated for any particular audience, one measure to go by is the "Magic Number 7 ± 2"; that is, any visual presentation of material is limited in its usefulness by the number of units the human brain can deal with simultaneously, often thought to be five to nine units (the problem, of course, is defining what a "unit" is). According to this theory, when your visual has more than seven to nine parts, you should consider stairstepping it. Figures 10.23a and 10.23b show two stages of a stairstepped visual.

Figure 10.23a   A Stairstepped Visual
Here we see only the highest levels of the organization.

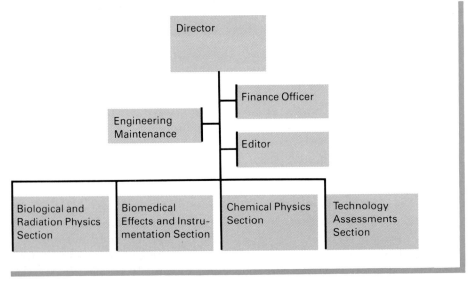

Figure 10.23b
Here we see the lower-level sections added.

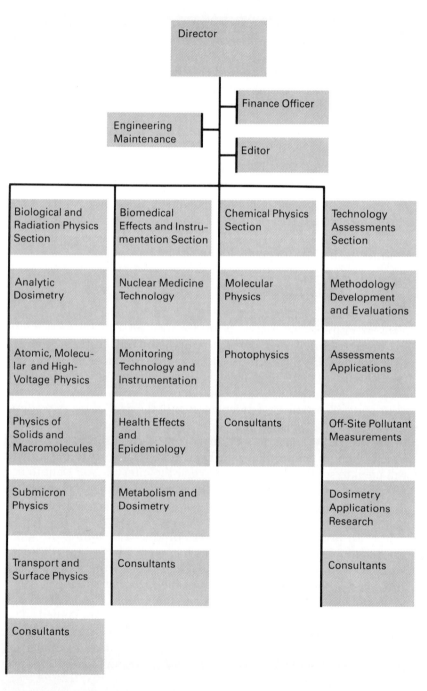

Stairstepping visuals also meshes well with another of the demands human beings, as information-processing systems, make on the structure of the information they process. Rather than presenting a mass of unsorted data (often called a laundry list), visuals that are stairstepped present a *hierarchically ordered* set or series of sets. Human beings, like most other information-processing systems, handle hierarchically ordered information better than unordered information. (This is also true for textual information; see Figures 9.4 and 9.5 and their accompanying text.)

5.   *How Can Visuals Be Produced?*

Physical production of some kinds of visuals is simple for anyone; other visuals (such as exploded or cutaway drawings) can be done only by professionals or, in the case of isometric graphs and isometric views (see Figs. 10.4 and 10.24), by computers. In general, how you produce each visual depends on its category.

5.1   Tables

Tabular graphics are relatively simple to produce with most mainstream word processing systems or by hand with just a pen and a straight edge. It's not even necessary to draw all those lines that you might think a table would involve; often having a vertical line after every five columns and a horizontal line after every five rows is sufficient and will produce a readable visual in the eyes of many readers.

When you're constructing tables with a word processor, you will probably need to remember to turn off the right justification feature, which when turned on aligns the right edge of your writing. And you will need to set the tabs on that page at the intervals you want to use to start each entry. On most machines, simply hitting the space bar to get from one entry to the next will produce a printed page that looks quite different from the nicely aligned screen you had made. The word processor reads the spaces produced by hitting the space bar as variable character spaces, which it uses to make pages of words look more like typeset pages by making some of the spaces bigger and some of them smaller. While variable character spacing is a plus in producing text, it can make the printout version of your table show zig-zag columns even when the screen showed them aligned. The solution is to get from one entry to the next with tabs, not the space bar, and to turn off the right justification. There are also numerous spreadsheet programs available for word processors that set up neat tables automatically.

Figure 10.24 An Isometric View Chart
This kind of visual requires a computer to produce.

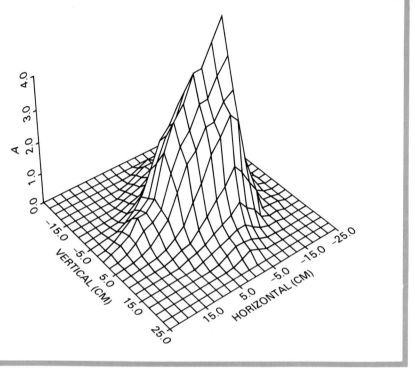

Source: Oak Ridge National Laboratory—Fusion Energy Division Annual Report, Period Ending Dec. 31, 1977, p. 15.

5.2 Graphs

Line graphs, bar graphs, column graphs, and pie graphs are becoming more common features of mainstream word processing programs and are also easy to do by hand. The most critical factor when you produce such visuals yourself is to keep them simple. Make sure the lines are crisp and clean and that the visuals are not crowded on the page. Few dot-matrix printers can produce a smooth enough curved line to make a good pie graph. In fact, most computer-generated visuals need a laser-quality printer to look their best.

Other problems can arise as these visuals become more complicated. Producing a good multiple line graph or segmented bar or column graph may require you to make more differentiation among different kinds of lines than a basic word processing program allows. Again, there are many different graphics software packages available that can handle this chore, and none of these kinds of graphics is hard to do by hand with colored pens and a straight edge. The most complex graphics, such as pictographs, probably should be left to professionals.

5.3 Structural Visuals

Two of the three types of structural visuals, flowcharts and organizational charts, are simple to generate by computer or by hand. The key points are to keep them simple and to make the entries clear. In the case of flowcharts, make sure to use arrows to indicate the direction of flow. The third type of structural visual, schematics, should be left to professionals to produce.

5.4 Representational Visuals

Line drawings generally cannot be done without the help of a professional artist (or a computer with an extensive clip-art file). Not only are the drawings themselves difficult for most of us to do, but the physical nature of the drawing itself—for example, the kind of ink it is done with—can defeat most laypeople who try to do their own. And the most useful line drawings, the cutaway and exploded views, are beyond the skill of laypeople. Photographs, on the other hand, can be relatively simple to add to your documents if you remember to leave time to have them screened by a printer and turned into halftones or have access to a reproduction-quality scanner that will allow you to digitize the photographs and enter them into your text via the computer. (These options were discussed in more detail in Section 3.4.)

EXERCISES

1. Choose three short factual articles from your local newspaper and produce sketches of visuals to illustrate them. You can assume the visuals are done for the same audience as the newspaper is written for. Try to use various kinds of visuals. Provide copies of the articles with the suggested placement of the visuals marked on them.

2. Zeno, the Greek philosopher, is said to have represented the way people learn in four stages of visuals: perception is an open hand, the mind's response is a partly closed hand, the mind's grasping of an idea is a fist, and knowledge is that fist enclosed in another hand. Divide one of the following processes into four to nine stages and produce sketches for conceptual visuals to represent each stage. Try to capture the same sense of

---

A ▶ indicates a case study exercise.

rightness that Zeno's example shows.

Pollution	Writing a	Environmental
Conservation	paper	restoration
Gravity	Import/export	Computer
Consumer	Hysteresis	virus
advocacy	Inertia	infection
Corporate	Centrifugal	Negative
responsibility	force	symmetry

3. Based on Figure 10.25, produce sketches for five different visuals representing the information the table contains. Assume an audience of taxpayers in general—people who are simply curious about how government
money is spent.

Figure 10.25   House of Representatives Research Budget Requests ($ millions)

Agency	1982 House appropriation	Administration request appropriation	1981
NASA	$6133.9	$6122.2	$5522.7
R&D	4938.1	4903.1	4336.3
Research and program management	1100.0	1114.3	1071.4
Construction of facilities	95.8	104.8	115.0
EPA	1201.5	1191.4	1351.0
Salaries and expenses	583.7	582.8	553.7
Abatement, control and compliance	422.5	413.9	540.2
R&D	191.2	190.6	253.0
Buildings and facilities	4.1	4.1	4.1
NSF	1103.5	1034.5	1022.4
Research and related activities	1065.0	1020.1	946.7
Science and engineering education	35.0	9.9	70.7
Special foreign currency program	3.5	3.5	5.0
TOTAL	$8438.9	$8347.1	$7896.1

Source: Chemical and Engineering News, Aug. 3, 1981, p. 14. Reprinted by permission.

Figure 10.26   State University Contributions to the Capitol City Economy

No statistics are regularly available to show the monetary influence of State University upon Capitol City. The purpose of this report is to provide such statistics. Table 1 shows the estimated direct and indirect impact of University expenditures calculated for the 1990–1991 fiscal year. Column 1 of the table lists total University expenditures according to specific category. They are adjusted by estimated proportions of spending occurring in Capitol City (Column 2) and by the personal-income factor (Column 3). The resulting estimate of total local wages and salaries attributable to the University's direct expenditures appears in Column 4.

University expenditures increase not only local wages and salaries but also profits made by local suppliers of goods and services. Estimates of retained profits in the Capitol City Standard Metropolitan Statistical Area (SMSA) were calculated using a uniform 10 percent rate of return on all sales, and assuming that one-half of the resulting profits remain in Capitol City. Column 5 lists updated retained profit estimates. Total local income (sum of Columns 4 and 5) is listed in Column 6. Total 1990–1991 expenditures of $341.2 million translate into a personal income increase of $188.4 million. Total direct and indirect impact of University spending is calculated by multiplying the figures in Column 6 by a personal-income multiplier of 1.75; resulting figures are listed in Column 7.

Student expenditures also contribute to the local economy and are thus listed here. The Office of Student Affairs provided a total 1990–1991 figure for direct student expenditures of $132.8 million. University-related visitor expenditures provide another source of local income. The best estimate of these expenditures comes from a local Chamber of Commerce study; their figure of $25 million is used here.

Total direct and indirect impacts of University-related expenditures may thus be seen to be $423.1 million for the 1990–1991 fiscal year. Of that total, payroll expenditures made 62 percent, other expenditures 16 percent, student expenditures 19 percent, and visitor expenditures 3 percent.

Figure 10.26 (continued)

TABLE 1 Estimated Direct and Indirect Impact of Expenditures of the University on the Capitol City SMSA Economy, Fiscal Year 1990–1991

Activity	Total Spending Column 1	Proportion of Spending in the SMSA Column 2	Personal Income Factor Column 3	SMSA Wages and Salaries Column 4	SMSA Retained Profits Column 5	SMSA Personal Income Column 6	TOTAL Local Impact Column 7
Payroll:							
State U.	$108,810,273	.90	1.00	$97,929,246	0	$97,929,246	171,346,180
System	10,188,516	.61	1.00	6,214,995	0	6,214,995	10,876,241
Ex. Station & Ex. Service	27,789,801	.32	1.00	8,892,736	0	8,892,736	15,562,289
Hospital	41,576,141	.90	1.00	37,418,527	0	37,418,527	65,482,422
Other Ex.							
Travel	9,615,584	.41	.06	236,543	197,119	433,662	758,909
Utilities	18,865,453	.94	.11	1,950,688	886,676	2,837,364	4,965,387
Communication	9,324,362	.93	.08	693,733	433,583	1,127,316	1,972,803
Supplies	37,735,967	.53	.66	13,200,041	1,000,003	14,200,044	24,850,077
Stores	30,267,701	.68	.39	8,026,994	1,029,102	9,056,096	15,848,168
Construc.	29,936,178	.69	.24	4,957,431	1,032,798	5,990,229	10,482,901
Other	17,058,420	.74	.29	3,660,737	631,162	4,291,899	7,510,823
Total	341,168,396			183,181,670	5,210,443	188,392,114	329,686,200

Figure 10.27 A Conceptual Visual
The drawing illustrates the idea of economy.

"The development should maximize the use of available funds."

> 4. You are a student intern in your university's Information Office, and you have been handed a copy of the report in Figure 10.26. Your task is to come up with ideas (including sketches) for five or ten visuals to illustrate the report when it is given to members of the local city council. The purpose is to impress upon them the importance of the university's contribution to the local economy.

5. Conceptual visuals offer an interesting way to draw readers' attention to a particular piece of text. Figures 10.27 and 10.28 contain examples of conceptual visuals. As the examples show, conceptual visuals are especially good for helping people notice, visualize, and understand ideas. For this assignment, search the printed matter that surrounds you in your daily life—textbooks, newspapers, magazines, pamphlets, and so on—to find four conceptual visuals. Make a photocopy of each one, including enough of its source for your reader to understand its context, and write a short report explaining how each of the conceptual visuals works (or doesn't work) in its setting.

Figure 10.28 A Conceptual Visual Sequence
The first (a) illustrates all the different areas of knowledge and skill a chemical engineer needs. The document has sections that correspond to each one, and each section is introduced by its own visual (represented here by *b, c,* and *d*).

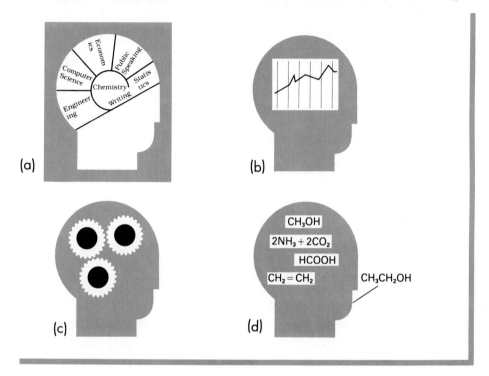

PART FOUR

Processes in Professional Writing

That a professional's writing consists of nothing more than reporting facts is a common misconception. It holds that the essence of good writing consists of keeping the words out of the way of the facts. Nothing could be further from the truth. If good writing is that simple—

- Why are our desks usually littered with three or four different versions of the user's manual for our computer and its software? Why do so many people purchase two or three manuals for each program in a vain attempt to find one that truly is "user friendly"?

- Why do articles with titles like this one, "If Johnny Can Read, Why Can't Johnny Write a Decent Memo?" keep turning up? (This one was in a recent MBA professional journal.)

- Why are scholarly, professional, and trade journals filled with so many poorly written, needlessly difficult-to-understand articles?

- Most critically, why does the student (or professional) who puts off writing until the last minute on the basis of "I've got all the data together; all I need to do now is write it up" so often meet with disaster?

Doing effective professional writing is more than just putting down the facts. From the simplest manual to the most complex multimillion-dollar proposal, there is an important abstract quality to effective writing. It's that quality that makes the words on the page "say" the same thing to the reader that the facts and data have "said" to the writer. In terms of the material presented in this book, that abstract quality expresses the way a document's purpose, message, and audience, along with the writer's role, work together to make a successful piece of writing.

What name should such an abstract quality of documents be given? If we were talking about how form and function, practical concerns and aesthetics, come together in an automobile, a building, or a chemical manufacturing process, we would call that quality *design*. Let us use the same word here and agree that *design is the abstract quality that unites the key intangible elements of the communication situation to produce a successful document*. A successful document, in this context, means one that goes beyond the bare presentation of facts to create effective communication:

- In a set of instructions, this quality of design may find its most concrete expression in the selection and placement of visuals.

- In a technical report, the quality of design may be most operative in decisions the writer makes about whether a particular passage gets placed as a major section or as a minor section, about what is more important and what is less important, and about what gets said in what order.

- In a proposal, that quality of design can be reflected in the outcomes of hundreds of decisions, both large and small, from the proposal's overall persuasive strategy down to specific word choices.

This quality of design becomes a critical factor in the changes that take place between the writing tasks students face and the writing tasks professionals face. In most school settings, writing tasks come neatly packaged. Your freshman English teacher may say, "Write a five-page analysis of your responses to *Hamlet* in the light of the last three days' class discussions," or, "Write a 10- to 15-page report on the effect of El Niño on Peruvian anchovy fisheries." In this kind of writing, the key intangibles (purpose, message, audience, and writer's role) are set, as are the document's formal requirements (length, margins, and so on). Thus, people whose primary experience with writing occurs in classroom settings might well be inclined to believe that just pouring the facts into the mold is writing's most important element—indeed, is all effective writing requires.

When you move into professional life, all of the key intangibles are unknowns, and thus their expression in terms of design becomes critical. That is, when all you know is that the Savannah plant, or the Oakland office, is not operating up to par, and when all you know of your writing task is that you need to analyze the problem and recommend a solution, you must be able to think of writing as something more than just reporting the facts. When you're given the task of explaining to employees that professional management depends on strict (but not mindless) adherence to policy and the only guideline you're given as to how to do that is that you must use words on paper, the single key you have as to where to begin lies in determining the values of the key intangibles so as to come up with some kind of overall design.

Design explains how all of the elements already discussed in this book add together with the processes discussed in this section to produce effective professional writing. If the earlier chapters were building blocks, these next six chapters explain the construction process. How those two sets of elements get put together is the element of design.

Chapter 11, Writing in a Professional Setting, shows how your audiences and purposes shift when you move from school- to work-related writing, how these changes require alterations in your writing, and how careful consideration of these elements will help you solve problems in design. Because it deals with intangible processes in writing as well as finished products of writing, Chapter 11 is somewhat more philosophical than Chapters 2–10. In that sense, Parts I, II, and III of this book are its first half (with Chapter 1 as that half's introduction), and Parts IV and V are the second half (with Chapter 11 as the second half's introduction).

Chapter 12, Writing Definitions and Descriptions, explains two of the most fundamental and most important skills any professional writer needs. Two examples of such writing, product descriptions and service descriptions, are examined in detail.

Chapter 13, Explaining Processes and Giving Instructions, presents the way to use narration to accomplish these important professional writing tasks. Writing user manuals, perhaps the premier instance of such writing, is explained in detail in this chapter.

Chapter 14, Solving Problems, addresses the professional writing situation that most requires the element of design: writing reports that effectively state solutions to hard-to-define problems. The processes described are those that can help you unite, for example, the definition and description techniques of Chapter 12 with the narrative techniques of Chapter 13. Although every professional field has its own approaches to problem solving—approaches requiring advanced knowledge of engineering, biology, chemistry, accounting, management, medicine, and so on—the processes presented here will be valuable additions to anyone's problem-solving skill, regardless of field.

Chapter 15, Making Recommendations, explains how to make the recommendations you include in your reports clear and convincing. Nothing hurts an otherwise well-written report more than poorly presented, unclear, or unconvincing recommendations.

A final writing process to master is library research, which is addressed in Chapter 16. Although students often think this is only a school-related skill, most professionals find that they need to know how to use research libraries often during their careers. The more writing professionals do, the more they use research libraries.

11. *Writing in a Professional Setting*

In school you write mostly to prove your knowledge; on the job you frequently write both to provide information and to guide the thinking of decision makers. In school the people in your audience—your teachers—are paid to read your writing, and they read it in order to measure your mastery of academic subjects. On the job your audience can (1) read your report, (2) skim it, (3) route it to someone else to read, (4) return it for revision, or (5) discard it!

Professional audiences read your reports primarily to make decisions about courses of action, to decide what to do or how to do it, and only indirectly to make decisions about *you*. And professional audiences are likely to be very different from your college instructors. Professional audiences come from all different levels in the organization, from outside the organization, from different educational backgrounds, and even from totally different cultures.

- Suppose, for example, that your company lands a big contract to supply machinery for 15 manufacturing plants in Central Europe. Assuming that the business people with whom you are dealing read English (generally they do), how will their different backgrounds require changes in your writing?

- Suppose your company has an opportunity to open up a market for financial consulting services in Japan. How will the letters you write to clients there need to be different?

- Or suppose your company sells a Middle Eastern state computer equipment. How will you need to revise the manuals and other documentation for that special audience?

As you move from college to professional life, these important changes in audience and purpose for your writing will require equally important changes in the way you design your reports.

Other features of professional writing will also require changes from the way you write in college. In college you may have a vague idea that one writing project you do is more important than another, requiring more preparation in terms of a special cover, professional art work, and so forth. On the job, most documents—letters, manuals, reports, brochures—are carefully categorized as belonging to one of several classes, depending on whether they are for internal or external audiences and on the level of formality they are felt to require. In college you are usually prohibited from collaborating with other students on the authorship of papers; in professional life joint authorship is the rule, not the exception. In college your report is turned in, and that's usually the end of it. In professional life you frequently have an editor, and possibly even a publication production process (professionally done graphics, typesetting, printing, distribution) to deal with. In college your writing almost always has about it the air of an exercise—essentially the same activity being done over and over by different students in order to build their mental muscles. On the job there's an air of reality about the writing assignments that gives new professionals a feeling that they may never have had about their writing as students. Here we will call that feeling *accountability*. When you write on the job you feel your writing matters; it matters to you, your business, your career, maybe even to the world. Be prepared for that feeling, because it can change your behavior as a writer in complex ways.

This chapter discusses five aspects of writing that are crucial for students making the transition to professional life:

1. Design decisions
2. Joint authorship

3. Editorial procedures
4. Computers and writing
5. Accountability and ethics

Each will have an effect on the process of your writing as a professional. You can begin to prepare for those effects and turn them to your advantage by reading this chapter carefully.

1. Design Decisions

Most writing in school operates on only one level: tracing the evolution of a student's understanding of some subject. For example, a typical lab report may well be a strictly chronological account of the activities in the lab: purpose, materials, procedures, results, and discussion. A typical English essay may trace the evolution of a relationship between two characters in a play, building out of that some idea of the relationship's significance. A typical business case analysis may trace the past, analyze the present, and speculate about the future of a company or market. Such writing operates mainly on the level of the writer's experience or understanding; it catalogs the writer's activities.

Although that kind of one-level (or *catalogical*) writing is both appropriate and traditional in schools, it is neither in professional life. As a professional person who writes, you will find that your readers expect more of you. Your readers want the material they read (including your experience and understanding of it) sorted out, analyzed, and discussed in terms of their understandings and their purposes. Rather than one-level, catalogical reports, your readers on the job will expect you to produce *multi-level, analytical* reports.

1.1 Choosing the Right Kind of Structure

Seen in their broadest dimensions, the questions of whether a document is designed to be analytical or catalogical and whether it is designed to be multilevel or one level are questions of structure. Although how you answer those questions may affect any number of other factors—typography, tone, style, placement and use of visuals, and so on—the primary difference is one of structure. The next two sections look at these structural differences in detail.

Catalogical Versus Analytical Reports. Did you ever try to figure out the organization of a catalog? Within individual groups of items, there may be an apparent pattern—men's clothing after women's, and children's after men's—but why then does it go to sporting goods? And why then to luggage, and then to housewares? You either have to know the organizational pattern or use the index. Such an organizational pattern, if there is one at all, is coming *not* from anything inherent in the way the reader uses the catalog but from some kind of (apparently hidden) quality of the subject matter. (In fact, one major department store chain organizes items in catalogs in the sequence one would see by walking around the inside of the store clockwise, which could be a useful organization if the readers understood that.)

The word *catalogical* applied to professional documents means any organization derived according to principles other than the reader's needs—that is, any structure that pays no heed to the reader's purposes in reading. An *analytical* structure is the opposite of a catalogical one; it usually implies that the writer analyzed who the readers would be and decided how their purposes in reading needed to be reflected in the way the document was written—especially in its structure.

Consider two different versions of the same proposal. (This example, like the one that follows, is adapted from an actual piece of writing presented to a professional editor for extensive revisions.) In its original, catalogical version, the unrevised proposal is eight pages long:

- Two pages describe the company proposing to do the work.
- Two pages describe similar other work done by this company.
- Two pages describe the credentials of the people in the company who will do the work.
- Two pages describe exactly what the company proposes to do.

This version requires readers to get to the end of the proposal before discovering what the company is proposing to do. The structure thus reflects the experience of the writer, who is working from her company outward toward the client and leading up to the proposed work, but it violates the needs of the reader, who is bound to ask *first* what is going on and only later who is doing it.

Now consider how a professional editor restructured the proposal into an *analytical* structure, asking from the start, "What will the readers want or need to see first, second, etc.?" This revised proposal is also eight pages long but considerably different in its overall design, especially its structure:

- A one-page list spells out what benefits will come to the client company if it accepts this proposal.
- Two pages spell out what the proposing company is seeking to do.
- Two pages describe the company's personnel.
- Two pages describe the company's previous related experience.
- One page describes the proposing company itself.

This analytical version, which is in some ways the reverse of the catalogical one, leads the readers from the project's positive impacts on them, to the project itself, and then back to the proposing company's qualifications. By thus starting in the reader's world and moving outward, this version is bound to get a better response from its readers because it was written with them firmly in mind. It illustrates clearly the source for the distinction between catalogical and analytical structures: the former is based on the writer's point of view, the writer's needs, the writer's experiences; the latter, on the reader.

The next example is a more complex illustration of the same distinction. This example is taken from before and after versions of an advanced user's manual for a general-purpose management information system called TETRA (not its real name). To make the presentation here clearer, the example has been adapted by omitting irrelevant sections. The context for this example is that at the end of TETRA's *Advanced User's Guide,* the author wants to show readers how to use TETRA to solve a typical complex management problem. In the examples, you will see the sample problem presented first in its original, catalogical form, and then in its restructured, analytical form.

In the catalogical approach (Fig. 11.1) the writer dives right into the process of *explaining* the technical content without adequately preparing the reader for the *significance* of the content (or, in this case in particular, preparing the reader for the fact that, just when the problem gets most interesting, the reader will be referred to another manual to see how to work out the rest of the problem). But these are exactly the things most readers will really want to know before they start into the problem. That is, here we have a fairly typical picture of a writer's presenting information in the manner of a catalog (here a series of steps) without first analyzing what higher-level generalizations or preparations the reader will need in order to be able to understand that series of steps most efficiently.

Compare that approach to presenting TETRA's sample problem with the one shown in Figure 11.2. Assuming a manager—a decision maker—as the reader, which would get a more favorable response? The second, analytical one would. It forecasts the structure of the upcoming problem, and it prepares the reader for the outcome. In the first (catalogical) version, the outcome seems to some readers to be a sort of halfway solution; that is, you know the project is in trouble, but this manual doesn't explain how to find the source of the problem. That frustrates some readers; so does not knowing the structure of the problem-solving process in advance. Because the analytical approach foresees and satisfies (in advance) these potentially frustrating points, it is the more desirable of the two. Because its underlying organization is based first and foremost on anticipation of the reader's needs, the analytical approach characterizes the kind of writing you will need to do for decision makers.

Figure 11.1   A Catalogical Approach to Writing About a Sample Problem

SAMPLE PROBLEM

TETRA can be used in many ways. One question managers frequently must answer is to determine how much is currently available in uncommitted funds. The equation for answering this question is: "Uncommitted funds equals Budget minus Cost-to-Date minus Uncosted Commitments minus the Projected Fixed Costs." Solving this equation is complicated by the fact that these data are held in different data bases. The advantage of the TETRA management information system is that it can access a number of different data bases as needed to solve this kind of complex problem.

The first problem is to figure out how much funding is allocated to the project. This information is available in two online data bases, OTTL and SUMMT. In order to access this information, instruct TETRA to . . .

[Two pages omitted here.]

Therefore the project's initial budget is $1,986,000.

Next, you need to determine Cost-to-Date. This information is available in the Ongoing Data Base (ODB). TETRA can access this data base through . . .

[Two pages omitted.]

Thus the Cost-to-Date is $1,087,781. These data can also be used in estimating the projected fixed cost . . .

[One page omitted.]

Thus the projected fixed cost is $80,889.

Next you need to take into account any other commitments outstanding. These commitments could be in any of three different data bases: direct outside purchases (DIROPS), subcontracts (SUBS), or outstanding work orders (OWORS) . . .

[Three pages omitted.]

Therefore the currently outstanding commitments total $1,023,505.

We now have values for all the unknowns in the equation. That is, . . .

[Calculations omitted here.]

. . . the uncommitted funds equals a *negative* $253,540. Obviously, the project is already in trouble. The TETRA management information system will help you isolate the problem. With TETRA, you can determine what specific accounts have problems and specifically where in each account the problem is. TETRA can be used to point out all the key data at these lower areas. This process is explained in the Sample Problem in the TETRA Expert User's Guide.

Figure 11.2   An Analytical Approach to Writing About a Sample Problem

SAMPLE PROBLEM

TETRA is a general-purpose management information system that can provide its users with information in many ways. This sample problem shows you how to use multiple data bases to answer a typical question managers have to deal with: "How much do I have available in uncommitted funds?"

Four variables must be determined in order to solve the problem: the budget, the cost-to-date, the uncosted commitments, and the projected fixed costs. Each of these pieces of information is in a different data base, and it is one of the particular strengths of the TETRA management information system that it can search a number of different data bases to enable you to solve just such a problem as this easily. This sample problem shows you how that is done.

At the end of this process, you will learn that the project in question is several hundred thousand dollars in the red. Just as TETRA can help you determine that important fact through the process shown here, so TETRA can also help you determine exactly where (in what accounts) the specific problems lie. Although this second-stage, lower-level use of TETRA is beyond the scope of this manual (it is explained in the TETRA Expert User's Manual), it is summarized here to give you some feel for how versatile TETRA can be.

The formula for finding uncommitted funds is . . .

[From here the two approaches are the same, until right at the end.]

The calculations show that the project is over $200,000 in the red. The TETRA management information system will help you isolate the problem. With TETRA you can determine what accounts have problems and where those specific problems are. Although this further application of TETRA is beyond the scope of this manual, it is explained in detail in the sample problem in the TETRA Expert User's Manual. But in case you are curious to know just a little more, here is a brief overview of this further application of TETRA.

[One page omitted here.]

Multilevel Documents. In the TETRA manual, the analytical element consisted of anticipating an obvious problem readers would have with the manual's first (catalogical) version (cf. Chapter 4, Section 2.7), and dealing with that possible problem by more carefully introducing the report—in this case, by more carefully framing the process to be described (cf. Chapter 13, Section 2.1). In other reports, the analytical element could be something quite different. It could, for example, be the presence of multiple levels of information presentation in the document.

If the professional journals in your field print abstracts or summaries at the beginnings of articles, you may already be used to seeing one form of two-level report. The presence of an analytical level, the summary or abstract, assists you, the prospective reader, in making the decision of whether to read each article.

Another example of the multilevel report is the multisection user's manual (described in Chapter 13). If you write a user's manual with three main sections—a how-to part for beginners, a description of the machine or system explaining each part's capabilities, and a troubleshooting section—you've written a three-level document, each part speaking to a slightly different audience or purpose.

Perhaps the most interesting and important use of the two- or multilevel report structure occurs in long professional reports of the sort described in Chapter 19. In such reports the multiple levels of design include not just the segmentation of the report proper but also the employment of multiple introductory devices. That is, in addition to the report's body being broken into sections corresponding to several levels, the opening pages of the report may include

- a short statement of authorization (explaining why the report was written).
- an abstract (summarizing the report's scope).
- an executive summary (tying the report's coverage to its significance).
- a formal (*cpo*) introduction to the entire report.

This use of multiple introductory devices, explained in detail in Chapter 8, exemplifies the use of multiple levels in report design. The practice is extremely useful for the decision makers for whom professional reports are designed, but it can be extremely frustrating for would-be writers who don't see why or how so many kinds of front-matter need to be used.

Subsequent chapters touch on the distinctions between catalogical and analytical and one-level and multilevel documents. The difference between writer-centered and reader-centered writing, explained in Chapter 12, Sections 4.1 and 4.2, is a more general form of this same idea. Compartmentalizing résumés (discussed in Chapter 6) and stair-stepping visuals (discussed in Chapter 10) are also examples of this principle on a lower level. More to the point, the recommendation report with the recommendation at the beginning rather than at the end (developed in Chapter 15) is a specific example of using two levels (one for the recommendation, another for the background to it) in a popular and successful application.

The following chart shows some of the distinctions between one- and two-level report writing.

One-level Report	*Two-level Report*
More writer centered	More reader centered
• Catalogs writer's experience	• Analyzes significance of writer's experience for reader
• Example: recommendation at the end	• Example: recommendation at the beginning

No particular difficulty or skill is involved in writing a two-level (as opposed to a one-level) report. You merely need to size up your purpose and audience accurately and to realize that the situation calls for a two-level design. Typically that situation is characterized by writing for a decision maker. The only additional process required is that of giving consideration to the psychological aspects of your report's design.

1.2   Choosing the Right Class of Document

In school you may put a little extra effort and finishing touches on an occasional piece of your writing, but typically most student papers reflect about the same amount of polish. On the job, there are many more possibilities, ranging from the memo you keyboard, print out, reread and revise, print out again, and circulate, all the way up to the annual report written by a professional writer, edited by a professional editor, with photography by a professional photographer, and printed by a professional printer. What you see on the job is many different levels of finish and many different levels of effort. The convergence of effort and finish is often expressed in terms of *classes of documents.*

There is no standard classification scheme for naming the degree of formality and finish, the amount of production and expense, needed for various kinds of documents. Although some reports, for example, are merely photocopies of typed pages, others are glossy, multicolor, bound documents. One popular scheme divides documents such as reports into three classes—A, B, and C.

Class A Reports. The Class A report (often called a "prestige" publication) is usually typeset. Its columns may be justified on both sides of the page. There is abundant artwork, and the art is integrated with the text. The report is printed and bound on high-quality paper, frequently in several colors. The format of Class A reports varies widely and is often quite creative. Class A reports are nearly always external communications (directed at people outside the writer's company). Typical Class A reports include annual reports to stockholders, prospectuses, and new product presentations. Figure 11.3 shows a two-page spread from a typical Class A report.

Class B Reports. The Class B report (the kind most frequently seen in business and industry) is usually word processed, with unjustified pages (although the use of word processing has made justified edges available in Class B reports as well). The illustrations and the text are on separate pages. Although the report is professionally reproduced and bound, it uses more economical materials than the Class A report. For example, the covers of Class B reports are usually standard, stock covers that the firm has printed in large quantities and uses on all its Class B reports. The format of the Class B report is usually determined by the company involved. Class B reports may be either internal or external communication. Typical Class B reports include technical memoranda, startup procedures, and training manuals. Figure 11.4 shows a two-page report from a typical Class B report.

Class C Reports. The Class C report is word processed, the illustrations and text are on separate pages (illustrations may be gathered together at the end), the copies are usually done on an office copier, and the format is strictly utilitarian. Class C reports are only for internal communication. Business-world examples include travel reports, planning reports, and personnel evaluations. The main elements of an authentic student report, another type of Class C report, are shown in Figure 19.8 in Chapter 19.

This classification system is intended to be representative only of the ways reports may vary. The situation you are in will obviously be the primary determiner of the class of report you produce.

2. Joint Authorship

Only rarely do students encounter a situation in which several people are encouraged to work together on a writing project; in professional life it happens all the time. Joint (or multiple) authorship can take many forms. For example, a large engineering firm, in bidding on a wastewater treatment project, may put together a project team of perhaps five to seven people, each with his or her own staff. This

Figure 11.3   A Two-Page Spread from a Typical Class A Report

Steam Turbines

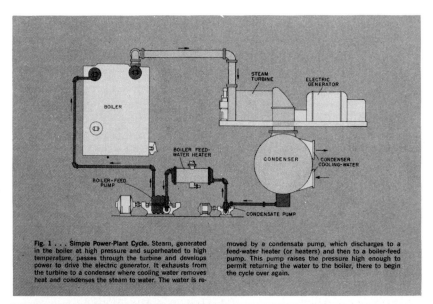

Fig. 1 . . . Simple Power-Plant Cycle. Steam, generated in the boiler at high pressure and superheated to high temperature, passes through the turbine and develops power to drive the electric generator. It exhausts from the turbine to a condenser where cooling water removes heat and condenses the steam to water. The water is removed by a condensate pump, which discharges to a feed-water heater (or heaters) and then to a boiler-feed pump. This pump raises the pressure high enough to permit returning the water to the boiler, there to begin the cycle over again.

Courtesy of Brown Boveri Corp.

Fig. 2 . . . Single-Cylinder Condensing Turbine. Initial steam conditions for this 2500-kw 3600-rpm turbine are 230-psi (16.2-kg/sq cm) pressure and 580 F (304 C) temperature. Letters refer to the following parts: A, journal and thrust bearing; B, balancing piston; C, impulse blading; D, reaction blading; E and G, journal bearings; F, coupling; H, governor; I, main oil pump; J, steam valve; K, steam-nozzle chamber; L, stop-valve flange; M, bleeder point for feed-water heating; and N, generator.

In a turbine, steam expands in stationary or moving nozzles from which it discharges at high velocity. The force of the high-velocity jets of steam causes the moving parts to rotate, thus making the energy in the steam available to do useful mechanical work— for example, to drive a compressor or generator.

After it has expanded in the turbine, the steam usually exhausts to a condenser, which serves two purposes. First, by maintaining a vacuum at the turbine exhaust, it increases the pressure range through which the steam expands. In this way, it materially increases the efficiency of power generation. Second, it causes the steam to condense, thus providing pure, clean water for the boilers to reconvert into steam. This simple cycle (Fig. 1)—water to steam, power generation, and steam to water—forms the basis on which most steam-power plants operate.

Construction . . . The parts of a steam turbine (Fig. 2) may be thought of as being in four groupings: (1) Stationary parts, (2) the rotor, or spindle, (3) governing mechanism, and (4) lubricating system.

The principal stationary parts are the steam-tight casing, or cylinder, the steam-admission valves, nozzles or stationary blading, shaft seals and bearings. The rotor, depending on turbine type, may consist of wheels mounted on a shaft or may be machined from a solid forging or a forging made up of welded sections. In either case, it carries securely fastened radial blades, or buckets. Turbine governors for small machines may be relatively simple mechanical mechanisms that directly operate a steam-admission valve. For larger machines, they may be very complex hydraulic systems which may not only control speed, by controlling the admission of steam, but also control the pressure of steam extracted for process purposes and the operation of valves or safety devices separate from the turbine. Similarly the lubrication system may be simple reservoirs in the pedestals of ring-oiled bearings or elaborate circulation systems, having pumps, coolers, filters, and devices that automatically shut down the turbine in case of low oil pressure or overspeeding.

Shaft seals at the ends of each casing are neces-

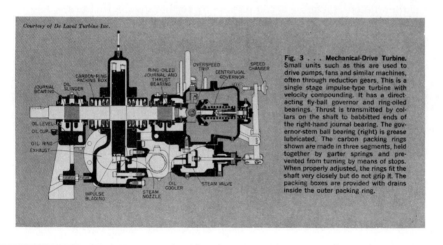

Courtesy of De Laval Turbine Inc.

Fig. 3 . . . Mechanical-Drive Turbine. Small units such as this are used to drive pumps, fans and similar machines, often through reduction gears. This is a single stage impulse-type turbine with velocity compounding. It has a direct-acting fly-ball governor and ring-oiled bearings. Thrust is transmitted by collars on the shaft to babbitted ends of the right-hand journal bearing. The governor-stem ball bearing (right) is grease lubricated. The carbon packing rings shown are made in three segments, held together by garter springs and prevented from turning by means of stops. When properly adjusted, the rings fit the shaft very closely but do not grip it. The packing boxes are provided with drains inside the outer packing ring.

Source: Steam Turbines and Their Lubrication, pp. 6–7. Reproduced by permission of Mobil Oil Corporation.

Figure 11.4   **A Two-Page Spread from a Typical Class B Report**

4

2.   COMPARISON OF EQUIVALENT ICE AND ELECTRIC CARS

The first tasks were the definition of a typical state-of-the-art electric vehicle and then the selection or development of characteristics of an equivalent ICE vehicle. (The opposite approach was rejected because it appears unlikely that electric vehicle performance can equal that of typical gasoline vehicles.)

The Electric and Hybrid Vehicle Act defines acceptable performance goals for electric vehicles (Table 1).[1] Typical values for today's electric car are compared with goals that were used to establish the performance required from the ICE vehicle.

A goal for the energy efficiency of the electric car is not set forth in the legislative act. This value depends on the car weight, battery/motor efficiency, and recharging rate and efficiency. The relation between weight, range, and efficiency is shown qualitatively in Fig. 1. Numerous tests indicate that the overall electric power use of a 950-kg (2100-lb) curb-weight electric car is at best 3 miles/kWh and more typically only 1.2 to 2.0 miles/kWh.[2,3]

Estimation of the fuel economy of an ICE vehicle of similar acceleration potential can be accomplished by correlating the fuel use and acceleration of today's typical automobiles (Fig. 2); a direct comparison cannot be made, because no present-day automobile performs as poorly as the goals set for the electric vehicle. An extrapola-

Table 1.   Summary of electric vehicle performance goals

Parameter	Goal	State of art
Acceleration to 50 km/h, s	<15	8-15
Forward speed for 5 min, km/h	80	65-80
Range,[a] km	50	45-70
Recharge time from 80% discharge, h	<10	<10

[a] Using driving cycle in Society of Automotive Engineers Standard J227a.

Source: Ref. 1.

ORNL–DWG 81–8335 ETD

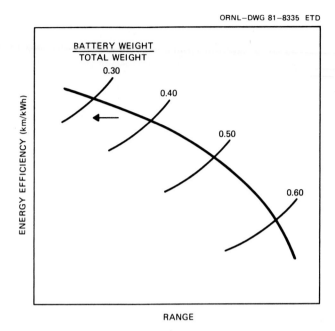

Fig. 1. Illustration of trade-off between range and fuel efficiency.

tion to very poor acceleration (0 to 50 km/h in about 12 s, which is still much better than the electric vehicle performance goal) shows that the corresponding fuel economy for a 950-kg (2100-lb) car would exceed 70 mpg over an Environmental Protection Agency (EPA) type driving mode. In fact, the equivalent ICE car weighs even less than this because it does not have the weight of the batteries. Nevertheless, a very conservative value of 60 mpg was used in this analysis. (Note that for a vehicle of this type, fuel consumption is already so low that further increases in efficiency make little difference to the running costs.)

The initial price of a mass-produced electric vehicle is difficult to estimate. Limited production two-passenger electric cars are selling

Source: R. L. Graves, C. D. West, and E. C. Fox, *The Electric Car—Is It Still the Vehicle of the Future?* (Oak Ridge, Tenn.: Oak Ridge National Laboratory, 1981), pp. 4–5.

group may include an architect, a mechanical engineer, a civil engineer, and an electrical engineer. Another engineer, one who has risen to the level of management, may well be made team leader. With all these people involved, who writes the reports? Everybody does, working together.

2.1   Working with Co-Authors

Usually the writing of a professional report is broken down into much the same kinds of divisions as are the technical tasks themselves. That is, each person writes the part of the project report that relates to his or her own specialty. With five to seven people involved, you can imagine what kind of report could result, so usually one person is responsible for making all the segments blend. Or, if the firm has a technical writer or editor, he or she may perform this function. But most companies choose to have their own technical people do the writing and editing. Sometimes, but not always, the team leader is the editor. More often, the editor is a team member who then passes the nearly finished report on to the team leader for final approval.

In other instances, two or three people may actually write a report, or a section of one, together. Even so, there are usually divisions of labor. Two team members may work together to make an outline, but then the actual writing is usually divided up between the two or three members of the team.

Certain problems and pointers apply to both kinds of joint authorship:

1. Most important, *all the people involved in any project must have the same mental picture of the project and the purpose of the report* being written. Depending on the size of your group and the nature of your project, it may well be indispensable to hold a first meeting at which everyone involved works to an agreement on the project, the report's purpose, and the report's audience. Any difference in understanding at this stage will be magnified each day it goes uncorrected.

2. *Everyone involved must agree on a division of labor and a schedule.* The timetable should specify preliminary-draft and finished-copy completion points for each member's contribution, and it should leave time at the end of editing and for a complete technical review by each key team member. Most important, once a schedule is arrived at, it must be followed.

3. *Staying within the timetable* is vital to the success of any group project. The kinds of problems that occur when work is late on a single-person project increase exponentially when work is late on a multiperson project. Many teams will specify that one member, usually the editor, check on each member's progress on a regular basis.

4. *Communication* among team members during the course of the project plays an important role in any project's success. You may want to schedule weekly meetings to review the progress of each member and to provide group support for solving any individual problems that may have come up. In case any members of the group object to regular meetings, remind them that the purpose is not just to have a meeting but to communicate and to draw on the strengths of the group to solve (and to prevent) problems.

As members complete their projects, they should review the technical content of each piece. Then they should funnel the reports to the editor. That person's responsibility (enlarged on in the following section) is to ensure that the various sections fit together into one integrated report. That means controlling style, correctness of grammar, treatment of visuals, and a number of other points, from the kind of typeface to the scope and coverage of the entire document. When these tasks have been performed, review copies of the entire report should be delivered to each team member, and a final meeting should be held to approve the report prior to its publication or delivery to the client.

Although actual practices may vary widely from those described here because of time pressures or other considerations, this process is representative of how joint authorship should function, whether there are two authors or ten. The flowchart in Figure 11.5 summarizes the process.

Figure 11.5 Flowchart for Joint Authorship of Reports

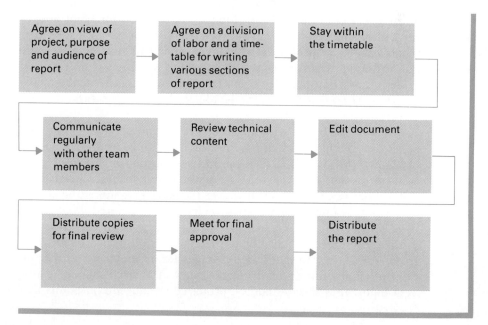

2.2 Being the Editor on a Joint-Authorship Team

Although there are too many complex processes involved in the profession of technical editing to explain here, any professional can profit from some knowledge of how technical editors work. The most important part of any editor's job is to maintain a positive relationship with the author(s). Occasionally an editor and an author will develop an adversary relationship, one in which they work against each other. As an editor, you should go out of your way to avoid conflict with the author(s). Both author and editor have the same goal: to produce the best possible document. Keep that in mind when it seems the authors you work with are making life difficult.

As the person in your group designated to edit the report, a number of functions are expected of you.

Functions of the Editor on a Joint-Authorship Team

1. Ensure that the various subprojects are completed on schedule.
2. Receive the individual reports and organize them into a coherent whole; take responsibility for the whole report's design.
3. Ensure the stylistic readability and grammatical correctness of the entire report.
4. Review any substantive changes with the appropriate authors.
5. Distribute review copies prior to the final approval meeting.
6. Be responsible for the final meeting; be responsible for the report's production in finished form and its appropriate distribution.

Items 1, 5, and 6 are really administrative functions, matters of efficient communication with other members of your team and of responsible record keeping on your part. Items 2 and 3 require writer's skills of the sort described in this text. Item 4 requires some skill at human relationships. Some authors become so ego involved in their work that they see every possible change in it as a personal threat. A good editor needs to be able to be assertive without being threatening.

You will do better with authors if you avoid altogether the kind of session in which you and an author sit down to go over a bundle of pages word by word. That kind of session not only is immensely (and needlessly) time-consuming but also promotes fatigue and frustration that can lead to needless bad feelings. You should be able to take minor grammatical and stylistic changes for granted. If you must review a portion of a report with its author, give the author a copy prior to the meeting, with the passages you want to discuss marked. Invite the author to mark any passages he or she wants to discuss. But do not get into a page-by-page review. Approach such a meeting as a negotiating session, and be prepared to concede on points that are matters of judgment on your part (rather than clear-cut decisions, such as grammar).

The functions of a professional editor are explained in the next section; obviously, they demand more time and expertise than can be afforded by most engineers, or accountants, or other professionals who happen to have been chosen their group's editors. If you remember the points discussed and listed here, you should be able to function effectively as the member of a project team chosen to do the editing. However, if the report you are working on is to be published (rather than merely word processed and copied), or if the artwork and illustrations are to be integrated into the text (rather than appearing on pages by themselves), you will probably need a professional editor's assistance.

3. *Editorial Procedures*

You can be a better author and a better editor of your peers' work if you know a little more about how professional editors work, what they do, and the sequence they do it in. Although the exact stages in the editorial process may differ a little from one situation to another (depending especially on the extent to which computers are used in the process), the flowchart in Figure 11.6 shows its basics.

Briefly, here's how the process works: Before any manuscript goes into production, peers should review it. After making revisions as required, the manuscript is turned over to production. Your editor will then copyedit the manuscript (more on this later), and the editor may review any important or technically questionable changes with the author. Then the edited draft is sent for composition, and the art sketches go to an artist to be drawn. The editor reviews the galleys and illustrations, and illustrations and text are combined into page masters. The editor makes a final check of the entire package, and turns it over to be printed. After it is printed, the editor will check it again.

When computers and word processors enter the picture, the process is compressed and accelerated. Suppose you start with on-line copy that has already been reviewed by peers and appropriately revised, and you send it (through the phone lines, computer to computer) to a publisher who uses computers for typesetting. Your editor will either edit a paper copy of your manuscript or edit it on-line. Once the editor's changes have been made and you have been consulted about any major changes, the copy is coded (on-line) for phototypesetting. The computer file is incorporated with the visuals (which are also computer generated), and the phototypesetter produces camera-ready copy (page masters), which go directly to the printer.

Figure 11.6   Flowchart for the Editorial Process

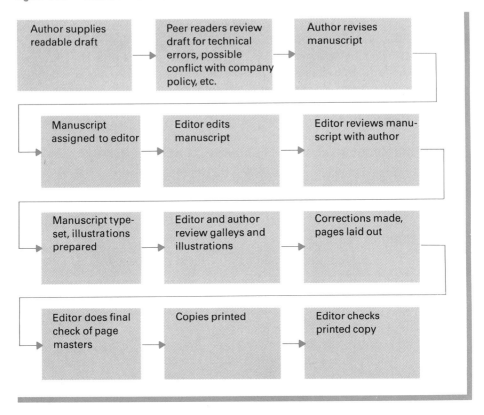

No matter which variation of the editing process your publisher uses, two important elements remain the same. Your editor will do essentially the same things, and you can help your editor in essentially the same ways. Your editor's job (very much simplified) can be divided into four stages, which are discussed in Sections 3.1 to 3.4.

3.1   Sizing Up the Manuscript

The first thing to do is size up the manuscript. That means reading the whole thing before ever starting to mark it. At this level the editor looks for the manuscript's completeness (Are important parts missing?) and overall structure. Does the large-scale arrangement of parts make sense, or should the order of the sections be rearranged? It is the editor's particular function to pay attention to the element of *design* described earlier in the introduction to Part IV. Only when your editor has a "feel" for the manuscript and what it is saying will he or she actually begin to mark it up.

3.2 Copyediting

The functions noted in the following bulleted list are performed by a professional copyeditor. If you are your group's editor, you need to know which of these you are expected to do.

Functions of a Professional Copyeditor

- Mark manuscript page numbers.
- Monitor project costs.
- Monitor schedule.
- Edit art sketches to conform to manuscript and design.
- Stay in contact with authors.
- Review manuscript changes with author where necessary.
- Ensure that all manuscript parts are present (cover, title page, etc.).
- Check the table of contents, references, and so on for accuracy.
- Ensure that the author's statements in the report do not contradict company policy or reflect unfavorably on the firm.
- Ensure that all sequences (numbered lists, etc.) are complete and in proper order.
- Check spelling.
- Check subject/verb agreement.
- Check completeness of sentences.
- Ensure that there are no inadvertent omissions.
- Ensure that figures are of appropriate quality and size.
- Mark mathematics for typesetting.
- Indicate margins, indentions, etc.
- Indicate appropriate type styles (*italics*, **boldface,** etc.).
- Indicate form and position of headings.
- Check capitalization, spelling, word compounding, proper use of abbreviations, numbers, bibliographic reference style, grammar, syntax, punctuation, usage, conciseness, and appropriateness of style and tone.
- Code all of the manuscript's parts that are not straight text (lists, headings, formulas, etc.) so that the compositor will know how to set them.

3.3 Author Review

If your editor has made any significant changes in your report, you will be given the opportunity to review them. Do not expect to be part of a line-by-line review. You will be expected only to review the specific sections your editor questions you about. The rest of the report should have been right before you sent it in.

3.4 Publication Production

When you and the editor have agreed on the final version of the manuscript, it goes into publication production. During production, the pages of text and the visuals are combined into final page proof, it is proofread one last time, and the whole package is sent to the printer.

Perhaps the most important thing you can do to help your editor is to have the manuscript complete and correct in neatly typed form before turning it over. Beyond that, there are a number of things you should *not* expect an editor to do.

What **Not** *to Expect of an Editor*

Do *not* expect your editor to

- research references.
- write any of the report.
- search for missing items.
- provide examples of other publications to help you decide how yours should be done.
- edit handwritten copy.
- handle multiple versions of a manuscript or incorporate more than one series of author changes into the manuscript.
- edit for technical content.
- deal with more than one author.

If you provide clean copy that is technically accurate, and you know what *not* to expect of your editor, your editor will work for you with a high degree of professionalism to make your report the best possible publication.

4. Computers and Writing

If you were to analyze the operation of any office and pare that analysis down to its essentials, you would almost always find that the office's primary purpose is to process information. Because computer technology so readily adapts to the processing of information and because today's professional world produces so much information to be processed, computers rapidly entered professional offices in every field during the 1980s. Now, in the 1990s, computers have also entered the lives of professionals in every field: however good computers are at handling information, they're even better at handling words. An office that uses computers fully is called an *automated* or an *electronic* office. It would be demeaning, though, to call a writer who uses computers fully an automated or electronic writer; rather, let us just call that person a *fully effective* writer.

Because the computer technologies that enable such writing and information processing are developing so rapidly, any specific hardware or software that might be named here would be obsolete by the time you read this book. Consequently, the following sections describe generally the current capabilities computer hardware and software offer to writers. If this text were to describe the capabilities of, say, the Aldus PhotoStyler, software that allows writers to scan images or photographs into their PCs, alter and enhance the images in a myriad of ways, and then insert the images into their PageMaker documents, then by the time you read this information, that product and service would probably be old news. But the needs of writers and the core capabilities of computers for helping writers will remain substantially the same, so that's what these sections focus on.

4.1 What the Possibilities Are

When *Modern Office Technology* polled 701 managers and professionals about how they would spend $10,000 given them by their companies to improve their productivity at work, 82 percent said they would spend additional money on automation; 41 percent specified they would use the money to buy a computer. In fact, most people who make the transition to word processing estimate that it doubles or triples their productivity; higher estimates are not uncommon. What is it computers can do for writers? The following list gives some examples.

A Short List of Work Computers Can Do for Writers

- Data base access
- Project planning
- Word processing (including spelling, grammar, and style checking)
- Graphics
- Spreadsheet budgeting and statistical modeling
- Printing
- Internal mail
- Mailing list management
- External mail
- Document transfer
- Record keeping

Data Base Access. The computer's modem can put any number of information storehouses at a writer's fingertips. Everything from the most up-to-the-minute stock market reports and news bulletins, to the archival information of major encyclopedias, is available on computers today. Many major periodical indexes of the sort used in research are also available in either online or read-only compact disc formats. Thus, even the smallest research libraries have—through PCs—access to vast amounts of information. And many professionals have almost equal access in their offices or in their homes, again through PCs. For you as a writer, this means that access to the kind of research information needed to write professional documents is no longer limited to what you can find on paper. Indeed, today you have to think of working in two media when you do research. Ask yourself both, "What can I find on paper?" and "What can I find via computer?"

Project Planning. For most writers, at the same time they are collecting information on a project, they are forming ideas of how it will be structured. Computers can help on two fronts here. First, there are programs that will help you build an outline for your project, automatically setting the indentations, capitalization, number, and so on, and then transferring those items into appropriate headings and subheadings in your document when you begin the actual writing. Second, there are "idea generator" programs that are designed to help you come up with ideas for writing projects and then to develop those ideas at greater length and in greater detail. With most PCs, it is easy to be simultaneously acquiring and collecting information on one screen and building an outline that you will subsequently insert that information into on another screen. And at any point you can print out a hard copy of your scratch outline or idea list to see how it looks.

Word Processing. Word processing today includes not just the ability to enter, delete, move, and copy text. It also includes commonly available software that does a reasonably good job of checking spelling, grammar, and writing style (these are discussed in detail in Chapter 3, especially Section 3.2). Today a large number of college students already are reasonably proficient with word processing. Many of those who aren't have never acquired basic typing skills; if you haven't had a typing class, *keyboarding* (as it is termed when typing is done on a computer) can be the most frustrating aspect of learning word processing. Of course, there is available software that can help you learn to keyboard, but a student who knows neither how to type nor how to use a computer can find trying to do even the simplest word processed assignment frustrating. Trying to compose even a short report while wrestling with both the keyboard and the computer will usually lead to serious disenchantment with the computer. An acceptable stopgap measure for such writers is to compose a first draft longhand, keeping the technology problems from interfering with the writing process. But as soon as possible, you need to transfer that rough draft to a text file on a word processor, and all subsequent revisions need to be done on the computer. Once something has been keyboarded, it can be manipulated, altered, cut up, scavenged, and used in multiple different ways, but it should never be keyboarded again; all the energy and time should go into revising and editing, not into retyping or rehandwriting.

Chapter 2 goes into much more detail about the processes of writing with specific reference to word processing, but one point needs to be made here strongly: Only the most experienced writers—people who write frequently and who always use word processors to do it—can rely on doing all of their revising and editing online. Most other people should keyboard the rough draft, print a copy, review it and mark the changes on the hard copy, keyboard the changes, print that copy, review it and mark the changes, keyboard the changes, and so on. Most people are better at seeing the need to make revisions and corrections when they're looking at hard copy than they are when looking at the computer screen. If you follow the process outlined here, which alternates between *seeing the need* for changes on the hard copy and *physically making* the changes on the computer, you will use more paper, certainly, but you will equally certainly do a better job of revising and editing.

Graphics. The combination of a computer and a laser printer brings within the realm of feasibility the practice of including graphics of acceptable (and in some cases, nearly professional) quality as an important part of what you write, as well as making it possible to prepare graphics such as transparencies for oral presentations directly from document files.

Chapter 10 discussed graphics in detail with some attention to computers, particularly with reference to using photographs and other images that are scanned into the computer and then merged appropriately with textual documents. The important point here is to come to know the graphics capabilities available for the computers (and especially the word processing packages) you use and then to employ those capabilities to their fullest. Most mainstream word processing packages have many more graphics capabilities than their users exploit; it may make more sense for you to work at becoming proficient at using the capabilities currently available to you than to try to learn a dedicated graphics package. Beyond that, the key point is to make sure that, whatever hardware and software you use, you have the option of seeing on the screen in front of you the pages with graphics incorporated into them *exactly* as the printer will print them. This WYSIWYG ("what you see is what you get") capability is crucial; otherwise you're trapped in trial and error.

Spreadsheet Budgeting and Statistical Modeling. Before the days of computers, complex budget tables and other tabular forms of statistical modeling of complex processes and business functions were a nightmare for everyone involved. Not only is there the difficulty of physically producing the tables, which frequently are turned sideways on the page and often are several pages long, there is also the difficulty of changing numbers in the tables when, at the last minute, a computational error is found in one of them. Often such an error will affect a third or more of the numbers on the page, so—in the precomputer days—the entire table would need to be done again. Today, with computers and dedicated spreadsheet software, spreadsheets and other related documents become easy to produce. Constructing the basic grid and printing the document sideways is not difficult at all for computers and laser printers. And most such programs today are interactive; if you change one number or set of numbers, the computer will (upon request) make the corresponding changes throughout the document file.

Printing. Printers for word processors generally appear in one of two types: the dot-matrix printer, which relies on the impact of varying numbers of pins (usually from 12 to 24) on a ribbon and a sheet of paper to produce letter shapes, and the *letter quality* or laser printer, which uses high technology such as lasers to produce fully formed letters. A few years ago it was common to say that dot-matrix printers were cheaper, faster, and more reliable than letter-quality printers, and hence they were the bread and butter of office or professional work. But today laser and laser-quality (such as bubble-jet) printers are nearly as inexpensive as dot matrix, operate much faster, run more quietly, and produce a much more attractive and readable image on the page. Thus, if you're looking for word processing equipment to help you with student writing, a dot-matrix printer will be very inexpensive and should (presuming you insert a fresh ribbon regularly) be sufficient for most assignments. If you need better-quality printing, you can probably find someone at your school with access to a laser jet printer. If, on the other hand, you're equipping an office, the laser (or laser-quality) printer is the best choice.

With the help of the laser printer and good graphics software, you can probably produce the body for Class C or B documents on your own machine. Then with a good copier, preprinted covers on heavier paper stock (these are usually standard within each company), and perhaps a plastic comb binding (which many offices can apply onsite), you can produce high-quality documents without using a professional printer. Class A documents are more difficult and usually require professional help, but with the help of the technology just described, you can provide the printer with a manuscript that is in fact a high-quality mock-up of the document as you eventually want to see it printed. A growing number of printers begin with your document files just as they come from your PC, building the files their computerized phototypesetters use directly from your own diskettes.

Internal Mail. One of the frustrations of any professional day is trying to contact people within your own company to talk with them. *Electronic mail systems* (generically called *E-mail*) use computers linked by networks or by telephone lines to make communicating within an organization much simpler. As soon as the person you're trying to reach returns to the office and turns the computer on, a message ("You have new mail," for example) comes on the screen. The receiver can scan a list of messages and respond appropriately. If the person you are trying to speak with is using the computer when you call, many systems allow you to signal the receiver that you are trying to get through. These E-mail systems allow you to send and receive messages and communicate with others in real time as well.

Mailing List Management. Computers allow you to send documents (such as letters) to as many people as you wish and to insert a different name and address (as well as other customizing features) on each one, as well as printing mailing labels or addressing envelopes, practically automatically. Usually all you need are two files: one of names and addresses and the other the core document. At the proper command, the computer merges them appropriately.

External Mail. Many worldwide computer networks allow you to send electronic mail to anyone whose computer address you know. These networks often give users the opportunity to converse in real time, although the more common use is still to keyboard a message in advance, save it as a file, access the network, and send the whole file to your recipient. Even without membership in such a network, you can use your computer with a modem and the appropriate software to send and receive messages electronically over telephone lines. In many industries, these kinds of facilities are used in the form of automated data exchange, which makes possible the ordering, shipping, and invoicing of goods without repetitiously filling out forms.

> *Caution:* Anyone who uses computer mail facilities needs to be aware of the ease with which computer viruses travel. If your computer is not fully protected and regularly checked for infection, use of *any* internal or external computer mail facilities is unwise.

Document Transfer. The same facilities that can be used to send and receive mail electronically can also be used to send and receive documents. The now-ubiquitous FAX machines should also be mentioned in the same regard. FAX machines carry the additional advantage of being able to transmit equally easily the parts of documents that might be difficult to keyboard—letterheads, signatures, drawings, and so on.

Record Keeping. A final use for computers that students don't usually consider is record keeping. One diskette can hold hundreds of pages of text; a hard drive can hold thousands. An increasing amount of corporate record keeping is being done using computer technology, especially on magnetic tape or on compact disks that hold information in a format that, for all practical purposes, will never wear out. For a writer, this means that projects from last year, the year before, even years and years ago, can be saved without having to archive stacks of yellowing paper. If the storage medium is anything other than magnetic tape or CD, it needs to be backed up on a second source and checked regularly.

4.2 How the Possibilities Can Work

Consider the case of Jack Johnson, an architect whose client has asked him to prepare a proposal seeking bank financing for the refurbishment of an aging downtown office building; the purpose of the project is to convert the building, now empty, into condominiums. A typical sequence of steps in his writing the proposal using computers would look like this:

1. Jack has stored in his computer at work the basic outline for a proposal as part of a *template* (a writing pattern specified all the way down to margins and line spacing). He brings that outline up onto his screen and begins thinking about the information he will need to fill in its parts.

2. To find recent information about office occupancy rates versus condominium sales for the nation and the state, he switches to a second screen and accesses several information data bases. While searching for that information, he runs into mention of two federal programs for supporting remodeling of abandoned downtown business properties that he decides he will also mention in his proposal ("Additional funding may also be available from . . ."). As he discovers information in the data bases that he can use, he marks it to be saved when he exits the data bases so he can import it into his document outline.

3. Exiting the data bases, he returns to his outline and inserts the information he saved from them at what look to be appropriate headings in the proposal. One section of the proposal needs to be

a description of his company and its capabilities; he enters another file from a proposal done last year and copies the corresponding section into his new proposal. Later he will go back over that section and update it.

4. Jack begins writing the sections describing his project and immediately encounters the need for corresponding visuals. He has several photographs of the building taken during its heyday that he scans into his computer. When he turns the document file over to the company's editor, she will work with the company's computer graphics specialist to enhance the images and make them of publication quality.

5. Jack has his notes from conversations with the client to keep him going during his composing. He likes to do the first very rough draft of this kind of project as fast as his fingers will fly; in fact he's often looking out the window of his office at the street below and listening to music on the radio as he composes. The point of this writing session is just to get some kind of rough draft down, not to agonize over fine points (that will come later). At a couple of points during the task, he uses his interoffice E-mail to ask technical questions of other specialists within the firm; as each of their answers comes in, a telephone icon appears on his screen. Each time that he clicks onto the icon with his mouse he sees the answer he wanted and copies it into his document.

6. Before he goes home that night, he runs his spell checker to pick up typos and then prints out a hard copy of the proposal draft. After dinner he will sit down for an hour and look it over, making notes to himself about changes. For him, these are usually either places where he needs to collect and insert more specific information or places where the writing is clumsy. The next morning he will give his company secretary some notes so the budget can begin being created while he continues to work on the body of the proposal.

7. After a few days more work on the project, including a review by his boss, Jack sends a FAX of a hard copy of the draft (labeled PRELIMINARY DRAFT FOR CONTENT REVIEW ONLY on every page) to the client and then has a long telephone conversation collecting feedback on that draft.

8. Once Jack has revised the report once more, assembled a few more visuals for the editor and computer graphics specialist to worry about, and inserted the budget, he turns the entire project over to the company's editor, urging her to do her best work but to be sure to get it back to him by next Wednesday at the latest.

9. The editor makes sure the proposal's page layout and overall style conform to the company's standards and reviews it for other relevant features (see Section 3.2 on page 291). She works with the computer graphics specialist to make sure the graphical elements

are clear and properly placed in the document. After a brief review of the manuscript with Jack to clear up a couple of small problems, the editor hands the proposal to her assistant for final printing on the company's own laser printer, copying, and binding. Her assistant works late Tuesday night so that he can hand the specified ten copies to Jack first thing Wednesday morning.

4.3 What the Problems Are

The processes that link computers and writing are wonderful when they work, but sometimes they don't. The biggest problem five years ago was *compatibility*—the problems that arise when a file originated on one kind of computer or with one kind of software needs to be moved to another kind—and it remains the biggest problem today. There are many kinds of customized software and special hardware packages available that assist writers in moving their files from one format to another, but the perfect transfer is still an exception. Usually any text file that has been transferred will need to be reviewed line by line to ensure that margins, indentations, underlines, and so on have not been garbled in the new format.

A second problem arises when the personnel involved, both managerial and support staff, lack the necessary training to get the most out of the computer equipment available, much less deal with the occasional problems that may arise. In particular, the full integration of graphics with text can cause headaches if the people trying to do it have to do these procedures only once or twice a year. Some of the software is still so complicated that familiarity with it is vital to proficiency.

5. Accountability and Ethics

Two more factors differentiate the writing college students do from the writing professionals do: accountability and ethics. The two are closely related to each other; the feeling of accountability that a professional who writes may have is especially strong in situations that may involve ethical problems, and the ethical dimension of professional writing can in fact be seen as one manifestation of its quality of accountability. Both factors have roles in the writing students do, but both play much stronger roles in professional writing.

Accountability. Writing that you do in college may have about it an air of unreality, a sense that "this is just an exercise; it's not *real.*" That sense is something that writers find hard to live with (it invites you sometimes not to do your best) and hard to live without (if you think the work is real, doing the work can be harder). As much as students and writing teachers dislike this quality of make-believe that haunts much classroom writing, many students find that when their writing really *does* matter, that realization itself seems to paralyze them. Often the first time a student writer feels this is when writing a job application, medical or law school application, or a report done as part of an internship. Some people never feel this accountability trying to block them until their first on-the-job writing task.

The following pointers will help you to deal with the accountability of on-the-job writing:

1. *Develop confidence in your ability to write effectively.* If you know your subject matter, adapt it to your audience in the light of their particular purposes for reading, and think about who or what kind of person you are supposed to be in this document, you can write effectively. Review the questions for audience analysis and the pointers for audience adaptation presented in Chapter 1, Section 7.

2. *Don't even try to make your first draft of any document perfect.* As Chapter 2 explained, plan to write at least two (and quite possibly more) drafts of anything that you produce as a professional. In the first draft, all you need to do is try to get the content into place. Don't worry about stylistic clarity, grammatical correctness, or readability until you get the content right.

3. *Recognize the source of the feeling that makes your writing seem difficult.* For those who possess reasonable basic writing skills, the biggest impediments to doing effective writing are nearly always fear and fatigue. The fear usually comes from inadequate knowledge of your audience; you're afraid you're doing the wrong thing or doing the right thing in the wrong way. An excellent way to deal with this fear is to get to know your audience better. The more time you spend with your prospective reader discussing the project in person, the less of this fear you will feel.

   Fatigue is an unfortunate fact of life in our society for many professionals; it usually comes from too little exercise and too little sleep. The cure for fatigue—more rest, more exercise, less stress—is beyond the scope of this book to present in detail. For writers, a good short-term solution for the problems fatigue brings to their writing is to be more careful when and where they write. Many professionals do their writing late in the day, after everything else they can possibly think of as an avoidance device has been exhausted. They also try to write at a time and place rich with opportunities for interruptions, such as in the middle of the workday, with the office door open, when everyone wants "just two minutes" of your time. If you find yourself in this predicament, try engineering your writing environment more directly. Try writing first thing in the morning, when you're fresh; you'll be amazed at how much easier it is. You may want to come into your office early in order to be able to attack your writing for an hour or two with the door closed before anyone else realizes you're there.

Ethics. There are certainly ethical situations that confront college students; many, such as plagiarism and collusion, apply to writing situations especially. In professional life, ethics—especially the possible ethical dimensions of writing tasks—can become much more important and much more troubling. Imagine, for example, a group of representatives from your company who go to tour a site where a hazardous waste spill has occurred. Your group's job is to assess the progress of the cleanup, and you are chosen to coordinate the writing of the final report. Does your report faithfully depict the extent of your group's examination of the site? Is it your responsibility to *volunteer* in that document the fact that your group's examination was superficial at best and that for the rest of the technical information you simply took the data from the on-site team and incorporated their findings into yours? If you want the report to say the ground was *treated* (to reflect the fact that there's still a sticky brown sludge on everyone's shoes after walking around, but the technicians say it's harmless), how do you deal with pressure from others to say the ground was *clean?* When you talk to the people who live around the site, do you tell them you've been sent to *monitor the progress* of the cleanup or to report to the company on its *completion?* Such issues are often faced in professional writing situations today, and perhaps no book can tell you how to deal with them all. Asking the questions presented in the following list of each such tough issue, however, may help you do a better job of dealing with each issue.

Three Questions to Help Untie Ethical Knots

- *What will the consequences of this action be?* The easiest answer to many ethical problems is in terms of consequences. If something will cause bodily harm to other people, for example, many of us would have problems doing it in most situations. If something you do will produce destructive effects, it's usually easy to decide not to do it.

- *To what extent and in what ways is the act itself wrong or right?* Many people respond to some actions automatically, not so much because they are forbidden by some legal code but by the individual's own, often unspoken, moral code. For example, many people feel that the taking of another human life is wrong, quite apart from whatever laws or other abstract notions are called into play. More to the point for writers, many feel that misstating information on paper—changing the numbers to make them say what you need them to say, for example, or affirming nothing is wrong when you know something is very wrong—is simply wrong, and that feeling exists quite apart from laws or consequences or other human responsibilities.

- *What are the rights of the human beings involved here?* Even in situations where there is no clear-cut adverse consequence and no clear feeling of right or wrong to the action under consideration, there are still human rights to consider (see also Chapter 4 on business goals versus human goals). You may find yourself worrying over something you're about to say on paper because it may hurt someone else. We all have the right to be treated fairly in life and to be respected by those with whom we do business; we expect business partners to continue to be business partners, strangers not to go out of their way to do us harm, and so forth. Similarly, each of us has our own rights, including the right not to be put into a compromising position by our employer, to be dealt with fairly and openly. This area of rights is the murkiest of the three, but it is one that must be at least listened to, especially by those who write as part of their jobs.

How would asking these questions of a difficult ethical tangle help the writer deal with whether to say in the site inspection report that the area had in fact been cleaned or to decide whether to point out that the survey team had accepted the on-site team's data and conclusions uncritically? The first two ethical questions have clear answers. The consequences of such a report (one "slightly slanted toward the positive side," as a co-worker put it) will inevitably be bad for the company (letting readers infer a site has been cleaned will ultimately mean the site will still need to be cleaned at some later date and probably cost much more), and the writer in question felt strongly that her own personal code of morality was being violated in being pressured into producing a misleading report. The third question caused the writer to see how to deal with the answers to the other two; she felt it was her right as a professional employee of that company not to be put into such a potentially compromising situation, just as it was the right of unaware members of the company's board of directors not to be misled by her report. That realization led to a long discussion with her supervisor, who explained that the company's best interest (as well as her own) was served not by a lightweight report that painted a rosier picture than the hazardous waste site really presented but by a more substantial report that clearly and honestly told the truth. Yes, her supervisor explained, everyone

around the office wanted to be done with the case, and so she probably did feel some pressure to make things look good in her report;
but if that meant getting into ethical problems, her supervisor assured
her that everyone, right up to the board of directors, would support
her being truthful, even if the truth hurt a little.

EXERCISES

1. Find a brief piece of writing—either one of your own or one from a textbook or professional journal—that you think fits the criteria established
   here for catalogical writing. Revise it into analytical writing. Turn both
   versions in to your instructor.

2. Go to a large administrative office of your university—the Liberal Arts
   College office, the president's office, the Placement Service, or the Testing
   Center, for example—and ask to be allowed to study a variety of the brochures, pamphlets, and booklets available there. Select three to five of them
   that seem to exemplify a variety of different levels of documents, and write
   a brief report for your instructor describing and analyzing them in terms
   of those levels. (This assignment can also be done with corporate publications if you have access to them.) Spell out the level of "gloss," the audience,
   the use of color, and whatever other qualities clue you in to the document's
   level.

➤ 3. With your class divided into groups of three to five students, all with identical (or similar) majors, each group is to produce a report of a length specified by your instructor entitled "Careers in [*fill in your major*]." The audience
   for the report should be college sophomores considering choosing that major. Each group should apportion the roles of the various people in it as
   the group thinks will work best. Each member of the group should keep
   very careful records of exactly what contribution *each* member makes. After
   the group report is submitted, each member should submit his or her own
   brief (one- or two-page) memo describing the way the group worked together.

4. Find a professional who has to write as part of his or her job. Interview
   that person to find out how group writing is handled in that person's company and what roles computers play in that company's writing. Write a brief
   report to your instructor explaining what you have learned.

➤ 5. (This may be done as a class project or as individual projects visiting various
   sites.) By prearrangement with the appropriate officials, visit an editorial
   office, either on or off campus, where you can see the process of editing
   taking place. Follow that with a visit to a printing shop, perhaps where your

---

A ➤ indicates a case study exercise.

student newspaper is printed, or whatever the material coming out of the editorial office you visited is printed. After touring the facilities, work individually or in groups to write this report: "How Computer Technology is Changing the World of Editing and Printing." The audience should be laypeople—the general public.

6. Recently there has been significant research done on measuring the dollar value of printed information of various types and the value of information in the U.S. economy. That is, researchers are now able to come up with relatively concrete answers to questions such as, "How much is this particular report worth to our company?" and "How much is information in general worth in this field of business?" With help from your instructor and librarians, find an article that discusses some feature of the dollar value of information. Write a brief report detailing what you learn from the article, being sure to discuss its significance for you as a writer. The audience should be your classmates.

7. Between the time the manuscript for this book was completed (December 1991) and the time you read this book, there will have been advances in computer technology not even hinted at in this chapter. Write a brief report, for laypeople, on recent advances in computer technology as it affects writers.

➤ 8. Do the research described in problems 4 and 7, and write a short piece for students hoping to major in your field. Your goal is to let them know what kind of computer skills they will need upon graduation. Your role is that of a junior professional, one who knows the requirements of the job market first hand.

12. *Writing Definitions and Descriptions*

Defining and describing may well be the most essential elements of professional communication. Even outside professional life, it's difficult to have so much as a five-minute conversation without involving yourself in some kind of defining or describing—of your emotional state, tomorrow's weather, the management exam you just took, or the problems you're having with your nuclear engineering lab's weekly 40-page report. As a professional, you'll find the twin processes of defining and describing at the heart of most of your writing.

This chapter presents 14 ways to define and describe whatever you need to write about. Doing professional writing usually involves explaining to your reader exactly what some object (or some abstraction) is; this chapter shows you how to do that.

Language establishes the meanings of words contextually; except for the most common nouns (like *tree* or *door*), the meaning of each word in a sentence is largely established by the sentence it is in. Consider this sentence:

> I threw the gilhickie out the window.

Few people know with any certainty what a gilhickie is, but we can tell from the sentence that it is something (a) found indoors and (b) light enough to throw out the window. We may also be able to speculate a little about what kind of thing one might throw out a window.

The more information contained in the sentence, the more of a definition of a gilhickie we can construct:

> Because it was beginning to smoke, I threw the gilhickie out the window.

Now we can construct a picture of a gilhickie as something that smokes when it isn't working properly. Add still more information, and we can construct a more precise idea of what a gilhickie is:

> Because it was beginning to smoke, I ripped the gilhickie from its mount and threw it out the window.

Now we're sure it's a piece of machinery bolted to the floor (but not too securely), something that runs, smokes when it isn't working properly, and can be thrown out the window.

Actually, *gilhickie* is a sailor's term for anything one can't think of the right name for. Rather than saying, "Hand me the thingamajig," a sailor might say "Pass me that gilhickie." This example shows that the richer a language context a term is embedded in, the clearer its meaning will be. Thus, a good working knowledge of techniques for placing whatever you're writing about into a language-rich context will help you make your meaning clear in any kind of writing task.

The 14 techniques examined in this chapter are divided into three groups: patterns in which the item being described is at rest, patterns in which the item being described is in motion, and patterns that describe the item indirectly.

The examples used within the body of this chapter come from a professional report, *Steam Turbines and Their Lubrication,* published by Mobil Oil Company; a few have been slightly rewritten to emphasize a particular pattern better. To see how all the techniques work together, you can find the first two pages of the actual report in Chapter 11, Section 2. Section 5 of this chapter contains a sample paper written by a student who was trying to use as many of these techniques as possible in one short report, a professionally written product description, and a technical description written for laypeople.

> ### *Fourteen Techniques for Using a Language-Rich Context*
>
> *Static Patterns*
> - The formal pattern
> - Explication
> - Analysis
> - Accumulation of detail
>
> *Moving Patterns*
> - Process
> - Cause and effect
> - History of the term
> - History of the object
>
> *Indirect Patterns*
> - Elimination
> - Comparison and contrast
> - Analogy
> - Naming examples
> - Pointing out examples
> - Showing examples

1. Static Patterns

These techniques treat the *X* being defined and decribed as it appears when it is standing still. Four techniques come under this heading: the formal pattern, explication, analysis (or division), and accumulation of detail.

1.1 The Formal Pattern

Usually seen as the fundamental definition technique, the formal (or logical) pattern has two stages:

1. First place *X* in a class.
2. Then differentiate *X* from the other members of the class.

You may already know this pattern, from courses in zoology or botany, as definition by genus and species.

The following passage uses the formal pattern to explain what a steam turbine is.

> The steam turbine is one of the world's most important prime movers. In a turbine, steam expands in stationary or moving nozzles from which it discharges at high velocity. The force of the high-velocity jets of steam causes the moving parts to rotate, thus making the energy in the steam available to do useful mechanical work—for example, to drive a compressor or generator. Steam turbines lead all other prime movers as the drive for electric-power generators. In addition, they are the versatile drives for many other machines, such as pumps, compressors, and mills of various kinds.

The first sentence places the steam turbine within a class (prime movers). The next two sentences differentiate it from other members of the class by explaining how it works. The last two sentences differentiate steam turbines from other prime movers by pointing out their leadership as drives for electric-power generators and their versatility as drives for other machines.

This process of logical definition is typically represented visually by placing X in a class and then establishing boundaries between X and the rest of the items in the class (see Fig. 12.1a).

Problems with the Formal Pattern. In placing X in a class and then differentiating it from the other members of its class, be careful not to violate the basic logic by which definitions work; that is, do not make the definition either too broad (including too many *not-X*'s in the class with X), too narrow (excluding some X's from the class you define), or circular (accidentally using X to explain X).

Too Broad. One way to go wrong with a definition or description is to include within X entities that are not X. For example, if you define an irrigation system as "anything that brings needed water to crops," you have accidentally included rain and snow as irrigation systems. The solution requires narrowing the explanation: "anything made by humans that brings needed water . . ."

Too Narrow. You can also err by excluding from X items that should be included within it. For example, if you define point-of-purchase displays as "any cardboard visual designed to be displayed with a product to increase its sales," you have accidentally excluded a number of types of construction (plastic, fiberglass) also common to P-O-P displays. Solve this problem by expanding the definition: "any visual media designed to be exhibited with a product . . ."

Circular. A circular definition accidentally presupposes a knowledge of X in the explanation of X. This occurs by using some form of the words that name X to define it—for example: "A magnetic bubble memory uses magnetic bubbles to store memory." People who

Figure 12.1a   The First Stage of Logical Definition
"First place X in a class."

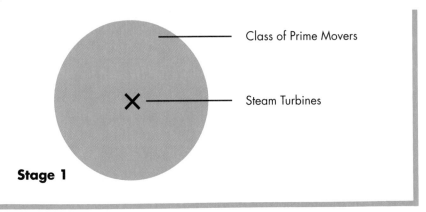

Class of Prime Movers

Steam Turbines

Stage 1

Figure 12.1b   The Second Stage of Logical Definition
"Then differentiate X from other members of the class."

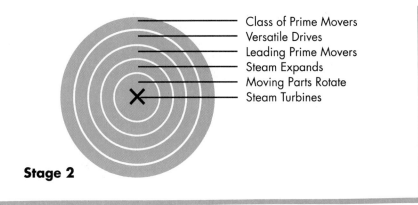

Class of Prime Movers
Versatile Drives
Leading Prime Movers
Steam Expands
Moving Parts Rotate
Steam Turbines

Stage 2

know what magnetic bubbles are probably do not need an explanation of magnetic bubble memory, but the explanation will not help someone who does not know what a magnetic bubble is. The solution is to add more information, being careful not to use forms of the words you're defining in the definition you're writing.

As Figure 12.1b suggests, the smaller an original circle (or class) we initially place X in, the easier the second stage will be. The second stage of the definitions uses a variety of techniques to draw the initial circle closer and closer—that is, define the class more and more narrowly, until it includes only the item or entity you are defining. For example, our definition of a steam turbine uses process as one of the ways to differentiate steam turbines from other prime movers. In an important sense, all of the following patterns are subcategories of this one: each pattern is a way to differentiate *X* from other objects and could therefore be considered a second stage of the formal pattern.

1.2   Explication

Explication consists of explaining the meaning of key words in a definition. As such, it nearly always occurs in combination with a formal definition. The following passage uses explication by expanding on the meaning of the terms *reliability, trends,* and *automatic data processing.*

> In all of these services, turbine reliability has always been of prime importance. This term embraces both availability for service and the ability to perform satisfactorily in service for long periods.
>
> Modern trends are placing greater and greater premiums on reliability. These are trends toward larger sizes, higher pressures and temperatures, and toward the use of automatic data processing equipment for:
>
> 1. logging and scanning of measured quantities and actuating alarms when necessary.
> 2. computing performance and adjusting operation accordingly.
> 3. full automatic operation of plants in some cases.

When you use explication, you may be tempted to grind through the several words pretty mechanically. Remember that giving in to such a temptation places your needs above those of the reader. Paragraphs 2–6 of "What Is a Robot?" (in Section 5 of this chapter) offer a particularly good example of explication.

1.3   Analysis

Analysis (or division) means dividing *X* into its component parts. These can be either *X*'s natural parts (properly called its *divisions*) or its artificial parts (called *distinctions*). Sometimes these parts will coincide with the words discussed in an explication, but sometimes they won't. Often this technique will be supported by the use of a visual.

The following paragraph is an example of a professionally written analysis:

> Among the most important of builders' efforts to achieve the reliability so essential to steam turbine applications is the provision of a lubrication system that will ensure an ample supply of lubricant to all moving parts. The size of the turbine generally determines whether the system is simple or extensive and complex. Small turbines, such as those used to drive auxiliary equipment, are generally provided with ring-oiled bearings, other moving parts being lubricated by hand. Moderate-sized units, particularly if driving through reduction gears, may have both ring-oiled bearings and a circulation system which not only sprays oil to the point of gear mesh but also supplies oil to the bearings of both the turbine and the gear set. Large turbines are invariably provided with circulation systems that supply oil to all parts of the unit requiring lubrication, to the turbine governor mechanism, and for the operation of valve gear and safety controls. Generators and other equipment driven by turbines are often lubricated from the turbine circulation system.

You can see the several layers of analysis or division used in the passage clearly through the following simple outline of it:

Builders' efforts to achieve reliability include provision of a lubrication system, and the lubrication system's relative complexity is determined by the turbine's size.

I.   Small turbines (such as those driving auxiliary equipment)
   A. generally provided with ring-oiled bearings
   B. other moving parts lubricated by hand
II. Moderate-sized units (especially driven through reduction gears)
   A. may have ring-oiled gears
   B. may also have circulation system
      1. circulation system sprays oil to point of gear mesh
      2. circulation system also supplies oil to bearings
III. Large turbines
   A. always provided with circulation systems
      1. systems supply oil to all parts of the unit
      2. systems supply oil to the turbine governor
      3. systems supply oil for operation of valve gear and safety controls
   B. often also lubricate generators and other equipment through turbine's own circulation system

1.4 Accumulation of Detail

Accumulating details may be the most common description technique. By adding detail to detail, the writer makes the reader's mental picture clearer. Determining which details to accumulate is critical; generally you should favor the ones that make *X* unique. Here is a paragraph built on accumulation of detail:

> Single-shaft, nuclear-powered turbines as large as 1200 mw are now in service. Cross-compound units (double shaft) as large as 1300 mw are also in service in nuclear-powered, electric generating stations. These large units operate at 1800 rpm. As a result of the tremendous impact on a system capacity when a 1200 or 1300 mw unit goes down for inspection or maintenance, the trend today is toward smaller, fossil-fueled units in the range of 400 to 700 mw operating at 3600 rpm.

2. *Moving Patterns*

Moving patterns place *X* in some kind of larger *sequence*—usually a sequence of process, cause and effect, or time. This pattern can happen four ways: explaining *X*'s process or operation, placing *X* in a sequence of cause and effect, explaining the history of *X*'s name (etymology), or explaining the history of *X* itself.

2.1 Process

If you define and describe *X* using process as your pattern, you tell how *X* works or what it does. If your subject is a machine lathe, for example, you explain its principles and methods of operation. If your subject is inflation, you explain what is occurring that we call inflation ("prices are going up, each dollar's buying power is going down . . ."). Here is an explanation of how a steam turbine works:

> In a turbine, steam expands in stationary or moving nozzles from which it discharges at high velocity. The force of the high-velocity jets of steam causes the moving parts to rotate, thus making the energy in the steam available to do useful mechanical work—for example, to drive a compressor or generator.
>
> After it has expanded in the turbine, the steam usually exhausts to a condenser, which serves two purposes. First, by maintaining a vacuum at the turbine exhaust, it increases the pressure range through which the steam expands. In this way, it materially increases the efficiency of power generation. Second, it causes the steam to

condense, thus providing pure, clean water for the boilers to reconvert into steam. This simple cycle—water to steam, power generation, and steam to water—forms the basis on which most steam-power plants operate.

Chapter 13 covers writing about processes in much greater detail.

2.2   Cause and Effect

Placing *X* in a causal sequence can help your reader understand what *X* is. What causes *X*? What is the effect of *X*? The following passage discusses the cause of turbine reliability—correct lubrication—and the effect of improper lubrication.

> Correct lubrication plays a major role in achieving reliability and low maintenance cost. High-quality Mobil turbine oils, including world-famous Mobil DTE 797 Oil, have unusual ability to protect against wear and to resist deposit formation. Deposits may cause faulty operation of hydraulic elements, lack of lubrication, and overheating. Downtime for repairs or cleaning of lubrication systems is kept to a minimum with proper lubrication.

2.3   History of the Term

Sometimes the word for *X* has an interesting and revealing history. For example, Smackover, Arkansas, is said to have gotten its name from the way the French settlers' name for the place sounded to ears unfamiliar with French; thus, the French name for the settlement, *Chemin Couvert* (covered road), became Smackover. Such information is more readily available than you might think; it can usually be found in the dictionary.

Remember that when you're discussing the word rather than using it, you need to mark that fact for the reader by underlining the word or printing it as *italic*. The paragraph below explains how a particular part of some steam turbines got its name.

> When shutting down large turbines, it is necessary to keep the rotors turning for sufficient time to ensure uniform cooling. If this is not done, uneven cooling may cause a distortion and bowing of the shaft. A *turning gear* is, therefore, provided on turbines of over about 10,000-kw size to rotate the spindle at 1 to 100 rpm during the cooling period.

2.4 History of the Object

Your aim in giving the history of *X* is to provide your reader with a richer and more readily meaningful context, thus enhancing the explanation. Although its history may not be crucial, many people will feel more comfortable with a new item if they know a little about its history.

> The actual "growing-up" era of the steam turbine took place after 1899. In 1900, turbines of 1000 and 1500 kw were built. By 1905, at least ten turbines of 5000 kw were in service, and between 1906 and 1909, turbines of 7500 to 14,000 kw were installed. After this period, sizes increased at a rapid rate: 20,000 kw in 1911; 30,000 in 1913; 50,000 in 1917; and 60,000 in 1924. In 1929, a 208,000-kw cross-compound turbine was installed, and this remained the largest turbine for many years.

3. *Indirect Patterns*

When you use indirect patterns of defining and describing, you explain *X* by describing or relating something else to it.

3.1 Elimination

Sometimes the easiest way to explain *X* is to tell what it is *not*. That is, you pick the other objects (*Y*, *Z*, and so on) closest to *X*, those with which it might be easily confused, and you clearly differentiate them from *X*. In this way, by paying close attention to the items that surround *X*, you help your reader understand *X*. The following passage explains what a "properly refined petroleum lubricating oil" does by explaining first what can happen if one does *not* use such an oil.

> Thick, surface-separating films cannot be maintained at all times in the lubricated elements of steam turbines and associated equipment. For example, during shutdown, the thick oil film in journal bearings is squeezed out of the pressure area, and some metal-to-metal contact is made. On starting (unless oil lifts are used), friction is high and some wear occurs. This condition of *boundary lubrication* continues at least momentarily until speed is high enough to develop a thick fluid film. Boundary lubrication may exist in other situations where speeds are low, loads are high, or lubricant supply is restricted.
>
> A properly refined petroleum lubricating oil has considerable natural antiwear value, or *film strength*, under boundary-lubrication conditions.

3.2 Comparison and Contrast

If you move from comparisons that are imaginative to comparisons that are literal, you move from analogy to comparison and contrast. The passage below compares several ways of reducing the rotating speed of turbine blades while maintaining efficiency and along the way makes a nice comparison between small and large turbines.

> To reduce rotative speed while maintaining efficiency, the high velocity may be absorbed in several steps, which is called *velocity compounding*. In some turbines this is accomplished by two rows of moving blades between which is placed a row of stationary blades that reverses the direction of steam flow as it passes from the first to the second row of moving blades. Other ways of accomplishing velocity compounding involve redirecting a steam jet so that it strikes the same row of blades several times with gradually diminishing velocity.
>
> Another method of reducing rotor speed while maintaining efficiency is to decrease the velocity of the jets by dividing the drop in steam pressure into a number of stages. This is called *pressure compounding*. Each stage consists of a single row of stationary nozzles and a single row of moving blades, and the method is equivalent to mounting several single-stage impulse turbines on a common shaft. In large units, velocity compounding is usually employed in the first pressure stage and pressure compounding in the remaining stages.

3.3 Analogy

Analogy, or the making of figurative comparisons, is one of the most neglected but potentially useful techniques in writing for business and industry. By creating one well-chosen figure of speech, you can suggest more and deeper meaning in fewer words than you can with almost any other pattern. For example, one business analyst wrote that the American automotive industry is like a sleeping dinosaur—either just resting, or hibernating through bad times, or on its way to total extinction. The problem, he said, is that we just don't know which of the three options is right. Although figurative language may be underused in some kinds of writing, that may only make the well-chosen analogy more effective when it does appear.

> The force of the high-velocity jets of steam issuing from the nozzles of steam turbines may be utilized in two ways. If the nozzle is fixed and the jet directed at a movable blade, the jet's *impulse* force pushes the blade forward. If the nozzle is free to move, the *reaction* of the jet pushes against the nozzle, causing it to move in the opposite direction. Both these actions can be observed when using the ordi-

nary garden hose. Many commercial turbines use both impulse and reaction principles.

3.4 Naming Examples

Few explanations are complete without examples. Many a writer of cloudy passages has learned to write reasonably clear explanations by learning to provide two or three good examples to support each important generalization. Here is an example:

> In many instances, efficient turbine speed is higher than that of the machine being driven. This may be the case, for example, where a turbine drives a direct-current generator, paper machine, centrifugal pump, blower, fan, or sugar-cane crusher. Under these circumstances, reduction gears are used to connect the high-speed turbine to the driven unit.

3.5 Pointing Out Examples

To make a really effective definition and description of *X*, you need to tell your reader where to find an example of *X*. That is, you point one out. In the example below, the writer gives a preliminary explanation of bearing lubrication and then tells the reader where to find a better explanation.

> Oil supply passages are provided in the housing and in the bearing shells through which a copious supply of oil is furnished from the circulation-oiling system. This oil is delivered to axial grooves and chamfers at the horizontal split, sometimes to both sides and sometimes only to one side. The grooves distribute the oil along the length of the bearing, and the chamfers assist in the formation of a film of oil between the journal and the bearing.*

> * A complete discussion of bearing lubrication will be found in the Mobil publication "Plain Bearings—Fluid-Film Lubrication."

3.6 Showing Examples

Despite the fact that it's such an obvious way to explain something to someone, using pictures and illustrations is as neglected a pattern as analogy. The illustrations in Figure 12.2 are from the page that faces the table of contents of *Steam Turbines and Their Lubrication;* anyone who looks at the table of contents also looks at these visuals.

Figure 12.2   Cross-Sections of Steam Turbines

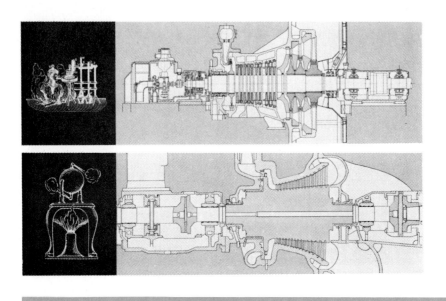

Source: Reproduced by permission. Copyright © 1965, 1981, Mobil Oil Corporation.

This chapter organizes 14 patterns according to a functional classification scheme, not necessarily a logical one. Depending on what you are writing about, one pattern may be pretty much like another, or one pattern may be exactly right and another totally inappropriate. Because of this, it is important that you go beyond looking at a particular pattern and debating whether it is comparison and contrast or analogy (to pick two patterns that often are hard to tell apart); the goal is for you to be able to use the patterns.

4. *Adapting Definitions and Descriptions*

At this point your instructor may be preparing to ask you to write a definition and description report in order for you to get intensive practice with using a variety of different developmental techniques. After college, although you will need to define or describe in nearly everything you write, the kind of piece you are asked to write will more likely be called an annual report, progress report, newsletter, sales promotion, design report, or procedures manual, to name only a few of the kinds of professional documents. Nonetheless, most of what you will be doing is defining and describing. The question then becomes how to adapt your defining and describing to the different settings and audiences professional life presents you with.

4.1 Reader's Purpose Versus Writer's Purpose

In all kinds of writing, from an academic exercise to a multimillion dollar proposal, you need to be very careful to balance your purpose with that of the reader. The balance you strike between reader's and writer's purposes will determine how you handle two key aspects of your report—its organization and its level of complexity (and abstraction). Before you can determine what method of organization or what level of detail and abstraction may be right, think about reader's purpose versus writer's purpose.

The conflicting demands of reader's versus writer's purposes always concern writers in business and industry. The tugs and tension between them are the forces you feel when you write. Your purpose as writer always must be balanced against your audience's purpose as reader(s). It is natural for any writer to want to tell everything he or she knows about a subject, just as it is natural for any reader to want to know only what directly concerns him or her. Writers are often not particularly aware of the amount of background a reader may need to be able to understand a document, or of the order of elements the reader needs to be able to understand with a minimum of difficulty. Readers, on the other hand, are especially aware of their need for background and particularly impatient with writers who have not provided it. Similarly, writers tend to use most those words with which they are familiar, while readers resent writing that contains too many unfamiliar words. A thoughtless writer may tend to deal with the topic at a level most comfortable to him or her, without regard for the reader's inability to handle the topic at that level of abstraction. Being aware of your reader's purpose in reading, including his or her level of interest and knowledge, can help you avoid these problems.

4.2 Organization

If your writing takes its reader sufficiently into consideration, it will be reader centered, not writer centered. The way many writers work, first drafts tend to be writer centered—in vocabulary, amount of detail, level of abstraction, and organization. In their first drafts, most professionals who write just try to get the content down on paper. At that point, the best way for the ideas to be organized is whatever way they happen to come out, just so long as they all wind up on paper. First drafts reflect the order of thought that allows the **writer** to put words on paper most easily. What the **reader** needs, however, is a report written following the order of thoughts by which he or she can most clearly understand it. Ideally, the later drafts of the same piece of writing are progressively more reader centered, as the writer gains enough control over the material to adjust it to someone else's needs. (This distinction parallels the catalogical/analytical distinction made in Chapter 11: writer-centered writing is often catalogical, and reader-centered writing is analytical.)

If you view the writing process as one of only two stages, such as creating and criticizing, it's easy to see the first—creating—as centering on the writer's need to get the subject down on paper somehow, and the second—criticizing—as centering on the reader's need to understand what has been written. The reader's need to understand, and in a broader sense, the document's overall effect on the reader, should be among your primary concerns when you criticize and revise your writing. Make sure the organization of your report builds from a base the reader will understand and that the order of facts and concepts will be logical to the reader.

Figures 12.3 and 12.4 show the first and second versions of the same simple memo. Between versions, the writer balances his initial concern with his own purpose against his growing awareness of what the reader wants, and will get, out of the memo.

As chapter 8, "Introductions and Conclusions," discussed, the temptation for writers is to jump into the middle of a topic, reciting facts and figures with no preliminaries. The problem with this approach is that most readers need to know the background before they can make anything out of those facts and figures. That background is so familiar to the writer that he or she may forget to include it, but the reader absolutely must have it. The rest of the report must build in a similar manner, taking into account the information the reader needs to have, and not simply following the order of items the subject had in the writer's mind when he or she first put the report together.

4.3   Level of Complexity and Abstraction

Decisions about the appropriateness of a passage's level of complexity or abstraction cannot be made sensibly without knowing your audience; language that is too abstract or complex for one audience, message, or purpose may be just right for another. Consider the defini-

Figure 12.3   A Writer-Centered First Draft

MEMORANDUM

To: Jim Hanson, Shipping Room Mgr.

From: J.R. Reed, V.P.

Subject: The shipping room mess

Date: 3/23/92

While you were on vacation I had occasion to come down and in-
spect the shipping room, and I was really shocked at what I saw.
There were uncoded packages and parts of shipments all around,
equipment not properly stored, not properly maintained. I couldn't
believe what I was seeing, or that you let the shipping operations be
so messy. I think that an operation like that needs to be neat and orderly with all equipment properly maintained and all shipments assembled in an orderly way. Not only does sloppiness increase the
chances of losing shipments, but it also is a considerable safety risk
to the people who work there. I shouldn't have to take my time to
tell you things like this. My desk is covered with my own work, and
it's a shame I have to do yours too.

tion in Figure 12.5: Is it right to ask whether it is too abstract or
complex without specifying what kind of audience it is aimed at?

Which sections of this definition are abstract? Some readers will
find the first sentence difficult because it has a quality (velocity)
changing according to another quality (time rate). The second sentence contains a concept (vector quantity) that may be unfamiliar to
some readers and make the passage difficult, and the logic of that
sentence (because velocity has both magnitude and direction, and
velocity is part of acceleration, then acceleration must also account
for changes in both magnitude and direction) certainly will give trouble to anyone who had problems with the concept of "vector quan-

Figure 12.4   **A Reader-Centered Second Draft**

MEMORANDUM

To: Jim Hanson, Shipping Room Mgr.

From: J.R. Reed, V.P.

Subject: Improving conditions in the shipping room

Date: 3/23/92

While you were on vacation I went down to the shipping room to
check on the status of a delivery. What I saw there wouldn't have
pleased you. There were packages and parts of shipments all around
the room in no apparent order, and much of the equipment apparently had been left wherever the last user had finished with it. The
equipment also didn't look like any maintenance had been done on
it since you left. That kind of sloppiness increases our chances of
confusing a shipment or having an on-the-job accident. I know
that's not the way you like to run your shop. When you get caught
up from being away, let's get together and review your policies on
neatness and safety and try to come up with a way to keep those
standards up even when you are not there to supervise them.

tity." The logic becomes more complex with the next sentence ("The
velocity of a point or object . . ."), which provides a specific example
of the previous sentence's point. The inference that can lead a reader
to the last sentence from the previous two is still more complex, the
result of an abstract reasoning process.

Some readers may have problems of the sorts detailed above; at
least in the view of the editors of *Encyclopedia Britannica*, most of their
readers will not. What is too abstract or complex for one reader may
suit another perfectly.

To adjust that definition of acceleration for a reader with less of
a background in physics would involve alterations in more than just

Figure 12.5   A Definition Written at a High Level of Abstraction and Complexity

Acceleration: the time rate at which a velocity is changing. Because velocity, a vector quantity, has both magnitude and direction, acceleration is also a vector quantity and must account for changes in both magnitude and direction of velocity. The velocity of a point or an object moving on a straight path can change in magnitude only; on a curved path, it may or may not change in magnitude, but it will always change in direction. This condition means that the acceleration of a point moving on a curved path can never be zero.

Source: From *Encyclopaedia Britannica,* 15th edition (1980), Micropaedia, Volume I, p. 48. Reprinted by permission.

the words used in the passage; the concepts themselves would have to be simplified or in some cases left out. The example would have to be more concrete and able to stand on its own without reference to the theory in the abstract explanation. But again, we cannot be specific about adaptation without specifying an audience and its purpose in reading the definition. If we can assume the audience for the *Britannica* is composed of college graduates, then we can see how concepts like "vector quantity" and abstract chains of logic may be appropriate. For an audience with only high school education, a better definition might well start out in this way: "*Acceleration* means any change in an object's speed along its current path." The definition may then go right into a specific example, such as a rocket taking off, and explain that its increase in speed along its path is its acceleration.

Determining the proper levels of complexity and abstraction for a document requires that you know its audience and their purpose in reading it. Most writers save adjusting the levels of complexity and abstraction in their writing for some stage of their writing beyond the first draft. As Chapters 2 and 3 showed, making your writing readable is most properly a concern during revision.

All aspects of the document, from sentence length to overall organization, come into play when you are measuring and adjusting its levels of complexity and abstraction. One of the most important such aspects is the kind of verbs you use. Especially when you write a passage that defines and describes something, it's tempting to make nearly every verb a form of *to be.* The few that are not are usually either forms of *to have* or they are in passive voice (the form "A was hit by B" rather than "B hit A"; see the Appendix, Sections 2.1 and

Figure 12.6   A Passage with a Chain of Weak Verbs

DISINTERMEDIATION

Almost everyone borrows money at one time or another. Sometimes qualified borrowers are unable to get a loan because funds just are not available. This is because of disintermediation. Many people do not know what disintermediation is or how it works. However, since it affects anyone who borrows money, it is useful to have an understanding of the subject. Disintermediation is an economic process. More specifically, it is a monetary decision. When investors withdraw their savings from financial intermediaries to make investments elsewhere because the ceiling interest rate is lower than the interest rate offered on securities, disintermediation has occurred.

2.2, for more explanation). Such a chain of weak verbs makes your passage needlessly abstract; nothing ever *does* anything to anything. Consider the example in Figure 12.6; the weak verbs are highlighted and connected.

The connected, highlighted words make a chain of weak verbs. The effect of that chain is to flatten the writing, taking away both life and precision of meaning. Three or four weak verbs in a row is bad enough, but you can easily find such sequences that go on for pages in many reports. Only the most determined reader will hack through this kind of deadly dull language, and it's deadly dull mostly because of the weak verbs.

The way to deal with such strings of needlessly weak verbs is to revise them out. Chapter 3 presents much more detail on revising weak verbs. The same passage, this time revised to strengthen the verbs, is shown in Figure 12.7.

Successfully dealing with issues such as organization, levels of complexity, and abstraction is crucial if you are to write effective definitions and descriptions. The combination of the right variety of techniques with the proper balance between reader's purpose and writer's purpose puts your definition or description on the track toward success. Any writer involved in writing an extended definition may be tempted to let his or her purpose as a writer—to get everything necessary into the definition and description by using the words and ideas that seem most convenient—prevent the fulfillment of the reader's purpose. In this kind of writing, as in any other kind that

DISINTERMEDIATION

Almost everyone borrows money at one time or another. Sometimes—because of disintermediation—qualified borrowers cannot get loans because funds just are not available. Many people do not recognize disintermediation or know how it works. However, since it affects anyone who borrows money, an understanding of the subject can help anyone who borrows money. Disintermediation is an economic process that occurs when investors make the monetary decision to withdraw their savings from financial intermediaries and to invest their money elsewhere. This usually occurs when the interest rate offered on securities exceeds the ceiling interest rate offered by financial intermediaries.

professionals do, remember that purpose and audience go together; in adapting your message to your reader, you must be sensitive to fulfilling the reader's purpose. Combining the reader's purpose and the writer's purpose establishes the document's purpose.

5. *Examples of Definition-and-Description Writing*

The following three examples show a few of the ways that defining and describing appear in the kinds of writing professionals do.

5.1 A Student Report

The report below was written by a business student for an advanced technical writing class. The assignment required the student to use as many of the different techniques presented here as possible but to work them into a readable paper.

A Definition of Selling Short

The term "selling short" indicates a specific type of trade made in a financial marketplace. Although selling short can be done in either the stock market or the commodity futures market, it will be explained here only in the context of commodity futures. One only sells short in a

market that is trending down, so the first part of selling
short is accurately forecasting the price of a commodities
market. To sell short successfully, one sells a given commodity at the current market price and then buys it back
later at a lower price.

The origins of the term "selling short" are obscure.
Since the action involves selling, the first word's sense
is obvious. "Short" is used in the expression in the same
sense as "getting caught short." When one sells short, one
runs the risk of getting caught without enough of the product to deliver——one is short of the product.

To understand selling short, one must first understand
certain particulars about the commodity futures market. It
is made up of 65 different commodities, which may be divided into four basic categories: metals, meats, grains,
and financials. Some of the commodities which make up
these four groups are gold, silver, platinum, paladium,
zinc, pork bellies, hogs, cattle, corn, oats, soybeans,
U.S. T-bills, U.S. T-bonds, British pounds, and the Standard and Poor 500 Stock Index. Each commodity is traded in
a specific contract size. For example, silver is traded in
either 1000 oz. or 5000 oz. contracts. Each contract
trades in specific increments. Silver trades in cents per
pound, so a one cent move in silver is worth either $10 or
$50, respectively.

All these markets are traded at a certain fixed time in
the future. Each contract is deliverable during a certain
specified month. For example, if one is trading December
1994 silver, that means one has control over a contract of
silver which is to be delivered in December 1994. By selling short a contract of December silver, the seller com-
mits to deliver either 1000 or 5000 oz. of silver in December. That commitment is cancelled when the contract is
bought back. Buying back means purchasing the same contract of the given commodity at the prevailing market
price and at the same exchange on which the short was
sold.

The profit in selling short occurs when the commodity
is bought back at a lower price than the one for which it
was sold. A loss occurs if the commodity is bought back at
a higher price. So if one sells short 1000 oz. of December
silver at ten dollars an ounce and buys it back at nine
dollars an ounce, the end result is a profit of $1000. On
the other hand, if the commodity is sold at ten dollars
and bought back at eleven, the result is a $1000 loss. Fig-
ures 1 and 2 demonstrate the two trades.

Figure 1.
Going short in a down-
trending market.
X = point where contract
was sold-short
O = point where contract
was bought back.

Figure 2.
Going short in an up-
trending market.
X = point where contract
was sold-short
O = point where contract
was bought back.

Selling short is the exact opposite of the more popular
type of trading--buying long. "Buy low, sell high" is the
advice every new kid gets on Wall Street. Yet the amount
of money to be made on the short side is in direct proportion to the amount made on the long side. Despite this
fact, Wall Street traders and the American public are much
more reluctant to sell short than they are to buy long.
Selling short is like betting against the man shooting
dice who has just had five successful passes: the odds are
he is not going to have a sixth. There is money to be made
betting the shooter's luck runs out.

One particular type of trader tends to go short more
than anything else--the Hedger. As distinguished from a
Speculator, the Hedger has some interest in the commodity
beyond that of making money from fluctuations in its
price. For example, a soybean farmer will use the commodities markets to hedge his soybeans, thereby insulating himself against the risk of price fluctuations during the
growing season.

Suppose the farmer anticipates that in September of 1993 he will harvest fifty thousand bushels of soybeans. His cost per bushel is three dollars, so anything more than three dollars per bushel is a profit. The current market price, in November of 1992, is six dollars a bushel. The farmer would like somehow to lock in that six dollar price for his next year's crop. So the farmer goes to the Commodity Markets and sells short 50,000 bushels (or ten contracts) of September 1993 Soybeans. He has committed himself to deliver 50,000 bushels of soybeans at six dollars a bushel in September of 1993 to an unknown destination for an unknown buyer.

In September of 1993 the market price of soybeans is four dollars a bushel. The farmer buys back his 50,000 bushels and sells them to a specific buyer in his part of the country. He executes both trades at four dollars a bushel. So by selling short in the commodity markets the farmer has increased his profit by two dollars a bushel. He has made $100,000 that otherwise would have been lost in price fluctuations.

The practice of selling short first appeared in the U.S. during the early 1700s. It was introduced by European merchants engaged in the cotton trade. If they were happy with the price they paid for a shipload of cotton they would offer the farmer an option on his next crop. For a small deposit they would have the right to buy the farmer's next crop at current prices. If prices had declined by the time the merchants returned, then they forfeited their deposit and paid market prices for their cotton. If prices had gone up, then the merchants exercised their option and bought their cotton at last year's prices. Since the farmer was supplying the complementary side of the option, he was essentially going short. If the price declined he made a little extra because he got to keep the merchants' deposit. If the price went up he lost a little, but he still sold his cotton at a price that was acceptable to him.

There are many commodity exchanges throughout the world on which one may sell short. The most important domestic exchanges are the Chicago Board of Trade, the Chicago Mercantile Exchange, and the New York Mercantile Exchange. Internationally, important exchanges are located in Zurich, London, and Hong Kong.*

---

* Used with the permission of Robert Vogel.

5.2 A Product Description

The following is from the *Quick Reference* manual for the IBM Personal System/2 Model 50:

What Is the Reference Diskette?

The Reference Diskette is a permanently write-protected diskette. You can *read* information from the diskette, but you cannot *write* (record) information onto the diskette. Make a backup copy of the diskette as soon as possible. Once the copy is made, put the original diskette in a safe place and always use the backup copy.

The Reference Diskette contains the following programs:

1. *Learn about the computer* provides information about:
   —Hardware: identifying features, using options, and handling diskettes
   —Software: practicing with different kinds of software
   —Communications: using telecommunications, local area networks, online information services, and linking to a mainframe
   —Troubleshooting: solving computer problems, testing the computer, and getting service.
   An index is also provided, so topics can be looked up quickly.

2. *Backup the Reference Diskette* makes a copy of the original Reference Diskette onto another diskette, but not onto a fixed disk drive. To make the backup copy, you will need a blank 2.0 MB capacity diskette that is not write-protected (see "Write-Protecting Diskettes" on page 40).
   Whenever you need to use the Reference Diskette, always use the backup copy. This ensures that the backup copy contains your computer's current configuration information and testing programs.

3. *Set configuration* is used to view, change, back up, or restore the computer's configuration.

The computer automatically sets its configuration when the Reference Diskette is started or when "Run automatic configuration" (see below) is selected. During automatic configuration, the computer makes a list of what it sees as being installed and assigns those items to operate a certain way. Then this configuration information is stored in the computer's memory and is kept current by the battery even when the computer is turned off.

The configuration lists the computer's:

—Installed memory size

—Built-in connectors and their assignments

—Installed *IBM* options with their location and assignments.

Whenever you change, remove, or install an *IBM* option or the battery, you must start the backup copy of the Reference Diskette so the computer can automatically configure itself.

View configuration shows you the present configuration stored in the computer's memory.

Change configuration is used to make changes to the configuration stored in memory. Changing the configuration lets you tailor the computer's operation to your needs.

Backup configuration copies the configuration stored in memory onto the backup copy of the Reference Diskette. If you have made changes to the configuration then you should back up (copy) the configuration in the event the battery is removed or replaced.

Remember, the computer's configuration is stored in memory and kept current by the battery. If the battery is removed or replaced, the configuration information is lost.

Restore configuration retrieves the configuration copied by "Backup configuration" and restores it back into the computer's memory. Use "Restore configuration" after the battery is removed or replaced.

Run automatic configuration is used when you want the computer to automatically configure itself. During automatic configuration, the computer makes a list of what it sees as being installed and assigns those items to operate a certain way.

4. *Set features* is used to:

Set date and time so that you have the convenience of recording the date and time of your computer activities. Once the date and time are set, the computer's battery keeps both current, even when the computer is turned off.

Set passwords to help restrict the use of the computer by unauthorized persons. Three passwords are available: a power-on password, a keyboard password, and a network server password.

Set keyboard speed to change the speed at which the keyboard responds when you type.

5. *Copy an option diskette* is used if you are installing an *IBM* option that comes with a diskette and instructions to update the backup copy of the Reference Diskette.

   This option diskette contains the option's testing program and configuration information. Be sure to follow the instructions supplied with the option.

6. *Move the computer* prepares and protects the fixed disk drive for a move.

7. *Test the computer* tests the computer hardware. If a problem occurs during this testing, an error message appears with the cause of the problem and the action to take.*

---

* IBM, Introducing the Personal System/2 Model 50 Quick Reference, pp. 10–12. Reprinted with permission.

5.3   A Technical Description for Laypeople

The following passage comes from the introduction to a book designed to introduce laypeople to the subject of robotics.

What is a Robot?

If you walk into the Nissan truck factory in New Smyrna, Tennessee, expecting to see shiny androids like C3PO assembling parts, you will be severely disappointed. The modern, industrial robot has far more in common with an ordinary piece of machinery than with a human. This only makes sense, since robots are machines. An example of a robot at work is shown in Figure 1-1 [Figure not included here.]. This industrial robot is trimming plastic dashboard components by moving them under a high-power laser. What is the difference between this robot and any other piece of automated machinery? Why is this machine called a "robot," not just an "automatic dashboard trimmer"? The robot is a special kind of automated machine. A robot can do not only this particular job of trimming, but it can be programmed and retooled to do many different jobs. This programmability and versatility is why all robots are automated machines, but all automated machines are not robots.

   There is only one definition of an industrial robot that is internationally accepted. It was developed by a group of industrial scientists from the Robotics Industries Association (formerly the Robotics

Institute of America) in 1979. They defined the industrial robot as ". . . a reprogrammable, multifunctional manipulator designed to move material, parts, tools, or specialized devices through various programmed motions for the performance of a variety of tasks." Let's take a close look at this definition to see just what it implies.

The first key word is *reprogrammable*. This implies that a robot is a machine that can not only be programmed once, but can be programmed as many times as one likes. Many electronic devices we use every day contain computer chips that are programmable. Programs are written on the chips of digital watches, for instance, that instruct them to do such things as play "Dixie" as an alarm. These programs cannot be easily changed, however. There is no allowance for input by the owner. You cannot, for example, put a song of your own into the watch when you get tired of waking up to "Dixie." The programs are "burned in" by the manufacturer. A robot, however, contains a program that is accessible, that can be changed, added to, or deleted, as the user chooses. A robot can have many programs to do different things in any sequence whatever. And, of course, to be programmable, a robot must have a computer that can be fed new instructions and information. The computer can be either "on board," which means the computer console is mounted on the robot itself, or it can be "remote," which means the computer that controls the robot can be anywhere you like as long as it can communicate with the robot.

The next key word in the RIA definition is *multifunctional*, which implies that the robot is versatile, that is, can perform more than one task. The same industrial robot used for laser cutting in Figure 1-1 could, with a simple change of end tooling, also perform welding, painting, or assembly operations.

The third key word is *manipulator*, which implies that a robot has a mechanism of some sort for moving objects for the performance of its work. It's the manipulator that separates a robot from a computer, just as it's the reprogrammability and versatility that separate a robot from other kinds of automated machines.

Finally, let's consider the meaning of the phrase *various programmed motions*. This implies that the robot is dynamic; that is, it is characterized by continuous, productive activity.

Although this definition may seem very broad and somewhat ambiguous, it does serve to separate industrial robots from, for instance, fixed-sequence automated machinery, or from multifunctional machines, such as food processors, that are equipped with interchangeable parts to perform various tasks, from blending sauces to grinding beef. It also removes robots far from the realms of science fiction, since any anthropomorphic (humanlike) characteristics a robot may or may not possess are merely a matter of efficacy.

From this perspective, then, the robot can be considered a major advance in the logical progression in the development of automated machines. We have moved from building machines that can do one job with human control to machines that can do many different jobs without any human control. The first industrial revolution has been said to be the start of an era of general industrial use of power-driven machines. The modern industrial renaissance may be called an era in which we are building machines capable not only of building other machines, but also of repairing and "reproducing" themselves.*

---

* From *Robotics: A User-Friendly Introduction*, by Ernest L. Hall and Bettie C. Hall. Copyright © 1985 CBS College Publishing. Reprinted by permission of CBS College Publishing.

EXERCISES

1. For practice in the techniques presented in this chapter, choose one of the following terms and explain it in each of the 14 different ways discussed in this chapter. Your explanations should be written on a level your classmates can understand, but you do not need to work the explanations into one continuous piece of writing.

amortization	gravity	prime
bluegrass music	inertia	interest rate
concrete	lathe	protein
conspiracy	libel	slag
containerization	liberal	software
credibility	education	space shuttle
enzyme	limit slope	spring
fusion	Noah principle	steel
generally accepted	optical molasses	supercomputers
accounting principles	petri dish	torque

2. For practice in the different techniques of defining and describing, choose one item from the list in problem 1 and explain it in seven different ways, with your classmates as your audience. Work the seven different patterns into one well-written short, informative report (300–500 words) designed to familiarize readers with important basic information about that term.

3. Choose seven separate paragraphs from the examples of definition and description writing presented in Section 5 of this chapter, and explain the

---

A ➤ indicates a case study exercise.

structural pattern(s) in each. (Do not use paragraphs discussed as examples in this chapter.) Use the patterns decribed in this chapter as much as you can; where they don't fit, explain the pattern you do see. Write the results of your analysis into a short (300–500 word) report. Be sure to identify carefully which paragraphs you are discussing.

4. In consultation with your writing teacher, choose an object or a concept in your major field of study and define and describe it at length (500–750 words). Your audience is an executive/layperson—someone who may not know anything about your subject but who has some decision-making power over your career. If you need to consult other printed materials to provide additional information, be sure to document your sources. Use a *cpo* introduction (described in Chapter 8), and make the reader benefits explicit. Use at least one visual and as many of the different techniques described in this chapter as you can. When you consciously use a particular technique, name it in the margin. Remember to try to strike a balance between your purpose (to explain your subject and to use a number of different techniques in doing so) and your reader's purpose (to understand something new and different). Thus, while you are trying to use as many different techniques as possible, concentrate on producing a unified and clear piece of writing (rather than mechanically grinding through the various techniques).

5. One common use of defining and describing in professional life is explaining new products (as in the example in Section 5.2 of this chapter). Find a description of a new product related to the profession you hope to enter upon graduation, and write a short (300–500 words) analysis of its use with the different techniques described in this chapter.

➤ 6. You are in your first year on a new job and, in casual conversation with your boss, have mentioned a new concept or piece of equipment in your field. Your boss asks you to write a short (300–500 words) description of it; the purpose is to help your boss understand this new item. Your boss has 20 years of experience in the field but generally does not keep up with new developments, choosing to leave that to new employees such as you. You want your report to be as professional as possible, so be sure to make full and frequent use of visuals.

➤ 7. Another common use of definition in professional writing is to describe a situation, such as a need to close down an unsuccessful venture your company has been involved in or a problem with the manufacture or marketing of a product. In such situations, the writer's task generally is to get to the heart of this relatively intangible subject and describe it in such a way as to give readers a good understanding of its complexities. This exercise asks you to give your classmates, in written form, the description of a situation. Put yourself in the position of being a student summer intern in either your city government or your college's administrative offices. You have been asked to provide a short background report on a situation that will assist decision makers in coming to agreement on how to deal with it. (For this exercise, you can choose the situation; this assignment works best when students choose something currently going on within or near the campus— something you can dig into a little yourself without having to do much library research, such as an ongoing parking problem, disputes over dormitory rules, problems with increasing fees, or policy about releasing campus crime information.) Find some such current local situation, and write a definition and description of it; assume an audience who is interested but not particularly well informed about the situation. Do not try to argue for one or another solution; what you are doing is presenting a 500-word background report. Describe how this situation came about (its history, its cause, and so on) and what its component parts are. Try to use a variety of definition and description techniques, but do not give in to the temptation to grind through them one after another.

13. *Explaining Processes and Giving Instructions*

1. **Basics**
   1.1 **The Opening**
   1.2 **The Body**
   1.3 **The Conclusion**

2. **Varieties**
   2.1 **Explaining Processes**
   2.2 **Writing Simple Sets of Instructions**
   2.3 **Writing Manuals and User's Guides**

One of the most common situations that working writers face is the need to describe a process. Chapter 12 presented *process* as one of a number of techniques that might be used in the larger context of explaining something. This chapter explains how to write about a process: not what something is but how something works or how something is to be *done*. Consider two examples:

- Writing for an executive/layperson audience, you may need to explain how a process works in general terms. For example, an engineer may need to explain to a prospective client—perhaps a banker or a lawyer—the particular method that engineer's firm proposes to use to solve a difficult technical problem on a project for that client.

- Writing for a mixed audience—laypeople, executives, technicians, the educated public—you may need to write a set of very specific instructions or procedures. For example, a computer specialist may need to write instructions for users of a customized software package that her company is selling on the open market.

These are just two of the many kinds of situations that require professionals to write about processes. While there are important similarities that connect all instances of writing about processes, there are also important differences— many of them triggered by differences in purpose and audience.

Because certain characteristics underlie all process writing and because there are so many different kinds of process descriptions that may need to be written, this chapter has two main divisions: *basics* (elements common to all process descriptions) and *varieties* (explaining processes, writing simple sets of instructions, and writing user's manuals). Three examples of process writing are provided within this chapter: a generalized process description written by a student ("The Process of Seining"), a simple set of instructions written by a professional ("Instructions for Modular Jack Converter"), and a short piece of user documentation also written by a professional ("How to Set Passwords").

1. Basics

Here is a list summarizing the key points of nearly any process description:

- Important principles that underlie the entire process, such as safety warnings, need to be placed *first*.

- The best process writing is always done by a writer who knows the process intimately. If you as a process writer are not in that position, it becomes doubly important to have good access to the process itself as a source of information while you're writing about it.

- The level of specificity in your process writing is a key to its general effectiveness and to your reader's feeling about the extent to which the writing was done with that particular reader in mind. When you are doing process writing, carefully monitor your level of specificity to make sure it meets the needs of that situation and those readers.

- As in writing definitions and descriptions, the tension between writer's purpose and reader's purpose needs to be observed carefully. It's easy for process writers to serve their own needs at the expense of frustrating the reader's needs. When you revise your process writing, check to see that your reader's interests have been honored.

- Visuals are especially important in process writing (see Chapter 10, Sections 3.3 and 4.1). Be sure to make full and frequent use of visuals when you are writing about processes.

Structurally, any piece of process writing will probably have an opening, a body, and a conclusion. Sections 1.1 through 1.3 describe each of these structural elements in detail.

1.1 The Opening

The opening of any process description should provide a brief overview of the process. What is it, who does it, and when, how, and where does it occur? If the process description is not part of some larger document, it also needs an introduction and a specific adaptation to that particular audience. Here is an example of an introduction specifically aimed at a layperson:

Home Auto Body Repair

It finally happened. After guarding your new car from scratches and dents for months, you accidentally tried to park just a little too close to that steel post. Can you still avoid the high cost of your local auto body repair shop? Yes! This article explains the process of minor automotive body repair and demonstrates how easy it is to perform at your own home.

Following the introductory paragraph, make a clear statement of the main steps in your process; usually this also forecasts the structure of your report. If you use headings, the main steps of the process should correspond to the main headings. This following extract forecasts the main steps in a process; subsequent major headings in the paper correspond exactly to the steps forecast here:

The primary steps of automotive body repair are preparing the surface, filling the dent, and repainting the body.

You may want to use a simple flowchart to tell the reader the steps of the process and simultaneously convey the process's dynamic qualities. Each part of the flowchart in Figure 13.1, from a report on

Figure 13.1   A Simple Flowchart

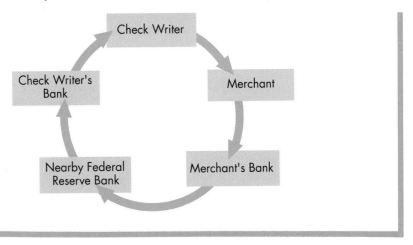

the check clearing process, corresponds to a stage in the process the report describes; each stage in the process has its own heading and section.

The next part of the opening should list all the parts, tools, and supplies needed for the process. This list will vary in complexity and specificity, depending on the kind of description you're writing. The following sample is for a generalized description of the process of making claw hammer heads; a set of directions for the same process would need a much more detailed and specific list.

> The following materials, machinery, and tools are needed in the making of claw-hammer heads:
>
> Round blanks of plain carbon steel (ASTM spec. 1078)
> ½ ton (453 kg.) drop-hammer forge
> 50 ton (45,000 kg.) trimming press
> Induction heat hardener/temperer
> Grinder
>
> The process also uses a Rockwell Hardness Tester. This is used between the various steps in the hammer's manufacture to assure that the surface hardness of the hammer-head meets the required specifications.

Just as important as listing parts, supplies, and tools is explaining whatever principles or special conditions apply, *especially safety warnings*. Even in professionally written process descriptions, the failure to indicate principles and warnings at the beginning may be the most common error. A writer may assume that the reader will read the entire set of instructions before starting to perform the process, but experience has proved that assumption false. Put *anything* your reader needs to know throughout the process at the *beginning* of the process description. If there are safety warnings, put them at the beginning, preferably surrounded by some sort of border to call attention to them, as in the following example:

DANGER!

Once X-rays are being produced, do not open the X-ray machine. Do not attempt to look at the sample during tests with units running. Do not stand on either side of the X-ray machine.

1.2   The Body

The parts of your process description will usually correspond to the
steps in the process. Because the process itself dictates the structure,
writing the body of process descriptions is usually fairly simple—as
long as you are sufficiently familiar with the process. You need to
have access to the process while you're writing, especially if (as too
often happens) it is one with which you are not as familiar as you
should be. Keep in mind the key points listed at the beginning of this
chapter, especially your audience and its purpose. The same points
about adapting definitions and descriptions according to the reader's
needs that were made in Chapter 12, Section 4, also apply to writing
about processes. Notice the different ways in which each of the following examples refers to its audience:

> *Example 1:* Once you have acquired the proper materials, you
> are ready to begin.
>
> *Example 2:* The purpose of this paper is to familiarize the
> reader with one particular type of manual inventory control
> process.
>
> *Example 3:* The specimen is loaded into the hydraulic test unit
> and the extensiometer is attached.

The writing in each example makes particular assumptions about
the kind of audience involved and their purpose in reading. In Example 1, the reader needs to know how to *do* the process, whereas the
readers of Examples 2 and 3 need to know only how the process
works. In the first example there's a close relationship between writer
and reader. In the second, the relationship is more distant. The third
example deals with its audience by not referring to it directly at all.
Your specific audience and its purposes must be firmly in your mind
while you're writing.

1.3   The Conclusion

At the end of your process description you need at least a brief conclusion. Here are five techniques for conclusions that are especially
appropriate for process writing. These and other conclusion techniques are discussed in more detail in Chapter 8, Introductions,
Transitions, and Conclusions.

1. Repeat the major points.
2. Emphasize the importance of the process.
3. Restate cautions and safety warnings.
4. Tell the reader how to evaluate the process—how to evaluate
   whether it has been done properly.
5. Describe alternative steps or troubleshooting procedures.

2. Varieties

The many different kinds of situations that require a professional to write about a process tend to fall into three categories:

1. *Process explanations*—how a process takes place or how someone does a process (as opposed to how to do a process).
2. *Simple sets of instructions*—a brief description of how to do something.
3. *Manuals, especially user's guides*—an entire document dedicated to helping someone perform a fairly complicated activity.

2.1 Explaining Processes

Many of the situations that require a writer to explain a process involve addressing an executive/layperson audience. You may be explaining a process that does not involve human agency at all—how a chemical reaction takes place, for example, or what goes on when pollutants leach out of a mine shaft and into a watershed. A research-and-development specialist may need to explain a complicated sequence of purchasing orders to an accountant; an accountant may need to explain a financial transaction to a customer. Or you may be explaining in general how someone does something—not for an audience who will ever need to perform the process themselves but for an audience who simply need a basic understanding of that process and that person's role in it. An engineer may need to explain to a banker the safety-shutdown procedures followed by computer technicians in the event of a power outage. In all such situations, you should explain the process specifically and accurately but not too technically.

Discovering the right level of detail to use in such situations means striking a balance between what you as a writer want to include and what you know the reader will accept (see Chapter 12, Section 4). The key elements are to put the process into a frame that will make the piece of writing appeal to the reader and to be appropriately specific without losing your audience.

Framing the Structure. Explaining a process to an executive/layperson presents particular problems for you as a writer. You cannot assume the reader has either the knowledge or the interest in the subject that an ideal reader might have. The first question an executive/layperson reader is likely to ask is, "Why should I read this?" You should answer that question in the introductory section of your explanation (see "Reader Benefits," Chapter 4, Section 2.4). The following example shows a writer framing the structure: leading the discussion from the reader's world to the writer's (in this case, a world in which inventory control is a matter of legitimate concern).

> Most businesses in the United States have a substantial portion of their capital invested in inventory. Hence, inventory control has become a necessity for all business sectors of our economy. Inventory and the controls associated with it affect not only the business sector but also the private sector of our economy. Every individual who purchases and consumes goods is affected by the inventory-control policy implemented by the retail firm the individual purchases from. If the business has an efficient policy, then the consumer will find that a substantial number of the items he or she seeks will be in stock.

All the introductory devices listed in Chapter 8, "Introductions, Transitions, and Conclusions," are especially important in this kind of situation.

Once again, making special conditions explicit at the outset is especially important. Writing for an executive/layperson you should explain not only the special condition but also *why* it is special (which an expert would probably know). The following example shows a special condition that often occurs in student papers: a specific (perhaps hypothetical) case is used to explain a general principle, and the writer needs to explain that fact.

> So that you will better understand the design process as landscape architects employ it, this report explains how the process would typically be used to create a specific design: a vacant downtown lot is to be turned into an urban park. The lot's location, dimensions, and surroundings are explained in detail, and the project's design criteria (including proposed budget) are also presented.

In this report a general concept—the process of design—is explained by describing the creation of one very specific design. Without such a specific case, an explanation of an abstract process (such as the design process) is likely to be too general, too abstract, difficult to write, and even more difficult to read.

Keeping It Concrete. In trying to simplify a process explanation for an executive/layperson, beware of becoming too abstract. However abstract a process you are describing, it's best to tie your explanation to a concrete example (or examples) throughout. As the previous example shows, if you are an architect, don't just describe the "design process," but rather describe how you design some specific project, drawing generalizations about the process when it's appropriate. If you are in management, don't just explain the "problem-solving process," but rather demonstrate it through a specific example. If you are an engineer, don't just explain the process of developing engineered standards, but rather use a description of the development of engineered standards for, say, a knitting mill as the vehicle for explaining the process in general.

Figure 13.2 gives an example of a generalized explanation of a process, "The Process of Seining," written by a student for a lay audience (his classmates).

2.2 Writing Simple Sets of Instructions

Because instructions may be written for any kind of audience or combination of audiences, it is especially important for the writer to know what kind of audience is involved. Often, though, you write a set of instructions for a totally anonymous audience, for *anyone*. In that situation you have to write with the assumption that anything that can be misunderstood will be. You must be especially clear and rely especially heavily on visuals. Whatever the audience, two aspects of the process can present particular problems: dealing with principles and warnings and maintaining a consistent point of view.

Principles and Warnings. Some elements that should go into a set of instructions are not strictly "instructions." Often these are general statements about the idea behind the instructions—something like, "Because of its difficulty, this maintenance is only performed once every 200 hours of operation. Because of that, you must be especially careful to perform it exactly as described here. Any variation from these instructions can lead to costly and time-consuming unscheduled repairs." Often the general statements you need to make are safety warnings, such as, "For your own protection, you must wear safety goggles any time this machine is in operation." In both cases, the important thing is to make such statements right at the beginning. Do not assume the reader will read through the entire procedure before beginning to perform it. And in the case of safety warnings, have them printed in such a way as to draw the reader's attention to them without fail. As the following example shows, using a contrasting color to set them off is wise.

> Although the laboratory process of tensile testing may seem routine, catastrophic failure of bridges and walkways may result from a designer's not giving careful consideration to a material's properties derived from tensile tests. In view of the potential for human and material losses in such failures, an appreciation of the nature and usefulness of tensile testing is especially important for designers.

Figure 13.2   A Generalized Explanation of a Process

THE PROCESS OF SEINING

As you motor down the lake in your boat, you notice a pecu-
liar sight on the shoreline. One person is standing on the
shoreline with a pole in his hand, and about thirty feet from
the shoreline another person is pulling a pole through the
water. As the person in the water reaches the shoreline, you
notice there is some sort of net between the two poles, and
now the people are picking fish from the net. This is the process of <u>seining</u> (see Figure One).

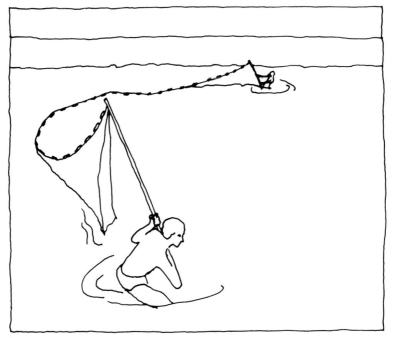

Figure One:
THE PROCESS OF SEINING.
Diagram of Seining a large area.

The purpose of this report is to describe the process of seining. Safety precautions, types of seines, and the details of the seining process will be described. When you understand how fisheries biologists do seining, you will have learned about one of the most valuable techniques available for measuring (and thus maintaining and improving) the fish population in our lakes and rivers. Maintaining and improving the fish population (and thus the water quality) of our lakes and rivers is important to everyone who enjoys the out-of-doors.

Safety Precautions

As discussed here, seining is a fish-management technique used in collecting small fish. Although this is a relatively safe procedure, a few safety precautions should be observed:

1. The person who pulls the seine through the water must wear some type of flotation device, such as a life jacket. Pulling the seine can take a lot of energy, and it is also possible to become tangled in the seine.

2. When transporting the seine from one site to the next, put the two poles together and drape the net over them (see Figure Two).

3. When seining, always wear a pair of old shoes, preferably tennis shoes. This will prevent cuts from broken glass or jagged rocks.

4. If you plan to seine, contact the State Game and Fish Commission concerning the laws and regulations concerning seining in your state.

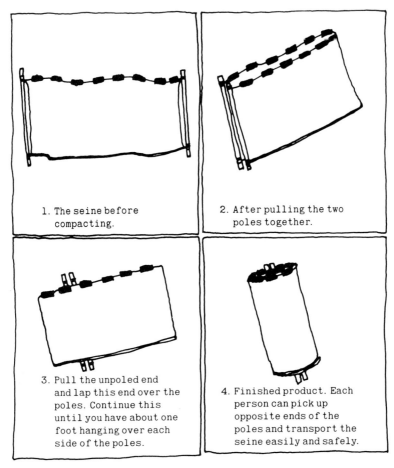

1. The seine before compacting.

2. After pulling the two poles together.

3. Pull the unpoled end and lap this end over the poles. Continue this until you have about one foot hanging over each side of the poles.

4. Finished product. Each person can pick up opposite ends of the poles and transport the seine easily and safely.

Figure Two:

COMPACTING A SEINE.

Steps involved in compacting a seine for easy transport.

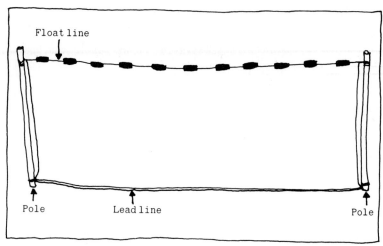

Figure Three:
A SEINE.
Diagram of a seine used for collecting small fish.

Required Equipment and Personnel
Seining can be accomplished by two people, although three are
preferable. The materials needed are:

1. A seine--a net connected at each end to a sturdy six-foot
   pole (see Figure Three).

2. Two or three people

3. A bucket

Types of Seines
There is basically only one type of seine--i.e., a seine by
definition consists of two poles and a net between them. But
the mesh size and the length of the seine will vary, depending on the type of sampling the seining is to accomplish. The
mesh size is the diameter of the holes in the net. Mesh size
varies from $\frac{1}{8}$" to 1".

Areas to Seine

Seining is a procedure that is applicable to all types of
aquatic habitats, but it is especially successful in shallow
areas, such as streams and ponds.

Steps in Seining

The process of seining may be divided into four stages:

1. placing the net in the water

2. pulling the water end of the net to the shoreline

3. keeping the lead line of the seine on the bottom

4. picking the fish from the seine

THE PROCESS OF SEINING

Seining is one of the sampling techniques of fisheries management that requires a minimum of materials and can be done by
anyone. The following discussion will describe in detail the
steps of a successful seining expedition.

Step One

Before you begin, it may be important to estimate the area
covered in one seine haul. That way, you will know how many
hauls you will need to cover the area assigned to you. Figure
Four illustrates the calculations involved in estimating the
area covered in one haul.

   The initial step is putting the seine in the water. If you
are the person who will pull the seine through the water,
then while wearing a flotation device, swim a straight line
from the bank. You can swim a sidestroke and carry your end
of the seine in one hand. Or you can fasten a light line to
your end of the seine, take the free end of the line out into
the water with you, and tow the seine out once you're in position. Make sure you have the seine's edge that is weighted
with lead weights at the bottom, then pull the seine taut.

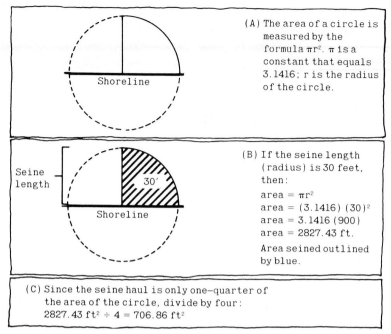

(A) The area of a circle is measured by the formula πr^2. π is a constant that equals 3.1416; r is the radius of the circle.

(B) If the seine length (radius) is 30 feet, then:

area = πr^2
area = $(3.1416)(30)^2$
area = $3.1416(900)$
area = 2827.43 ft.

Area seined outlined by blue.

(C) Since the seine haul is only one-quarter of the area of the circle, divide by four:

2827.43 ft^2 ÷ 4 = 706.86 ft^2

Figure Four:

ESTIMATING THE AREA OF A QUARTER HAUL.

Steps involved in calculating the area of a seine haul.

When the seine is taut, begin the quarter-circle back to the shoreline. Maintaining the tautness of the seine is important for covering the area estimated. Because it covers one-fourth of a complete circle, this type of seine haul is called a quarter-haul.

Step Two

When you begin the quarter-circle you usually cannot touch bottom, so it may take time for you to swim the quarter-circle around toward the shoreline. When the quarter-circle is nearly completed, the person standing on the bank should begin to pull gradually on the pole to maintain the tautness of the seine. This will also keep the lead line on the bottom of the seine in contact with the bottom of the stream or pond.

Step Three

It is very important to keep the lead line in contact with the bottom of the stream or pond. If the lead line comes off the mud, fish may escape under the seine. To prevent this, you and your partner should angle the tops of the poles back toward the water (bringing the bottoms of the poles forward toward shore) as the swimmer approaches the shoreline. When you reach the shoreline, one of the two of you should lay the pole on the ground by bringing the bottom forward (trapping the fish in the seine, not under it). This person then begins to pull the lead line, making sure it remains in contact with the bottom because fish are now trapped in the seine and are looking for a way out. If three people are working the seine, the third person would pull the lead line.

Step Four

While the lead line is being pulled, the person on the opposite end should carefully pull the pole up the bank. When the lead line is completely out of the water, both poles may be picked up and held parallel to the ground. Then the seine should be moved farther up the bank. This is a precaution against fish flopping back into the water. It is also easier to collect the fish from the seine when you are on surer footing. Finally, the fish collected from the seine are placed in the bucket for easy transport.

CONCLUSION

The seining process is an important technique for collecting fish. Other possible techniques include electroshocking or gill netting, but both of those are biased toward larger fish. If the steps outlined in this report are followed, seining is an especially successful technique for sampling small fish.

Point of View. Grammarians call the way you refer to yourself and to your reader in your writing the *point of view*. The most commonly used points of view are

Singular

First Person: *I* next connect *a* to *b*.

Second Person: You next connect . . .

Third Person: *He* next connects . . .

                        She next connects . . .

                        One next connects . . .

                        The technician next connects . . .

Plural

First Person: *We* next connect . . .

Second Person: *You* next connect . . .

Third Person: *They* next connect . . .

The grammatical point of view you use says a great deal to your reader about the assumptions you make about your role as writer and your audience's purposes as reader(s). Earlier in this century many writers preferred either the anonymous third person ("One next connects . . .") or the anonymity of passive voice ("Next *b* is connected to *a*"), which avoids point of view entirely. In our time, however, most writers and readers prefer either the second-person ("You connect . . ."), the second person with the personal pronoun left out ("Next connect . . ."), or a combination of the two. Unless you want your writing to be as formal and stiff as a late-nineteenth-century drawing room or you are in a specific situation in which you are certain you cannot use personal pronouns (thus usually requiring passive voice), stay with some form of second person for your set of instructions.

Figure 13.3 presents an example of a very well-written set of instructions from Western Electric, "Instructions for Modular Jack Converter."

2.3   Writing Manuals and User's Guides

A manual is a document used to instruct or remind people how to do something. Because manuals must be replaced or have sections updated frequently, they are usually done in a looseleaf-notebook format. There are many kinds of manuals; we will discuss four of the most common: personnel manuals, hardware manuals, repair manuals, and user's guides. All of these types of manuals are, to some extent, process descriptions.

Figure 13.3   A Well-Written Set of Technical Instructions

INSTRUCTIONS FOR
MODULAR JACK CONVERTER SWINGER

1. CAUTION:

Your telephone connecting block may have varying amounts of electricity in the wires and screws. Therefore, to avoid the possibility of electrical shock follow the instructions below:

If you have a telephone at a location other than the one you are converting, take the handset off the hook. (This will keep the phone from ringing and reduce the possibility of your contacting electricity. While you're doing this, ignore messages coming from the handset which ask you to hang up the phone.) If you have only one telephone, take the handset of that telephone off the hook.

– AVOID HAND CONTACT WITH BARE WIRES OR SCREWS.
– USE TOOLS WITH INSULATED HANDLES OR USE RUBBER GLOVES.
– DO NOT PLACE THIS DEVICE WHILE A THUNDERSTORM IS IN THE VICINITY.

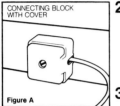

CONNECTING BLOCK WITH COVER

Figure A

2. **STOP:** Before you go any further, make sure you take the handset of one of your telephones off the hook according to instructions under "CAUTION" above. Then follow steps 3-10 below.

Tools needed: Screwdriver with insulated handle, wire cutters with insulated handles (or scissors with insulated handles).

3. Find the plastic connecting block in the room where the telephone is to go **(Figure A)**. Loosen the screw in the center of the connecting block and take off the cover. Discard the cover and screw.

THESE CORDS ARE EXCESS.

WALL

FLOOR

Figure B

One end of cord is lying unattached on floor.

WALL

Figure C

Cord runs from connecting block to old telephone.

4. Follow 3 steps below to cut off any excess cords which may be attached to the connecting block:

● Refer to **Figures B-F** to identify which cords are excess and which are not. You may have one or more of these at your connecting block.
● Take each excess cord and gently bend it away from the connecting block so that the wires which make up the cord are clear of all other wires on the connecting block **(Figure G).**
● CUT each of the wires in the cord, one at a time, at the base of the spade tip which connects the wire to a screw. (See enlarged drawing.) Be careful not to cut any wire other than those in the cord being removed.

NOTE: DO NOT UNSCREW THE SCREWS ON THE BLOCK. Leave the spade tips under the screws.

THESE CORDS ARE NOT EXCESS. DO NOT CUT THEM.

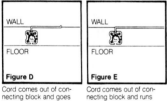

WALL

FLOOR

Figure D

Cord comes out of connecting block and goes into wall.

WALL

FLOOR

Figure E

Cord comes out of connecting block and runs along wall to another connecting block.

WALL

FLOOR

Figure F

Cord comes out of connecting block and runs along wall to a small box which is plugged into an electrical outlet.

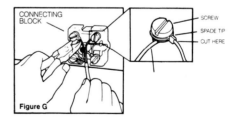

CONNECTING BLOCK

SCREW
SPADE TIP
CUT HERE

Figure G

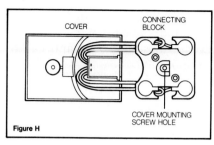

Figure H

5. Snap the four colored buttons of the converter onto the four screws in the corners of the connecting block. **(Figure H)**

- Snap the red button onto the screw which already has a red wire attached to it.
- Snap the green button onto the screw which already has a green wire attached to it.
- Snap the yellow button onto the screw which already has a yellow wire attached to it.
- Snap the black button onto the screw which already has a black wire attached to it.

If the colors of the wires cannot be determined,

- Snap the red button onto the screw marked **R**.
- Snap the green button onto the screw marked **G**.
- Snap the yellow button onto the screw marked **Y**.
- Snap the black button onto the screw marked **B**.

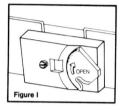

Figure I

6. Attach the converter to the connecting block by fastening the screw in the converter to the hole in the center of the connecting block. Be careful not to pinch any wires between the converter and the connecting block. **(Figure I)**

Give the converter one turn with the buttons in place. This will keep the wires from being pinched.

Figure J

7. Plug the end of the cord from the *new* telephone into the jack on the converter. **(Figure J)**

Rotate the cover in a clockwise direction to expose the jack opening.

INSERT FREE END
OF CORD INTO
OPENING UNTIL
CLIP LOCKS

8. Hang up the handset which you took off the hook to keep your phones from ringing.

9. Lift the handset on the new phone and listen for a dial tone. Dial a phone number to make sure the phone works. **If the dial tone does not go off when you dial:**

- Again, take the handset off the hook to keep your phone from ringing.
- Unscrew the converter.
- Switch the red and green buttons.
- Screw the converter back on.
- Hang up the other phone.
- Check the new phone again.

10. If you are attaching more than one converter, follow these instructions again, leaving the handset of the first phone off the hook.

IF YOU NEED ASSISTANCE, CALL THE TELEPHONE NUMBER YOU WERE GIVEN.

Ⓑ Western Electric

Source: Reproduced by permission of AT&T.

Personnel, Hardware, and Repair Manuals. In 1971, Stello Jordan's *Handbook of Technical Writing Practices* presented a general outline for manuals that can still be adapted to serve for most kinds of manuals. The handbook suggests the following organization:

Front matter
Introduction
General description of product
Theory of operation
Description of the operation controls
Adjustments and operator's activities prior to actual operation
    of equipment
Actual operation instructions and operation technique
Troubleshooting section
Emergency or standard servicing
Index

Although many manuals may not need quite so full a treatment, this list—with appropriate modifications—can serve as an outline for the content and structure of most such manuals. Beyond that, the particular content of any manual you may be involved in producing will probably dictate its structure to a large extent.

Another area of special interest to a technical writer producing any kind of manual is the way the words and pictures are laid out on the pages and the way the pages are bound together. This textbook has stressed that the purpose, message, audience, and situation in which any technical document will be used should have a strong influence on how that document is written. Manuals provide particularly vivid examples of the ways in which those factors can affect the way a technical document is done. Some critical questions for the writer or editor of a manual are

• What kind of person will be using this manual?
• What environment will this manual be used in?
• What will be the goal of the person using this manual?

Answers to these and other similar questions must be taken into account if a manual is to function effectively.

For example, consider the binding of the manual. If the manual is glued, stapled, or stitched, it will be difficult to use with no hands; such bindings do not allow the document to lie flat. If, on the other hand, it is bound with a spiral binding or three-hole punched to be put into a looseleaf notebook, the manual can be laid flat, freeing the operator's hands for technical work. The looseleaf format also allows the operator to replace pages that carry out-of-date or no-longer-accurate information with new pages, without replacing the whole volume.

Figure 13.4   A Page from a Typical Personnel Manual

Group Insurance Plans

Basic Plan

All full-time faculty and staff members and regular part-time employees who are scheduled to work at least 30 hours a week are eligible to participate in the basic group hospitalization plan. Academic appointments must be for at least 75% time and for a minimum of one semester or 2 consecutive quarters in order to meet eligibility requirement. If both the employee and spouse work for the state, both must elect to be insured under separate policies. No person eligible for coverage as an employee may be included as a dependent. Coverage under the Group Program will become effective the first day of the calendar month coinciding with or following completion of one month of active service provided the employee has completed the necessary enrollment forms. Under this plan, the individual pays 40% of the premium and the University pays 60%. (This is the only insurance coverage for which the University may legally contribute toward premiums.) The monthly contribution of the individual varies, depending on salary and coverage of dependents. The basic group-insurance package contains hospital and surgical insurance, major medical insurance, emergency accident expense benefits, term life insurance, and special accident insurance. The plan also has an additional optional special accident insurance (payable upon death or dismemberment), which can be elected by the individual. Details of all benefits under this program are set forth in the Group Insurance Booklet issued to all participating employees.

For another example, consider the way words and illustrations are laid out on the page. Figure 13.4 shows a sample page from a typical personnel manual.

Figure 13.5 shows the same information, reworked to take more into consideration that the document is a manual and will be used differently from other kinds of reports.

These kinds of variations in the way each page looks—use of typefaces, chunking of information, careful cross referencing, and so on—are characteristic of the kind of thought and work that needs to go into many different kinds of manuals. By enhancing the text's

Figure 13.5   Reworked Version of the Page in Figure 13.4

Basic Group Hospitalization Plan

Who's Eligible?	The following University employees are eligible for the Basic Group Hospitalization Plan: • All full-time faculty • All full-time staff members • Regular part-time employees who work a minimum of 30 hours per week
Restriction on Eligibility	Academic appointments must be for at least 75% time and for a minimum of one semester or 2 consecutive quarters.
Restriction on Spouse Coverage	If both employee and spouse work for state, both must be insured under *separate* policies.
When Does Coverage Begin?	Upon applicant's submission of all necessary forms and the completion of one month of active service, COVERAGE BECOMES EFFECTIVE THE FIRST DAY OF THE CALENDAR MONTH.
Premiums	Under the Basic Group Hospitalization Plan, • you pay 40% of the premium, • the university pays 60% of the premium. • (Your monthly contribution varies, depending on salary and coverage of dependents.)
What Do You Get?	Under the Basic Plan, you get the following coverage: • hospital and surgical insurance • major medical insurance • emergency accident expense benefits • term life insurance • special accident insurance • additional special accident insurance (optional). This insurance is payable upon death or dismemberment.
Comments	Details of all benefits under this program are set forth in the Group Insurance Booklet issued to all participating employees.

Source: Michael E. Hall, "Improving the Readability of Training Manuals" (Master's thesis, University of Tennessee, 1984).

visual qualities, the writer makes the material within it more accessible to the reader.

User's Guides. A special instance of process writing occurs when you must write procedures for complex systems. This could be routine shutdown procedures for any kind of power plant, an operations manual for an airplane, or a user's guide to a particular piece of computer hardware or software. All of these (and many similar) cases present problems for you as the writer—problems that occur both because of the complexity of the system being operated and because of the operator's attitude toward the process. Because writing computer documentation of one form or another—usually a user's guide—occupies the attention of growing numbers of writers from whatever background, we will use the writing of user's guides here as an example of writing instructions for complex systems.

One of the most rapidly growing areas of writing is software manuals. These manuals typically take several different forms:

New user's guides
Advanced user's guides
Programmer documentation

These three types of software documentation provide perfect examples of the importance of audience analysis and adaptation. Although the three types of documents cover many of the same areas of a program, the way each does it is completely different from the way the others do it. For example, a **new user's guide** has to focus to a great extent on the user's lack of experience with the system, and this limitation affects everything the guide does and how it does it. The writer of such a guide has to resist the temptation to be catalogical (or encyclopedic) about what the system is and what it can do and realize that the reader wants to know first how to access the relevant programs (this may even include something as basic as turning the computer on and logging on), and then how to get started in a simple routine. Similarly, a new user will feel considerable discomfort if something does not go just the way the instructions suggest. Discussion of more complex procedures has to wait until the reader has learned the basics.

An **advanced user's guide** can be more theoretical and comprehensive. Advanced users have the time, inclination, and courage to try new procedures, to understand considerations beyond those currently being discussed and to experiment on their own.

Programmer's manuals work in almost the reverse manner to the way user's manuals work. Where the user's manual is taking an incremental approach, trying to build up an understanding one piece at a time, the programmer's manual takes a catalogical (or encyclopedic) approach, laying out the whole system, not according to the logic of a user (who looks at the system from the bottom up) but according to the logic of a programmer or systems analyst (who looks at the system from the top down).

Let's assume you are the manager of a small business, and you have just purchased a software package called Paymaster to serve as your business accounting system. What will its structure look like? The following listing offers a sample outline of a typical new user's manual; you may want to compare it to the Stello Jordan manual outline given earlier in this chapter in Section 2.3.

Outline of a Typical New User's Manual

- Clipped to the inside front cover is a "ready reference" card, for quick reminders of frequently used commands and routines.
- Before the title page is a page listing the numbers and titles of all the chapters, in big print.
- The title page has a copyright notice and a "proprietary information" notice on it.
- The Table of Contents is quite detailed, using different typefaces, perhaps even different colors, to give readers quick access to particular parts of the manual.
- A one-page piece, "About the Paymaster Manual," tells who it is for, what its parts are, and how to use it. It also reminds the reader that there is a reader-response card at the end of the manual to fill out and send in with comments about the manual.

Chapter One: Before Starting   This chapter introduces the system; gives a one-page overview of the whole system, possibly in the form of a flowchart; describes a shortcut that experienced users who may happen to get this manual can take; and gives a quick overview of what the system will do for its users.

Chapter Two: Getting Started   This chapter explains what the system needs in terms of software, hardware, and supplies; how to load the disks and copy them; and how to set up Paymaster.

Chapter Three: A Practice Run   This chapter sets up a simple trial problem and explains how to solve it using Paymaster.

Chapter Four: Setting Up Paymaster   This chapter explains the different options available and procedures used to set up Paymaster in its basic form.

Chapter Five: Running Paymaster   This is the most detailed chapter, explaining each function of the software in detail.

Appendix A   This section explains possible error messages and how to deal with them.

Appendix B   This section gives paper copies of the various formats used to enter data into Paymaster, so that users can marshal the appropriate data in proper form before they start keying them into the computer.

Appendix C This section gives technical information on Paymaster.

Glossary

Index

Although this structure or format will vary considerably from one user's manual to another, it is typical of what you will see. Notice that the user is introduced to the system with the briefest of all overviews, instructions on how to use the manual, and a short tutorial on a simple problem. Notice, too, that technical information of any kind of detail or complexity at all is placed at the back of the manual.

Four features of writing for user's guides deserve specific mention: the document's overall orientation, its structure, the user's attitude, and the testing of the document.

Orientation. There are two alternatives for the way a user's guide (or any set of instructions for operating a complex system) approaches its subject: the guide can be system (or machine) oriented or task (or function) oriented.

A *system-oriented* description explains the system's characteristics in an encyclopedic manner and describes how the system works using the system as the focal point. Only incidentally does such a document tell a person how to use or operate the system; that information must usually be extrapolated from the encyclopedic description of the system.

A *task-oriented* description takes the *user* as its focal point and explains in a step-by-step manner each interaction between the user and the system. Only as support for those steps will a task-oriented user's guide explain the system's characteristics or how the system itself works. In writing instructions on complex systems, a task-oriented approach works better.

A simplified hypothetical example makes this distinction clearer. Suppose you have come to a world in which elevators had never been invented. Now having invented the elevator, you must write its documentation. Your brochure can look like either Figure 13.6 or Figure 13.7.

It may seem to you that no one would ever write a user's guide with the kind of machine- or system-oriented approach described in Figure 13.6. Unfortunately, it seems the larger the system being described is, the greater the tendency is to do just that. Perhaps the tendency of an unsure writer to begin any document by telling everything he or she knows about the subject is another manifestation of the encyclopedic tendency described in Chapter 11. If you are writing *instructions*, take the user as the focal point; write task-oriented instructions, even for complex systems.

Figure 13.6   A System-Oriented Description

```
                        ELEVATOR

The Orbis elevator is a rectangular chamber 6 feet deep,
10 feet tall, and 8 feet wide. The chamber's face has dou-
ble sliding doors to allow entry and exit. The doors have
pressure-sensitive rubber bumpers and electric eyes to pre-
vent injury to people entering and leaving. The chamber
will carry as many as 10 to 12 people (1900 lb.) from one
floor to another of the building with a minimum of effort.
The chamber carries its own fluorescent lighting system,
its own ventilator fan, and a control panel with which to
select floors (as well as an emergency stop button and
emergency phone), and it comes equipped with a roof-
mounted escape hatch. All Orbis elevators are licensed by
the state and inspected yearly. Certificates of in-
spection . . .
```

Structure. Many times, the most usable and thorough way to explain the operation of a complex system is to produce a three-part
user's guide. The first section is the user-oriented "how-to-do-it" set
of instructions. The second section is a troubleshooting "what-to-
do-if" section. The third section is a brief description of the system's
technical characteristics (possibly the system-oriented description)
and an indexed reference section.

Suppose, for example, that you are writing the user's guide for a
word-processing program your company is marketing. It would be
useful to write it in this way:

1. Section 1 would be aimed at beginners. It would have to include
   even the most basic elements, such as how to turn the computer
   on and off and how to enter and exit various programs, as well as
   how to perform basic functions.
2. Section 2 would explain how to solve common problems the user
   faces—how to copy files, for example. It would also explain errors
   users often make, such as what to do if you exceed a disk's memory
   capacity.
3. Section 3 would be the system-oriented description and reference
   section. It would catalog the program's characteristics and provide
   an index section as well.

Figure 13.7   A Task-Oriented Description

```
              Tired of walking from floor to floor?
               Try taking an elevator instead.
                    Here's how you do it:

Walk to the end of the hall, and push one of the arrow-
shaped buttons—either the one pointing up or the one
pointing down (depending on which way you want to go).

When an elevator arrives at your floor, the door will
open, and the lighted arrow over the door will indicate
the direction the car is headed. If that's the direction
you want to go, step aboard. (If not, wait for one that is
headed your way.)

When you step aboard the elevator, push the button on its
control panel that corresponds to the floor you want.

When the elevator reaches the floor you want (as indicated
by the number over the inside door), the door will open
and you simply step out . . .
```

Sections 2 and 3 would have to be cross-referenced to each other, and all sections would have to be organized so that the right pages could be found quickly. Such a guide would be useful to beginners and also give them all the information they would need as their skill levels increase.

Because a three-part structure like this can produce a bulky document, it's good to put it in a format that helps make items within it accessible. A looseleaf notebook with pages reinforced on the inside edges and color-coded index-tabs (for major sections) on the outer edges makes a good format for such a document. The notebook format also allows readers to insert new or replacement pages as the need arises.

User's Attitude. If you are writing instructions for the operation of a complex system, and especially if the anticipated user is a beginner, be particularly aware of the user's state of mind during the operation of the system. Even more than average users, beginners are almost exclusively dependent on your instructions for their actions. Anything that occurs which your instructions do not explicitly mention may greatly distress the user, even if that occurrence is so totally routine you chose not to mention it. Remember that little things that experienced users deal with routinely can baffle beginners.

For example, if you are writing the word-processing user's guide mentioned earlier, be sure to tell the user how to exit the program. Contrary to the apparent beliefs of many authors of computer manuals, the nearly universal exit codes, such as Control-K, are not widely known outside of computer circles. An inexperienced user, carefully following your step-by-step instructions, can be distressed far out of proportion to the actual seriousness of the situation by being unable to get out of a program, or to solve a gridlock (the situation in which no key you push seems to stop what the computer is doing), or by any other unforeseen occurrence. Even experienced users of instructions for complicated systems can have tunnel vision when it comes to failing to think of even simple alternatives to an instruction that didn't work the way the guide indicates that it should.

Testing. The more complicated the system is that you write instructions for, the more important it is to test the instructions. To do this, find a person with the same knowledge and experience level as your document's intended audience, and persuade him or her to try out your instructions. To make the test even more useful, ask your trial user to tape-record his or her thoughts while trying to use your instructions. Then you can alter your instructions as required, based on the trial user's experiences.

Figure 13.8 shows the pages from a *Quick Reference Manual* for the IBM Personal System/2 Model 50.

Figure 13.8   Sample Pages From a User's Manual

How to Set Passwords

These instructions are also on the Reference Diskette. Because the computer must be turned off, then on again to follow some of the password instructions, the information is provided here for your convenience.

Set a Power-On Password
Setting a power-on password helps restrict the use of the computer by unauthorized persons. The password can be no more than seven characters and does not appear when typed. Once the power-on password is set, whenever you turn on the computer, the password prompt (a small key) appears. To use the computer, you must type the correct password and press Enter. When the password is entered correctly, an OK appears momentarily. If the password is entered incorrectly, a key with an X over it appears. If you have not entered

the password correctly after three tries, you must turn off the computer, then on, and try again.

To set a power-on password, start the backup copy of the Reference Diskette (see page 13). Go to the main menu and select "Set features," then "Set passwords." And then select "Set power-on password" and follow the instructions on the screen.

Change a Power-On Password

To change your power-on password, turn on the computer. When the password prompt appears, type your current password followed by a slash (/), then your new password, and press Enter.
Type: *Current Password/New Password* and press Enter.

Remove a Power-On Password

To remove your power-on password, turn on the computer. When the password prompt appears, type your current password followed by a slash (/), and press Enter.
Type: *Current Password/* and press Enter.

Forget the Power-On Password?

To eliminate your power-on password because you forgot it, you must remove the battery from inside the system unit for at least **20 minutes** to erase the memory containing your password. (See page 29 for instructions on removing the battery.)

Set a Keyboard Password

Setting a keyboard password lets you lock the keyboard without turning the computer off.

How to install the keyboard program:

To set a keyboard password, you must first install the keyboard program from the Reference Diskette onto the fixed disk drive or diskette that contains your disk operating system (DOS). To do this, start the backup copy of the Reference Diskette (see page 13). Go to the main menu and select "Set features," then "Set passwords." And then select "Set keyboard password" and follow the instructions on the screen. After installing the keyboard program, continue with the instructions for how to set and use a keyboard password.

How to set and use a keyboard password:

1. Start the computer using the fixed disk drive or diskette that contains DOS and the keyboard program.
2. Go to the DOS prompt (usually "A>" or "C>"), type *KP* and press Enter. Continue with step 3.

Source: IBM. Reprinted with permission.

EXERCISES

1. Write a brief critique of the excerpt from *Instructions for Modular Jack Converter* on pages 352–353. How does the piece employ the principles discussed in this chapter? Can you suggest ways to make the piece a better set of instructions without making it significantly longer?

2. Choose a simple process—changing a tire, tying a knot, building a birdhouse, getting the best bargain while buying new luggage, formatting a diskette, registering for summer school, and so on—and write a 200-word set of instructions for it. Assume your reader has *no* knowledge of the process. To accompany your instructions, write a brief sketch of your intended audience: their age, purpose in reading, education level, and so on.

3. Write a description of a process—how it takes place or how someone does it but *not* a set of instructions—with your instructor as the audience. Choose an everyday process (baking bread, balancing a checkbook, a process of the sort listed in question 2), and try to make your description about half visuals and half words. The description should be about 300 words long.

4. Find a set of instructions (perhaps part of a computer user's manual, the instructions for a laboratory exercise, the instructions for programming a VCR, and so on) and write a two- or three-page critique of the way they are written. At the end of the critique, rewrite two pages of the instructions, incorporating the kinds of improvements you have suggested. Be sure to attach a copy of the relevant pages of the original to your report.

5. Choose a process that is important within your field of study—one either students or professionals must do. In order for your writing to be effective, the process needs to be one you know very well or can gain good access to. Write a description of the process (how someone does it or how it takes place) or a set of instructions for performing it, with your instructor and your classmates as the audience. Frame the discussion appropriately, and make it specific and concrete. Use a *cpo* introduction, headings and subheadings, abundant visuals, and a conclusion. Your finished piece should be 500–750 words long.

➤ 6. It's the summer before your senior year in college, and you've taken a job in your university's student counseling center for the summer. You've been assigned the project of writing a guide for freshmen to use during their first weeks on campus. It should cover such topics as dorm life, finding your way around campus, making the best use of your time, dealing with your new freedom, and so on. Write the text for that guide (500–1000 words).

A ➤ indicates a case study exercise.

➤ 7. You've just landed an entry-level job in your chosen field, and the boss has asked you to do a small special project. The last time the company purchased word processing and spreadsheet programs for the main office, many employees were unhappy with the documentation that went with the software, and the boss was especially unhappy with it. Your task is to write a short analysis and critique of the *documentation* that accompanies what you estimate are the three most popular word processing and spreadsheet programs available. As the boss said, "All of the programs do pretty much the same thing; the big difference is in the manuals. Most of them are completely inscrutable. What I want you to do is to find one that is scrutable." Your report should be 500–750 words long. *Hint:* There are two ways to find this information: look at the documentation itself and look at published reviews of it in computer journals. To do the best possible job, you should probably do both.

14. *Solving Problems*

1. Exploring the Problem
   1.1 Define the Problem
   1.2 Place the Problem in a Larger Context
   1.3 Make Your Definition More Concrete
   1.4 Assign Priorities to Your Goals
   1.5 Make Sure You Are Aware of All the Facets
       of the Problem

2. Finding a Rich Array of Solutions
   2.1 Brainstorming
   2.2 Visual Thinking
   2.3 Asking Questions
   2.4 Linear Analysis

3. Testing for the Best Solutions
   3.1 Explanatory Power
   3.2 Prior Probability
   3.3 Predictive Power
   3.4 Clarity
   3.5 Provocative Power
   3.6 Falsifiability
   3.7 The Crucial Test

4. Making Your Choice
   4.1 Check Your Work
   4.2 Rank Your Alternatives
   4.3 Get Advice
   4.4 Make Your Choice and Document it

5. Doing the Writing

The most troublesome writing tasks are often those that involve one-of-a-kind situations: once in a while you will find yourself writing about a problem that is difficult not just to *solve* but even to *define*. For this reason, one of the fundamental skills that any writer in business and industry needs in order to be effective is the ability to solve problems. As it is used here, "problem solving" does not refer to a concrete element of writing (in the sense that introductions, conclusions, definitions, descriptions, and process explanations are concrete elements of a piece of writing). Rather, "problem solving" here means a *process* we go through when we write; it's a process that is part of writing, just as the process of sensing when a reader will need more information is a part of writing. Problem solving can be an underlying element in any of the types of writing described in this book. This chapter discusses how to do the kind of problem solving that typically goes on when a professional person is using writing to create solutions on paper to problems like these.

Examples of classroom problems:

- Design a particular robot arm end so that the robot can change its own "hands."
- Analyze an airline's last few years of financial ratios to see whether forecasting its bankruptcy should have been possible.
- Derive a formula to determine limit slope for various shapes of concrete drainage channels.
- Determine whether radial tires make a big enough energy saving in agricultural use to justify their initially greater expense.

Examples of professional problems:

- Design your company's exhibit for a trade fair.
- Reduce your unit's operating costs by 15 percent without lowering production quality or quantity.
- Explain how your city can successfully operate a mass-transit system with insufficient funds and aging equipment.
- Research how to save energy costs in the operation of a particular building.
- Determine why a particular model of fan brings complaints from so many people who use it (and suggest how to improve its design).

When you face assignments like these, you need to go through the steps listed here. The basic procedure for analyzing problems has four steps:

1. Explore the Problem.
2. Find a Rich Array of Solutions.
3. Test the Best Solutions.
4. Make Your Choice.

Note: Every professional field has its own particular, technical approaches to solving problems. That approach may range from carefully conceived and controlled laboratory experimentation, to computer modeling, to regression analysis, to philosophical inquiry. This chapter is not intended to replace the education the professors in your own field are giving you in problem solving, but rather to add techniques to those they cover. By demonstrating how these problem-solving techniques underlie professional writing in general, this chapter will help you learn how to work the techniques characteristic of your own field into your own writing more effectively.

1. Exploring the Problem

Any experienced problem solver will tell you that it's critical not to get locked into thinking in depth about particular solutions before you have made sure that you fully understand the problem. That is, the more you know about the inside of the problem—its nature, depth, and complexity—and about the outside of the problem—its context, background, and limits—the better you will be able to come up with a good solution. Thus, your goal should be to try to understand the problem *before* you let your mind fasten on any one solution. The five steps for exploring any problem thoroughly are found in the following list.

Steps in Exploring Any Problem

1. *Define the problem:* What is the conflict or key issue?
2. *Place the problem in a larger context:* Why is it a problem?
3. *Make your definition of the problem concrete and operational:* What specific goals need to be reached?
4. *Assign priorities to your goals:* Which come first in terms of their importance? Which come first in terms of when they must be solved?
5. *Make sure you are aware of all the facets of your problem:* Are there any important features of it that you've failed to consider?

1.1   Define the Problem

The first priority is to gather facts. For example, if people are involved, who did what, where, when, how, why, and to whom? Focus on the problem's internal characteristics, and dig into the problem as deeply as you can. Figure 14.1 illustrates this stage.

1.2   Place the Problem in a Larger Context

In the second stage, work around the edges of the problem, collect background and peripheral information on it, and try to fit the problem into perspective. At this stage it's often useful to talk with other people who may view the problem differently, know more about it,

Figure 14.1   Define the Problem
Discover as much information about it as you can.

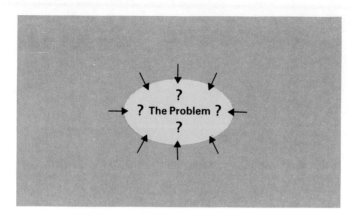

Figure 14.2   Place the Problem in a Larger Context
Ask questions, questions, and more questions.

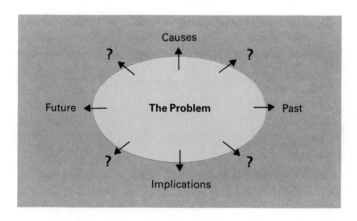

or feel differently about it than you do. Ask questions, questions, and more questions. Figure 14.2 illustrates this stage.

1.3   Make Your Definition More Concrete

By this point you should have a clear understanding of the problem. Now you need to envision exactly what the desired state of affairs would be. You're not searching for what the problem's solution will

Figure 14.3   Make Your Goal Concrete
Visualize exactly what the desired state of affairs is.

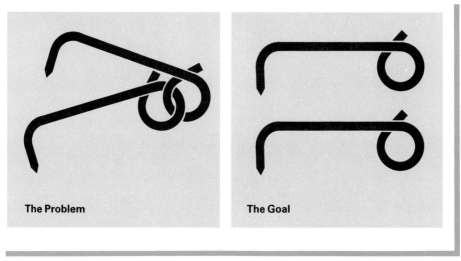

Source: Adapted from Henry Boettinger, *Moving Mountains.* Reproduced by permission.

look like, but what the problem area will look like when it's solved. Figures 14.3 and 14.4 show these two steps.

Making your definition of the problem more concrete and determining what goals need to be reached are especially important. Until you have a clear idea of exactly what you want, it will be difficult to be satisfied with what you get or even to know when you've found the best answer. Observe how each of the vague problem statements in the following list is revised to make it more accurate.

Clarifying Vaguely Conceptualized Problems

General Problem: How can we solve the campus parking problem?
Concrete Goal: How can we ensure that students and faculty have easy access to campus buildings?

General Problem: How can we improve the marketing of cattle?
Concrete Goal: Will telemarketing improve the marketing of cattle by bringing buyer and seller closer together?

General Problem: What will we do when the federal government reduces the funds they pay our state for highway construction?
Concrete Goal: How can we fund highway construction on the state level?

[List continues on page 372]

Figure 14.4   Visualize the Desired Result

Imagine that all you know is that you need to tie the line in **A** into an "eye splice." Without a picture of some sort of what the problem territory will look like after the problem is solved, you simply cannot go on. With some sort of visualization, you can begin to figure out how to get from **A** to **B**. Until you have a very clear idea of what you want, it's hard to know what direction to go.

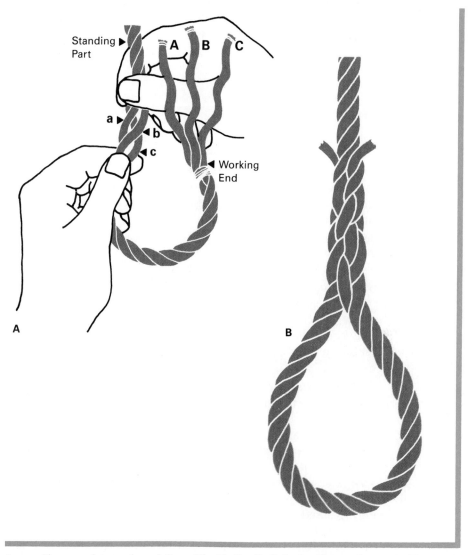

Source: Figure and text adapted from *The Marlinspike Sailor* by Hervey Garrett Smith, *Rudder* Magazine, 1952. Reprinted with permission of Baker-Glazen Publications, Inc.

General Problem: How do incoming industries affect a town's economy?
Concrete Goal: Given a specified town, how will a specified industry affect that town's unemployment rate?

General Problem: What's wrong with the undergraduate biochemistry major?
Concrete Goal: What employment opportunities are there for a person with a B.S. in biochemistry?

1.4   Assign Priorities to Your Goals

Often your larger goal will have several smaller parts; one goal may be composed of any number of subgoals. When that is the case, you need to rank the several subgoals in order of their priority. You can do that two ways—by urgency or by importance. For instance, if you are trying to solve the campus parking problem, you may have the following concrete subgoals:

• Students and faculty need easy access to classes.
• Parking should be inexpensive.
• The plan should be feasible.
• Delivery, repair, and emergency vehicles need access.
• The parking scheme should work in all weather conditions.
• The parking scheme should be reasonably noise free.
• The parking scheme should allow people to get to and from their cars safely at night.

The way you order these subgoals—which is most important, next most, next most, etc.—will control the way you solve the problem. Once you have formulated a detailed and ordered list of subgoals, you may find you are well on your way to solving your problem.

If you organize the subgoals in terms of which one is most important in an absolute sense, you might choose "Students and faculty need easy access to classes" as the first point to tackle. That would lead your problem-solving process in a very different direction from an ordering based on which one needs to be solved *first*. For example, if you are working in a situation in which there have been recent nighttime muggings of people going to and from their cars, you may be pressured to come up with a quick response to "The parking scheme should allow people to get to and from their cars safely at night," and that would lead your problem-solving procedure in a very different direction.

1.5   Make Sure You Are Aware of All the Facets of the Problem

Listing the goals and subgoals will often also mean that you are at least indirectly listing the various facets of the problem, but many times making a separate, deliberate list will help you. Here is a partial list of several facets of the campus parking problem:

- Dorm parking lots are overflowing.
- Commuters have no parking spaces.
- Faculty and staff cars crowd the campus.
- Exhaust fumes are increasing air pollution.
- Traffic noise is increasing.
- Safety for pedestrians and bicyclers is questionable.
- In bad weather campus traffic is a hopeless snarl.
- People waste too much fuel getting to class.

Notice that a new element, fuel waste, has appeared in the list. If that is to be a part of the problem, then energy efficiency must be added to the ranked list of goals. Inclusion of this new item is an example of how stopping to consider all of the facets of your problem will help you deal with it more effectively. Another way that identifying the parts of your problem can help solve it is the process called *reduction.* You may find that only one part of the problem really must be solved. Or you may find that if you solve one crucial part of the problem, solutions to other parts come much more easily. In this way, reducing the problem to its component parts can help you solve it.

While you are exploring your problem, it is important not to get prematurely locked into a solution. Each of us feels the temptation to seize the first solution that comes up or at least to filter all subsequent exploration through that solution. But to analyze the problem thoroughly and accurately, do not think about solutions until you have completed the exploration phase. Certainly you should jot down possible solutions as they occur to you, but then you should put them out of your mind until you finish fully exploring the problem.

Where does writing play a part in the problem-exploration process? Students often have developed the habit of trying to do this stage only in their minds—writing down the output of the mental operations but not the input or the operations themselves. That may work at some level for problems of the size and complexity students face, especially early in their educations. But as the problems you face become larger and more complex, and especially as you move into professional life, you will do well to write down the steps you go through (on paper or online) from start to finish. That way, you free your mind to do creative and critical thinking without carrying the added memory burden of the various parts of each step. You also have a record of the steps you went through, a "paper trail," that will often prove useful when you turn to writing your actual report that results from the problem solving.

2.   *Finding a Rich Array of Solutions*

Just as you should explore all sides of the problem, so also should you examine a variety of solutions. Just as you should not allow your first understanding of the problem to be your only understanding of the problem, you should also not allow the first solution that comes into your mind to become the *only* solution that you consider.

We each have our own best ways of coming up with a rich array of solutions. The methods presented in the following pages are among the best used by professional communicators.

2.1   Brainstorming

Brainstorming may be the most commonly taught idea-generation technique. Brainstorming means producing ideas (in this context, solutions to a problem) absolutely freely, not doing any screening, refining, or selecting until some later stage. The idea is to separate the positive, imaginative thinking required to *produce* the ideas from the negative, critical thinking required to *choose* the best idea.

Any number of people can brainstorm effectively. It's helpful to tape record the process so you will not have to stop to write down ideas. The first principle of brainstorming is not to criticize ideas while they are being generated. The second principle is to feel free to piggyback one idea on another; if someone says "paint it green," you piggyback onto that idea with "paint it green with white stripes." In this way, one idea actually helps to produce the next.

2.2   Visual Thinking

Many people find they think much more clearly and creatively when they think visually. One such person typically says, "I just can't visualize what you're telling me—can you put it on paper?" For him, thinking visually means seeing lists, sketches, charts, and numbers. Another person, an architect, sketches everything she talks about, from sailboats, to houses, to the way a foot should hit a soccer ball, to the way a properly smoked turkey should look. You don't need to be an artist to think visually; all it takes is an active imagination and the willingness to try.

What does your problem look like? Give it a shape, a color, a size in your mind. You may find that actually sketching or doodling on a pad while you visualize will be helpful. Or you may just need to lean back, close your eyes, and conjure up images in your mind. Whatever it takes for you to think visually, you may find that it enhances your problem solving, making it well worth the time and effort.

2.3 Asking Questions

The solutions you find to your problem generally will be no better than the questions you ask. Working to improve the quality and variety of the questions you ask can thus improve the quality of the solutions you find. There are very few—if any—unsolvable problems, but there are many problems about which the right question has not yet been asked. The following is a partial list of a few of the most important kinds of questions to remember.

> ### Basic Types of Questions
>
> *Basic Questions:* Who, What, When, Where, Why, How?
>
> *Reminding Questions:* "Have we forgotten that . . . ?"
>
> *Challenging Questions:* "If we look at this, being as specific and hard-nosed as possible, . . . ?"
>
> *Eliciting Questions:* "What extra information could we collect that will help . . . ?" or "Can we find three specific examples of . . . ?"
>
> *Furthering Questions:* "If we follow *X* course of action, what will be the effect of . . . ?"
>
> *Clarifying Questions:* "What is the evidence that supports this generalization? Does the evidence really connect with that generalization?"
>
> *Deflecting Questions:* "Can we go around this problem—move in some other direction to avoid it entirely?"
>
> *Structural Questions:* "What is it that holds the parts of this problem together?"
>
> *Testing Questions:* "What conditions would have to occur to falsify our current understanding . . . ?"

If you will remember not to allow the first questions you ask of the problem to be the *only* questions you ask of it, if you can take the time and trouble to ask a variety of different kinds of questions, you will do much better at problem solving.

2.4 Linear Analysis

One commonly used technique involves flowcharting (a variant of visual thinking). One special kind of flowcharting, *linear analysis*, leads you through a problem using a series of yes/no decisions. To perform a linear analysis, you take apart a problem, divide it into its various parts, subdivide the parts, and finally make decisions as to which parts can be dealt with simply and which need more attention. The technique has a double advantage: linear analysis is very systematic, and it makes the problem-solving process easy to visualize. It is also especially useful in the kind of situation that leaves you paralyzed by its complexity. Systematically breaking the problem down into visualizable, manageable parts may leave you more able to solve it. Figure 14.5 shows a generic flowchart for decision making using linear analysis.

Consider the following simple example, solved here first using words and then using linear analysis. Suppose you find yourself at one end of a strange city with limited funds and no car, and you need to get to the other end of the city to attend a meeting. What are the possible solutions? You brainstorm this list:

Taxi	Bus	Combination
Subway	Walk	Hitchhike
Call a friend	Rent a car	Do nothing

Then you exclude the totally impractical alternatives; here hitchhiking is not safe, and doing nothing is not feasible. Renting a car is too expensive, and you have no friends in this city. That leaves five alternatives, in descending order of preference:

Taxi
Bus
Subway
Walk
Combinations

You rank Combinations last because it's the most complicated solution. Then you realize the distance is too far to walk. Now you have the following list:

Taxi
Bus
Subway
Combinations

The taxi is too expensive to take clear across town. The bus runs through the part of town you're in but stops at downtown, and the subway is not in your part of town. This moves you to look at Combinations for a solution. And indeed you discover you can take the bus to the center of town, take the subway from there nearly to your destination, then (depending on how much time is left) either walk or taxi the last ten or twelve blocks from the subway to your destination. Problem solved!

Now consider the linear analysis of the problem (see Figure 14.6). Notice that the choices (to rank-order by preference and to test by expense) are specific to this problem. One could have rank-ordered

Figure 14.5 Linear Analysis
A Flowchart for Decision Making

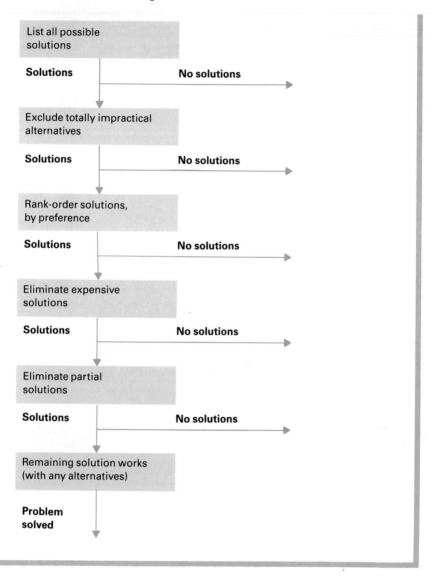

and tested according to any number of different criteria (see Section 4, "Making Your Choice").

The role of writing in finding a rich array of solutions is somewhat bigger than it was in exploring the problem. Having someone writing down the words that pop up in a brainstorming session—on a flip-chart or a blackboard, for example—can help the brainstorming

Figure 14.6   Decision-Making Flowchart
The choices to rank-order by preference and to test by expense are specific to this problem.

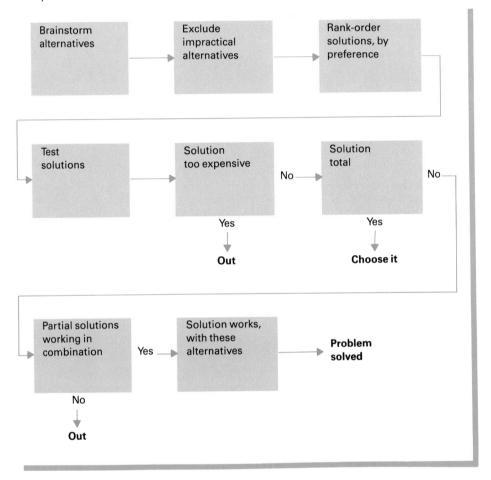

session be productive. Because the participants in the session have a visual reminder of the items that have already come up without having to tax their memories to do so, they get an ever-growing supply of fuel for new ideas. Specific attempts to visualize problems result in manifold sketches and doodles that feed writing directly and productively. The questions you ask of a problem and the answers you write to them can turn into sentences, paragraphs, or whole sections of your reports. And the linear analysis you do to help you solve the problem systematically can become part of the report's overall structure immediately. (See Sections 1.1, 1.2, and 1.3 of the following chapter for more on this.)

In fact, each of the four activities that comprise finding a rich array of solutions can lead to specific words, phrases, illustrations, and even structural elements in your report. The stage of exploring the problem may often be so internal for a writer, and thus so remote from the surface (and often somewhat later) activity of writing the report, that the process's elements never show up in the report in recognizable form. The stage of finding a rich array of solutions, on the other hand, begins to feed the actual report writing process in concrete and direct ways. We will see in the next two sections that each subsequent process feeds the writing process more and more.

3. Testing for the Best Solutions

Once you have explored the problem and found a rich array of solutions, you need to test to find the best ones. Much of the time, that testing will be empirical; that is, it involves work in the laboratory, field research, and calculations. More challenging situations occur when you cannot design empirical tests to evaluate the various solutions. Methods of making causal inferences (such as, "If we do this, we solve the problem") are not so well known as methods of empirical testing, nor are they as specific to each field. Yet they can be tremendously useful. Here are seven ways to test solutions, only one of which (number 7) requires empirical methods:

1. Explanatory power
2. Prior probability
3. Predictive power
4. Clarity
5. Provocative power
6. Falsifiability
7. The crucial test

3.1 Explanatory Power

When you are comparing several solutions, which one seems to work on the greatest number of facets of the problem? Which one seems to explain the problem? This is a good way to detect solutions that are actually smoke-screens, mirages that don't really deal with the problem.

3.2 Prior Probability

Which solution seems at first glance to be the best one? That is, if people just now looking at the problem saw your list of solutions, which would they choose at first glance? This often tells you which solution will be easiest to sell to someone else or which one(s) may be too farfetched to consider further without some good reason.

3.3   Predictive Power

Which solution allows you to plan for the future most effectively? That is, which one will enable you to predict post-solution conditions most effectively? Often you can pick a safe solution this way.

3.4   Clarity

Many times lists of solutions include some poorly defined ones, especially those whose limits are impossible to discern. For instance, deficit spending may be a solution to a government's need for capital, or loans may be a business's solution, but where are the limits of those solutions? A solution whose limits are undefinable may be a solution to avoid.

3.5   Provocative Power

In some situations you can choose a solution because it seems to open up a number of interesting and potentially attractive possibilities; you choose a solution because it leads you to ask a number of exciting questions. For instance, at the conclusion of the U.S. program to land a person on the moon, those in government who were interested in a continuing U.S. space program had to choose a direction in which to continue. One of the reasons that the reusable shuttle was chosen was the number of doors it would open, the number of interesting and exciting projects it could be used on. (Notice that this criterion is almost the opposite of Clarity.) In some situations, when a number of feasible solutions are competing, the one with the strongest provocative power may be the most attractive.

3.6   Falsifiability

One of the most useful nonempirical tests requires you as the problem solver to describe the conditions that would invalidate the solution you are testing. What would have to happen in order for that solution not to be the right one? Then you check to see whether those conditions do in fact exist or what their relative probabilities are. You may want to choose a mediocre solution that definitely *will* work over an excellent solution that *might* work.

3.7   The Crucial Test

This is the classical empirical test: can you design a crucial test situation that will allow you to select clearly among the alternatives? Can you design a test that only the right solution can pass?

Nearly anything you do to test the solutions you come up with can become part of a report on the subject. Few clients would be interested in an engineering report recommending solutions to a problem without any evidence as to how and with what results the solutions were tested. As this section has suggested, those tests are not always empirical. And the less empirical they are, the more they will have to rely on words on paper to substantiate them.

4. Making Your Choice

The quality of your choice depends mostly on the quality of your work in the previous stages of problem solving, but it is still important to be systematic in making your choice. Use this simple checklist to keep you on track:

1. Check your work.
2. Rank your alternatives.
3. Get advice.
4. Make your choice and document it.

4.1 Check Your Work

Too often people behave as though "checking your work" applies only in math classes. But successful professionals will tell you that making sure the earlier stages of problem solving have been done accurately is an important step toward success. Checking your work *before* you choose a solution takes much less time than solving the problems created by making an error that leads to a wrong solution.

4.2 Rank Your Alternatives

To be orderly about problem solving, list your possible (or recommended) solutions according to some clearly explained criterion (or set of criteria). This list is necessary to the success of the next two stages.

4.3 Get Advice

When you can, it's a good idea to get advice at this stage. Show a friend at work, your spouse, or maybe your boss your rank-ordered list of alternatives. They can help you catch any errors in your thinking and give you good practice explaining the decision-making process you used.

4.4 Make Your Choice and Document It

If you have followed all the preceding steps, your choice should usually be obvious by now. When you choose, you should prepare to document your choice. Whether your decision was good or bad, right or wrong, you can count at least on someone's asking how you did it and at most on writing up a full report. Collect all of your notes, brainstormed lists, data, tests, and so forth—any elements that contributed to your decision—and arrange them in order. At the very least, outline on paper the process you followed, for your own records. Often you will want or need to put together a short written account of how you made your decision. This account can then be keyed to the various lists, tests, and data you have gathered and the whole package either stored for future reference or used as the rough draft for your report. At this point, the process of testing your solutions, the problem-solving process, blends directly into the report writing process.

5. Doing the Writing

Solving the problem is one thing; writing up the solution in such a way as to fulfill a particular document's purpose can be quite another. If you take your reader through the process you used in solving the problem, you may be running the risk of doing exactly the wrong thing. Using the terms developed in Chapter 12, should your report have a *writer-based structure* (as it would if you duplicate the problem-solving process in the structure of your report) or a *reader-based structure* (perhaps beginning with the solution, and then explaining its aptness)?

In terms of the distinction between catalogical and analytical writing developed in Chapter 11, over and over again many writers produce catalogical reports (reviewing problem-solving processes) rather than analytical ones (presenting their solutions in the best possible way). It's not that people are not interested in how a problem's solution was discovered—*some* audiences, for *some* purposes, are—but you cannot assume that interest. For whatever writing situation you are in, you must analyze that specific situation—purpose, message, audience, and writer's role—and make your choices as to what to include and what to exclude, what to put first and what to put last, on the basis of that analysis alone. The examples in the following chapter, on making recommendations, illustrate these choices.

EXERCISES

1. Given a cigar box containing a claw hammer, five assorted nails, and a household utility candle, make a design that will fasten the candle to a wooden wall so that the candle will burn in an upright position. Write a report describing your solution and giving a full account of your problem-solving process. The audience is your classmates. (If you experiment, be careful with the lighted candle!)

2. Design a container that will hold a fresh egg and protect it well enough that you could drop the container ten feet onto concrete without breaking the egg. The container must be something that will travel with the egg. The optimum design criteria are simplicity, economy, and reusability. As an appendix to the report in which you describe your design, describe your problem-solving process.

3. Write a report to your classmates presenting the solution to the Tower of Hanoi puzzle (shown in Figure 14.7). The object of this simple child's game is to restack the rings on another post with the biggest again at the bottom and the others in order so that the smallest is at the top. You can move only one ring at a time, and you can never put a larger ring on top of a smaller one. Include in your report specific mention of any important principles involved. Could the puzzle be solved with seven rings and three posts? How?

Figure 14.7   The Tower of Hanoi Puzzle

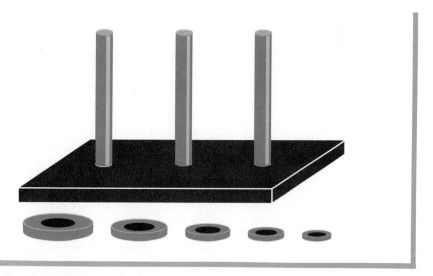

A ➤ indicates a case study exercise.

4. If your class has as its final requirement a major report, write a problem-solving report detailing your search for a topic to write about. Your report should end by persuasively recommending that your instructor approve the topic you suggest.

5. Do a short report on how problem solving (or decision making) is handled in your field of study. Consult textbooks, your instructors, library resources, and working professionals. How compatible are those methods of problem solving with the one this chapter presents? The audience for the report is entry-level professionals in your field.

➤ 6. In consultation with a faculty member within your major field of study, pick a professional problem within your field, and write a five-page report on different ways people are trying to solve it. For example, an accounting student might write about the 150-hour major requirement, a nuclear engineering student might write about work being done toward designing a safer source of nuclear power, or a transportation and logistics major might write about attempts to revive railroads in the United States. The focus of this paper should be on a short definition of the problem followed by summaries of several approaches being used to solve it, primarily taken from published literature. The audience should be an educated layperson, not an "insider."

7. Pick a campus problem (such as parking, class size, discrimination, and so on), and work with a small group of your classmates to come up with a good solution for it. Then write a report detailing how you came up with the solution. The thrust of the report should be to show that the process used to arrive at the solution is a valid one, not to argue for the solution in and of itself.

➤ 8. Sometimes problem solving involves deciding not only *what* to say but also *how* to say it. Do you need a report, a memo, a letter, or something else? The following case problem will test your ability to solve a problem, to decide whom you need to present the solution to, and to decide how you need to do it:

> You're in charge of the program committee for the local chapter of your professional society. Your committee's responsibility is to bring in guest speakers and create other interesting and informative programs once a month for the chapter members. Last week the program was a disaster. The speaker, a local lawyer who had been highly recommended, apparently had not put any time into preparing his presentation; he didn't even seem sure exactly who these people were he was addressing. While his topic, "The Law and You," seemed to have some good potential for your chapter's membership, in fact his comments were so disjointed, general, and apparently haphazard that the membership was notably unhappy by the end of the evening. When he got around to inviting questions, there was an awkward and stony silence.

Since the meeting, your colleagues at work, many of whom were there, have seemed to be unwilling to bring the subject up, and your boss has been heard to say, "Whoever is responsible for the program planning that produced that speaker needs to clean up their act or they won't last long in this business." The local chapter's president won't return your telephone calls. The chapter's immediate past-president, a friend of yours, recommends that you deal with the problem on paper to keep it from mushrooming. Now you need to figure out what, exactly, on paper you need to do. Here are just a few of the possibilities: A letter to your boss? A letter to the chapter president? An open letter to the membership? A column in the chapter newsletter? An information sheet for prospective speakers?

Your task is to select and write the document(s) you need to deal with this problem.*

---

* Adapted from *Business Communication Casebook,* American Business Communication Association, 1974. Reprinted with permission.

15. *Making Recommendations Persuasively*

The kinds of problem-solving processes that professionals use (discussed in Chapter 14) can take place over periods of time ranging from ten minutes to ten years (or more). One of the most common ways the results of such a process are reported is through a recommendation report. In fact, *any* kind of report can also have the presentation of recommendations as an essential element. The frequency with which reports have recommendations at their heart makes it even more striking that weak presentation of recommendations is the most common rhetorical flaw of reports.

What good is it to spend six weeks, six months, or a year collecting data and coming to an incisive understanding of a problem if, finally, no one acts on your recommendation?

A recommendation is only as effective as the extent to which a reader puts enough faith in it to act on it. Thus, the goal of this chapter is to describe ways to present recommendations not just clearly but also persuasively.

1. *Processes That Result in Recommendations*

Recommendations don't just come out of nowhere; they result from particular processes, such as those described in Chapter 14. Those processes that result in recommendations can be divided into three kinds:

Historical processes
Methodical processes
Logical processes

Each kind of process has its own implications for the structure of any recommendation the process leads to and for the problems its approval will meet with.

1.1 Historical Processes

Sometimes the historical flow of events merely needs to be observed to lead one to a recommendation. For example, as federal regulations surrounding cigarette smoking become stricter, tobacco companies recommend actions to their stockholders based on the flow of historical events. Similarly, a growing number of state wildlife departments face the problem of acid precipitation and have to come up with recommendations to deal with it, their recommendations again stemming from a historical process. In such cases, the historical pattern needs to be made clear in the recommendation. Do not assume your reader will be as familiar with the flow of events as you are. Frequently someone's *partial* familiarity with the flow of events leads to one conclusion about what should be done, whereas a *full* understanding of that flow of events can lead to another.

1.2 Methodical Processes

It may well be that most recommendations are arrived at methodically—by processes such as those described in Chapter 14. The scientific method is but one of a large number of methods designed to lead to recommendations. In some situations—such as those involving empirical tests—making the methodical process by which you arrived at your recommendation very clear and very convincing (in its apparent reliability and thoroughness) will be crucial to your recommendation's acceptance. In others, the method may be the least of the things the reader wants to know about.

1.3   Logical Processes

Some problems are "thought problems," requiring mainly an ordered set of mental operations for their solution. Einstein is said to have worked best when he worked most conceptually. Many times this kind of process leading to a recommendation poses the biggest problem for a writer. It may be very hard for you (or whoever solved the problem) to come up with a good, convincing explanation of how the problem was solved. And until that process is clear, some people will be reluctant to accept your recommendation. Section 2.1, "Argumentation Leading to Recommendation," gives specific advice on how to approach such a situation.

2.   *Patterns That Underlie Recommendations*

The internal patterns of recommendations can take a number of forms; two of the most common are argumentation and comparison.

2.1   Argumentation Leading to Recommendation

Suppose your company asks you to come up with a recommendation on whether they should continue to put money into a certain operation or abandon it. How would you write the report? Obviously you'd have certain facts at hand—a complete accounting of how much has been put into the project so far, how much the project is losing, and projections of its future profitability based on varying levels of new investment from your company. How do you make your report's contents logical? How do make the logic persuasive?

Persuasive logic is a little different from the formalities of induction and deduction that you may have been taught in the past. To be persuasive in writing for business and industry, you need to answer three questions:

1. What do you recommend?
2. What evidence supports your recommendation?
3. What connects your recommendation to your evidence?

Rephrased in general language, these questions become applicable to any situation requiring persuasion:

• What is your claim?
• What is your evidence?
• What connects your claim with your evidence?

Figure 15.1 shows the relationship among these three parts of the persuasive logic pattern.

Suppose you want to recommend that Acme Property Management (your employer) abandon its attempts to make the renovated

Figure 15.1 A Simple Persuasive Logic
The connections are the middle step(s) between the evidence and the claim.

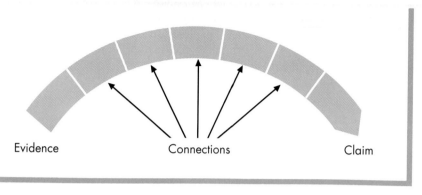

Evidence   Connections   Claim

James Hotel profitable and recoup whatever losses they can from the project by selling the building. How do you put that argument into this pattern of persuasive logic? Here's the persuasive analysis of your argument:

> *Evidence:* The cost/benefit projections show that it will cost more to operate than it brings in and that even with a massive infusion of new capital the project may well only break even for years to come.

> *Connections:* The projections are reliable. That is, all of the figures have been checked and double-checked, and they have been worked out under a variety of different economic assumptions (strong economy, recession, and so forth).

> *Claim:* We should abandon the James Hotel Project.

Depending on your decision about the report's larger structure (discussed in the next section), you can organize these as Claim/Evidence/Connections (the reader-based structure) or Evidence/Connections/Claim (the writer-based structure).

The important point is to make the connections explicit. The connections explain why, how, or under what conditions the evidence supports the claim. By performing this function, the connections help the reader bridge the gap between evidence and claim, which makes your argument more persuasive. At times your reader may simply not know that extra piece of information or may be too tired (or uninterested) to call that information to mind.

Making the connections explicit also serves to transfer some of the burden of the recommendation from you to the facts. For example, in terms of where the responsibility for the decision lies, there is a substantial difference between stating, "The project should be abandoned because the projections show the money won't work out" and

Figure 15.2   Using Persuasive Logic: Example One

First Argument

Claim: We can encourage people to ride the bus more by selling multistop tickets (which allow people to make several stops along one line without buying new tickets).

Evidence: The increased convenience and savings our riders get from multistop tickets will encourage more people to ride the bus.

Discussion

Assuming that customers are attracted by convenience and savings, the argument is plausible. It would be stronger if there were some evidence offered that increased savings and convenience really do result in more people riding the bus.

First Argument Revised

Claim: We can encourage people to ride the bus more by selling multistop tickets.

Evidence: People want increased convenience and savings.

Connection: Our local market research shows that multistop tickets will provide riders with increased convenience and savings.

---

stating, "The project should be abandoned because the projections show the money won't work out, and these are the most reliable projections available." The second way of putting it places at least some of the responsibility on the reliability of the projections. Figures 15.2 through 15.5 show examples of arguments, first without the connection, and then—after discussion—with the connection.

There are three other parts to this persuasive logic. The *backing* for the connection supports it in the same way that the evidence supports the claim. If someone challenges your statement of connec-- tion with "How do we know that is so?" the backing will answer that question. *Reservations* add an explicit statement of what circumstances would negate the claim. "If present economic trends continue" would be a statement of reservations. *Qualifiers* add an explicit statement of the force of the claim by bringing in such words as probably, perhaps,

Figure 15.3   Using Persuasive Logic: Example Two

Second Argument

Evidence: The length of time it takes our designers to produce drawings is holding back our firm's productivity.

Claim: We should begin using computer-controlled drafting.

Discussion

How will switching to computer-controlled drafting increase productivity? That is, will the computers work enough faster to justify their greater initial cost?

Second Argument Revised

Evidence: The length of time it takes our designers to produce drawings is holding back our firm's productivity.

Connections: Computer-controlled drafting is so much faster than manual drafting that we will be able to at least triple our productivity. This increased productivity should mean that the computers will pay for themselves in a year.

Claim: We should begin using computer-controlled drafting.

certainly, etc.) Figure 15.6 on page 393 shows how these parts fit into the basic persuasive logic model.

Adding these other parts to your recommendation continues the process begun by making the connection explicit. By further qualifying your claim, you make the argument tighter, more restricted, and less threatening.

If you will use this persuasive logic as an internal pattern, your recommendations will meet the criteria set out earlier in this chapter: they will be clear and persuasive.

2.2   Comparison Leading to Recommendation

One specialized kind of internal pattern leading to a recommendation occurs so often it merits special attention here. Over and over again, writers find themselves needing to compare two or more items and

Figure 15.4   Using Persuasive Logic: Example Three

Third Argument

Claim: Chrom-Ex does not induce in any way any type of hydrogen embrittlement (H_2 molecules trapped inside a chrome coating, caused in part by hydrogen being released from the surface to be coated by an acid in the coating).

Evidence: The surface preparation is done manually.

Discussion

Very few people would be able to see the connection between this evidence and the claim in this argument. The audience's lack of knowledge of this coating process makes it necessary to state the connection explicitly: when such coatings are applied manually, acid is not used; therefore, no hydrogen is released from the surface to be trapped under the coating; therefore, no flaws in the coating occur.

Third Argument Revised

Claim: Chrom-Ex does not induce in any way any type of hydrogen embrittlement (caused by acid).

Evidence: The surface preparation is done manually.

Connection: Manual preparation of the surface means no acid is used.

---

come up with a recommendation. There are two basic ways to do this: a two-part pattern and an alternating pattern.

The Two-Part Pattern. Suppose you are making a recommendation about a choice between A and B, and you want to compare them on qualities 1–10. The two-part pattern devotes one part to all of A's qualities (A:1–10) and another to all of B's qualities (B:1–10). When A and B have qualities that do not correspond (such as A:11 and B:12), those qualities usually go in a separate section.

The Alternating Pattern. For longer comparisons, you may want to put the comparisons side by side. The pattern then goes A:1–B:1, A:2–B:2, A:3–B:3, etc. When there are characteristics that don't correspond, such as A:12 and B:13, they usually go in separate sections.

Figure 15.5   Using Persuasive Logic: Example Four

Fourth Argument

Claim: We can use these plants successfully in interior environments only if they are properly conditioned first.

Evidence: These plants are native to tropical and subtropical regions.

Discussion

Again, the necessity for making the connection explicit comes from the audience's probable lack of knowledge. Plants that are native to tropical or subtropical regions would have to be properly conditioned (slowly acclimated) to any kind of interior environment.

Fourth Argument Revised

Evidence: These plants are native to tropical and subtropical regions.

Connection: Plants native to tropical or subtropical environments must be properly conditioned (slowly acclimated) to any kind of interior environment.

Claim: We can use these plants successfully in interior environments only if they are properly conditioned first.

Figure 15.6   A Fully Developed Persuasive Logic

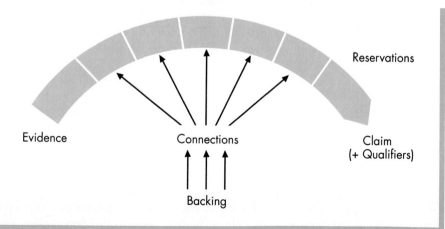

Evidence   Connections   Claim
                                       (+ Qualifiers)

Reservations

Backing

Guidelines for Comparison. Comparisons are relatively easy when A and B have mostly the same kinds of qualities. As long as you can compare A on qualities 1–10 with B on qualities 1–10, the comparison is fairly straightforward; you can often even use a chart or grid to make the comparison visual. The comparison becomes more difficult to write when A has qualities 1–5 and B has qualities 6–10. Of course, in this situation the alternating pattern is usually ruled out. But despite the necessity here for the two-part pattern, it's usually a good idea to add a part in which you do your best to compare the two head to head. Such situations, in which comparisons are being made on the basis of two different sets of characteristics, require special consideration on the writer's part. There are three guidelines to follow in this and any other kind of comparison leading to recommendations:

1. *List the Points of Comparison.* Before you get involved in the actual comparison, let your reader know what you're going to do. Be very explicit in the introduction about which points of comparison you will discuss. Bring up in the introduction any points your reader is likely to think of that you will *not* discuss, and explain in the body the reason for their omission. If you write a *cpo* introduction (as explained in Chapter 8), you will satisfy this criterion.

2. *Maintain a Neutral Stance.* The attitude you have toward your comparisons and eventual recommendations usually must be neutral. You want your reader to feel (in between the lines) and to see (in the lines themselves) that you have taken a completely fair and impartial attitude into the process of comparing and recommending. Be scientific in the attitude you show in your writing. In writing for business and industry, the most effective way to persuade is through a neutral stance. Of course, you will make your recommendation, and you will hope the reader endorses it, but the way to gain your reader's endorsement is through a nonjudgmental comparison.

   The psychologist Carl Rogers (among others) has shown convincingly that when one person explicitly chooses one side of a discussion, it invites other people to take the other side. That is, as soon as I see a report entitled "Whole- Versus Term-Life Insurance: The Whole-Life Ripoff Conspiracy," I immediately lean toward the other side of the issue. If you and I are discussing A versus B and you open by a strongly biased statement in A's favor, you force me psychologically toward the other side of the issue. The writer of the report who wants to recommend term-life insurance to me would be far better off titling the report "Whole-Versus Term-Life Insurance: A Comparison and Recommendation." Otherwise I suspect the writer has prejudged the issue. Put the facts in your report, and let *them* do the arguing for you.

In the "Recommendation" section of your report, you can make the results of the comparison clear; but again, be neutral—a judge, not an advocate. That means you present *the* recommendation, not *my* recommendation; demonstrate that it grows clearly out of the evidence, not out of your feelings.

3. *Make Your Comparison Complete.* It's always tempting to use only the points of comparison that will help you make the recommendation you prefer. But any points you fail to include invite your reader to suspect your work is incomplete, and your recommendation thus may not be fully persuasive. Certainly you can subdivide your "comparison" section into two large groups—one of the areas in which there are significant differences, and the other of the areas in which no important differences exist. You can often put that second section in an appendix so it won't get in a hurried reader's way. But one way or another, you need to account for all the points a complete comparison requires. At the least, if there are obvious points of comparison that you choose not to discuss, explain that fact and the reasons for it in your introduction.

3. *Patterns That Recommendations Show*

There are two goals you want your recommendations to reach:

1. The recommendation is *right* (whatever you are recommending will work).
2. The necessary people *endorse* it, *support* it, and *put it into action.*

You will have to pay the necessary attention to ensuring that your recommendation is right. But you also want to ensure that the necessary people endorse your recommendation, support it, and put it into action. Fulfilling this second goal can require some considerable thought on your part as a writer. The simplest pattern for a recommendation to follow is shown below:

- *Problem:* Why do something?
  ("Because Town Lake is polluted, property values are declining . . .")
- *Solution:* What to do?
  ("Continuing pollution of the lake needs to be stopped. This can be done by . . .")

All of the other patterns presented here are variations on that one. Like problem/solution, several ways of presenting recommendations just seem to make sense to writers. Other methods are less intuitively obvious to writers but may actually be more persuasive because they appeal more directly to readers. The following two sections describe these two kinds of patterns—writer based and reader based—in detail.

3.1   Writer-Based Patterns

The list below presents what may be, after the simple problem/solution pattern, the most popular writer-based pattern:

- What is the problem?
  ("Town Lake is polluted.")
- Why is it important to solve it?
  ("Property values along the lake are declining; new investment along the lake is nonexistent.")
- What caused the problem?
  ("Industrial waste and urban runoff, plus sewage spillover during heavy rains, have caused . . .")
- What are the alternative solutions, in rank-order?
  ("We have five options: . . .")
- What is the top-ranked alternative, in detail, and how can it be implemented?
  ("A Water Quality District should be established with the power to . . .")

This pattern has the strength of being methodical and predictable; each phase of the process seems to be the next logical step. The pattern's disadvantage is that the actual recommendation does not come until the very end of the document. An anxious or impatient reader may want to find out exactly what needs to be done earlier than the end of the report. If you want to include data (for example, facts and figures concerning the nature of the problem, its causes, and how and why that solution will work), those data come between the reader (who starts at the beginning) and the recommendation (placed at the end). Notice that this pattern allows the writer to present the report in exactly the same sequence as the one the problem was solved in. Figure 15.7 shows an example of this writer-based pattern.

Another writer-based pattern often used in recommendations is shown in the list below:

Introduction

- Purpose of the report
  ("This report examines the scope and significance of the pollution of Town Lake.")
- Nature of the problem
  ("As Town Lake's water quality has declined, parts per million (ppm) of 10 major carcinogenic or otherwise harmful chemicals . . .")
- Scope (criteria) of the solution
  ("By the year 1995 the lake should be able to pass all standards established by . . .")

[*List continues on page 399*]

Figure 15.7   A Writer-Based Recommendation Report

At the request of XYZ Insurance Company, we have evaluated your electrical-service entrance equipment to measure its ability to handle present and projected future loads. This report includes our analysis of your problem and our recommendations on how to solve it.

Here is our analysis of your present and future electrical-service loads:

	Connected	Demanded
Present load	4200 kVA	2207 kVA
Add new computer	86 kVA	86 kVA
Add new building	655 kVA	590 kVA
Total loading	4941 kVA	2883 kVA

As these figures show, a possible peak demand will result in overloading the utility company's 2500 kVA pad-mounted transformer. At the time your building was designed, that transformer was adequate. However, the combination of the new computer and the climate-control equipment it requires clearly means you have outgrown your electrical service.

Four methods of solving this problem were considered:

1. Installing capacitors to improve power factor and to release existing system capacity.

This solution would reduce the possible peak demand to 2550 kVA, which would still be an overload of the utility's transformer. In addition, such capacitors can often be difficult to properly maintain and when they are switched can cause damaging voltage transients. For these reasons, we do not recommend this approach.

2. Asking the utility company to install a one-point 4000 kVA service at 480 V.

The largest pad-mounted transformer your utility can supply is 2500 kVA. Therefore to handle your load would require a substation. You would be required by the utility to provide space (50' by 50'), plus the right-of-way for overhead lines, plus an expensive transition from the delivery point to your service equipment. The higher fault current available from such a substation would also require replacing many circuit breakers inside your existing building. Since this method is by far the most expensive and unsightly, we do not recommend it.

3. Asking the utility to install a new 1000 kVA pad-mounted transformer to serve only the new building.

This 1000 kVA transformer should provide adequate potential for the load growth of the new building. The cost for installing new service equipment to include trenching for underground utility conductors, pad, service switch, and entrance conductors we estimate at $12,500.00. We estimate the power billing for the new building would be $120,225.00. Added to the annual power billing for the existing building ($400,000.00), the total power billing for both buildings we estimate would be $520,225.00.

While this method would have the lowest initial cost, your existing building by itself would still be within 300 kVA of overloading the 2500 kVA transformer. That leaves you almost no potential for load growth in the existing building.

4. Asking the utility to install adjacent to the current 2500 kVA transformer a new 1500 kVA pad-mounted transformer, to divide the load between the two existing service entrances, and to totalize the metering.

Estimated cost of installing this new service equipment, including trenching for underground utility conductors, pad, new entrance conductors, and a feeder to the new building is $34,000.00. We estimate the annual power billing for the two buildings would then be $500,000.00.

We recommend this as the most desirable option for the following reasons:

(a) Although the initial cost is higher, there is a minimum annual savings of $20,000.00. This results from the energy being billed at a more favorable rate schedule the utility makes available for customers receiving such service.

(b) Splitting the service results in some load being removed from the existing substation, which would provide more flexibility for load growth in each building.

(c) With this solution the new building's service can begin immediately, without waiting for the utility to install the new transformer. By the time the peak summer load becomes possible (eight months from now), the new transformer will be in place.

(d) Should you decide in the future to add more equipment, having two separate sources of power will give you improved service flexibility. Having two separate sources of power will also mean more reliability: if one system goes down, the other can handle all but the peak loads.

For these reasons we recommend the fourth option. Should you have any questions about the information in this report, please call us. We will be happy to assist you in this construction in any way.

Body

- Presentation and interpretation of data
  ("Tests have been run. . . . As a result of the pollution problem, property values here . . . New investment has declined to the point that . . .")

Conclusion

- Recommendations and alternatives
  ("A Water Quality District needs to be established and given the powers to . . .")

This scheme also has a reassuring, right-at-first-glance appearance. It's predictable and easy to use. The disadvantage, again, is that the recommendation may seem to be buried at the end of the report. Once again the ordering of elements in this pattern of recommendation report has been determined primarily by the ordering of the writer's investigation of the subject. Figure 15.8 shows an example of this pattern.

3.2　Reader-Based Patterns

You may find that you need a structure that appeals more directly to your reader than the writer-based patterns. The preceding patterns reflect your need as a writer to produce a document whose structure mirrors the process by which you arrived at your recommendation more than they reflect the reader's need to learn your solution to his or her problem. When you want a structure that responds more directly to your reader's needs than the structures above, it usually means that you need a structure that begins with the recommendation, as shown below:

Recommendation: Briefly, what should be done?
  ("A Water Quality Board should be created . . .")
- Background: What is the nature of the problem?
  ("Increasing water pollution caused by . . . has led to declining property values to the extent that . . .")
- Detailed Description: What is the exact nature of the solution?
  ("Creation of a Water Quality Board will enable us to . . . and to . . .")
- Future: What methods will implement the solution?
  ("Three areas will be studied and, eventually, regulated by the board: industrial pollution, urban runoff, and sewage spillover.")
Appendix: Presentation and interpretation of data.

This structure can contain the same parts as the writer-based recommendation, but here those parts have been reordered significantly, with the reader's goals as the ordering principle. You can decide

Figure 15.8   **Another Writer-Based Recommendation Report**

This report investigates the availability of information on how to write computer user's manuals. Depending on that availability, I may request your permission to write my major report for English 4140, Advanced Technical Writing, on how to write user's manuals. Part of the report would then be my rewrite of the user's manual for the university's DEC-10.

User's manuals are currently the subject of much attention from both computer manufacturers and computer users. Manufacturers are realizing that many people who buy computers, especially micro-computers, are first-time computer users. Such people often make their decision about which computer to buy based in part on their response to the readability of the user's manual. The buyer's long-term satisfaction with the computer (and thus the manufacturer's chances for resale) also depends in part on the user's manual. Thus manufacturers are looking for people who can write readable user's manuals.

From the computer buyer's point of view, the user's manual can be the primary factor in determining which computer to buy and whether that purchase proves to be a satisfying one. Thus even people who have never seen a floppy disk before are learning to ask— and to test—whether the user's manual for the microcomputer they are considering buying is "user friendly."

A satisfactory search for literature on this subject should not take more than eight hours in the library, should yield a clear-cut answer, and should be conducted in such a way as to be reliable as a source for deciding whether to go ahead with this topic for a term paper.

I searched the Main Library's card catalog for books on the subject and found none. In the Main Library's periodicals, searching through three periodical indexes and going back five years, I found eight useful articles. A brief check of the Undergraduate Library's collection showed it has nothing on the subject that Main doesn't have. I worked with reference librarians in both libraries, so I'm reasonably certain these results are reliable. I also interviewed the owner of our local Computerland franchise, who said he knew of nothing devoted to the subject (but he wished there was something written about it).

Based on this search, which took approximately ten hours, I believe there is an even greater need for information on how to write good user's manuals than you and I previously believed. However, I do not think there is enough published literature on the subject to make a term paper on it feasible if the paper is mainly to be a review of the literature. There simply is not enough literature to review. While I remain interested in the subject, I recommend we look for ways to modify it, perhaps by expanding the length of the DEC-10 documentation rewrite, before we proceed with our planning.

between the two kinds of structures by determining your reader's needs and goals. A reader who is in a hurry, or who may not be especially interested in background information and technical data, may well respond more positively to the reader-based structure than to the writer-based one. However, if you have a reader who will (1) need convincing that there *is* a problem and (2) need to be shown in detail that your response is the *right* one, then for that reader you may well be better off using one of the writer-based patterns. Figure 15.9 presents the same recommendation as Figure 15.7, this time following the reader-based pattern.

Figure 15.9   A Reader-Based Recommendation Report

At the request of XYZ Insurance Company, we have evaluated your electrical service entrance equipment to measure its ability to handle present and projected future loads. This report includes our analysis of your problem and our recommendations on how to solve it. The Appendix to this report discusses alternative solutions and their drawbacks.

Recommendation
We recommend that you ask the utility to install adjacent to the current 2500 kVA transformer a new 1500 kVA pad-mounted transformer, to divide the load between the two existing service entrances, and to totalize the metering.

Estimated cost of installing this new service equipment, including trenching for underground utility conductors, pad, new entrance conductors, and a feeder to the new building is $34,000.00. We estimate the annual power billing for the two buildings would then be $500,000.00.

Analysis

Here is our analysis of your present and future electrical-service loads:

	Connected	Demanded
Present load	4200 kVA	2207 kVA
Add new computer	86 kVA	86 kVA
Add new building	655 kVA	590 kVA
Total loading	4941 kVA	2883 kVA

As these figures show, a possible peak demand will result in overloading the utility company's 2500 kVA pad-mounted transformer. At the time your building was designed, that transformer was adequate. However, the combination of the new computer and the climate control equipment it requires clearly means you have outgrown your electrical service.

Explanations

Here are the reasons we recommend this course of action:

(a) Although the initial cost is higher, there is a minimum annual savings of $20,000. This results from the energy being billed at a more favorable rate schedule, which the utility makes available for customers receiving such service.

(b) Splitting the service results in some load being removed from the existing substation, which would provide more flexibility for load growth in each building.

(c) With this solution the new building's service can begin immediately, without waiting for the utility to install the new transformer. By the time the peak summer load becomes possible (eight months from now), the new transformer will be in place.

(d) Should you decide in the future to add more equipment, having two separate sources of power will give you improved service flexibility. Having two separate sources of power will also mean more reliability: if one system goes down, the other can handle all but the peak loads.

Should you have any questions about the information in this report, please call us. We will be happy to assist you in this construction in any way.

Appendix
In addition to the solution recommended above, we considered three other solutions. Each has drawbacks, as described here:

1. Installing capacitors to improve power factor and to release existing system capacity.

This solution would reduce the possible peak demand to 2550 kVA, which would still be an overload of the utility's transformer. In addition, such capacitors can often be difficult to maintain properly and when they are switched can cause damaging voltage transients. For these reasons, we do not recommend this approach.

2. Asking the utility company to install a one-point 4000 kVA service at 480 V.

The largest pad-mounted transformer your utility can supply is 2500 kVA. Therefore to handle your load would require a substation. You would be required by the utility to provide space (50' by 50'), plus the right-of-way for overhead lines, plus an expensive transition from the delivery point to your service equipment. The higher fault current available from such a substation would also require replacing many circuit breakers inside your existing building. Since this method is by far the most expensive and unsightly, we do not recommend it.

3. Asking the utility to install a new 1000 kVA pad-mounted transformer to serve only the new building.

This 1000 kVA transformer should provide adequate potential for the load growth of the new building. The cost for installing new service equipment to include trenching for underground utility conductors, pad, service switch, and entrance conductors we estimate at $12,500.00. We estimate the power billing for the new building would be $120,225.00. Added to the annual power billing for the existing building ($400,000.00), the total power billing for both buildings we estimate would be $520,225.00.

While this method would have the lowest initial cost, your existing building by itself would still be within 300 kVA of overloading the 2500 kVA transformer. That leaves you almost no potential for load growth in the existing building.

4. *Checklist for Effective Recommendations*

For your recommendation to be effective, it needs to satisfy at least these three criteria:

1. The recommendation itself must be clear. What do you think needs to be done?
2. The reasons behind the recommendation must be clear. Why do you think this needs to be done?
3. The connections between the recommendations and the reasons must be clear. What connects your reasons to your recommendation?

4.1   Clear Recommendations

Be very explicit about exactly what needs to be done. Do not hesitate to make full and frequent use of charts and other kinds of visuals to spell out your plans. Be very clear with your audience about whether you've come up with a general solution, with details to be worked out later, or a very specific, itemized plan of action. If the latter is the case, review Chapter 13 on process descriptions, because part of your recommendation will need that kind of specificity.

4.2   Clear Reasons

You need to be very clear about the criteria used in choosing this solution: is it the most feasible, the most exciting, the best if cost is no consideration, the most economical, or something else? State clearly and explicitly the reasons behind your recommendation. Often those reasons will be grounded in history, in your problem-solving method, or in logic and can be explained in exactly those terms. But the persuasive reasons behind a recommendation may not always be directly or most clearly rooted in *how* the recommendation itself was discovered. Sometimes the most persuasive reason you can offer in support of your recommendation lies not in its past but in its future, not so much in the fact that it was arrived at in a systematic and acceptable way but in the fact that it will work. Because you wind up justifying your prediction by explaining in depth how it will work, your prediction really can be persuasive (if it is clear and detailed). It also can seriously endanger your career if your prediction is wrong. For that reason, it's often wise to qualify your prediction carefully ("if present trends continue") and to base it squarely on explicit reasoning. That is, don't just answer the question, "How will it work?" Also answer, "How do you *know* it will work this way?"

4.3   Clear Connections

It's good to make your recommendation and its reasons clear, but for your recommendation to be fully persuasive, you need to connect your reasons to your recommendations carefully and explicitly. Suppose your recommendation is to centralize all of your firm's purchasing in one department rather than to continue to allow each separate unit to do its own purchasing. The reasons may be that the current scheme's costs have risen astronomically in the last two years, and you suspect waste, pilferage, and duplication of orders are to blame. To be persuasive, make the connections clear. *How* will centralizing the purchasing functions solve those problems?

Figure 15.10 shows a sample recommendation report that meets the three criteria.

Figure 15.10   A Sample Recommendation Report

Energy Conservation Opportunities at Building 12

Significant opportunities for energy savings exist in Building 12, a 22,680-square-foot structure with seven air-conditioning systems and eleven exhaust air systems. We recommend modifying the three largest systems (AC 1, 2, and 4) to provide night setback controls. This will save $13,028 yearly, with a total payback period of 1.7 years.

Background

Acme Engineering uses Building 12 as a metallurgical and quality-assurance laboratory. To control hazardous contaminants, the work performed in this building requires continuous operation of selected exhaust systems within the building. Providing make-up air to the exhaust air systems requires the continuous operation of the building's heating and cooling equipment. Reheating system fan discharge air accomplishes temperature control of multiple zones served by the air handling system.

The existing make-up air systems are central-station types located in the attic. AC 1 and 4 are equipped with zone steam reheat and are located at the eastern and western extremes of the building. AC 2 is a single-zone system serving the central area of the building.

AC 4 handles 100% outside air. AC 1 and 2 mix 58% and 47% outside air by design, respectively, with system return air. All systems are fitted with manually operated outside and return air dampers. The fan discharge temperature is manually reset as the need arises to satisfy building comfort level and is controlled independently of zone reheat demands. The nominal air and tons of refrigeration for systems AC 1, 2, and 4 are 6000 cfm with 28 tons, 10000 cfm with 42 tons, and 3000 cfm with 20 tons, respectively.

Recommendation

By modifying the three largest systems (AC 1, 2, and 4) to provide night setback control, make-up air systems will cycle in response to dedicated night space thermostats. The thermostats should be set seasonally at a moderate temperature, such as 60°F in the winter. This will minimize the operating time and capacity level of these heating and cooling systems. Since at least one of the three systems will probably be operating at any given moment, the possibility of a total absence of make-up air for sustained periods is low.

   The major control elements for the proposed night setback mode for each system are electric space thermostats, automatic outside air control dampers, electric-pneumatic relays, or (as in the case of AC 1) a two-position steam control valve to isolate steam reheat. All these space thermostats are programmed by associated time clocks. We recommend that Acme Engineering modify each system as specified here, based on the following conceptual estimates of in-place costs and energy savings:

System	Installation Cost	Energy Cost Saving Per Year	Discounted Payback in Years
AC 1	$9600	$5556	1.5
AC 2	$7000	$3730	1.6
AC 4	$7600	$3742	1.7

Appendix

[The appendix explains how the costs and other data were figured. It consists of about one page of figures and is not reproduced here.]

EXERCISES

1. Choose a frequently discussed campus problem (parking, the price of athletic tickets, the rising costs of attending college, etc.), and write a recommendation report suggesting a solution to the problem. Write two versions of the report—one in a reader-based pattern and one in a writer-based pattern. Each report should be about 300 words long. The audience should be the provost, chancellor, or president of your institution.

2. Choose a research-oriented journal in your field of study and examine the structures of articles in two or three issues. Can you make generalizations about those structures, especially in terms of their being writer based or reader based? Write a recommendation report in which you compile your findings and recommend improvements in the structures that are frequently used in the articles you look at. Make the structure of your report reader based. The audience should be the people who write articles for those journals, especially entry-level professionals in your field.

3. Investigate the career opportunities that will be available to you when you graduate, then write a report recommending that students follow:
   (a) your course of study if they want good jobs upon graduation
   (b) a modified variation of your course of study if they want the best chances for jobs
   (c) some totally different course of study
   The report can be either reader or writer based. The audience should be other students who are in the process of selecting majors.

▶ 4. Assume you work for Smokey's Pharmacy, which runs a very busy delivery service in the Cedar Springs suburb. One of your first jobs as a new assistant manager is to recommend to your boss what kind of car to buy for the new fleet of delivery cars. Wanting to make a good decision, you're using as your primary information source the issue of *Consumer Reports* (available at most libraries) that each year compares new cars. You need to write a 300-word report containing your recommendation. The criteria you want to use are economy, reliability, and durability. Other information you might need: The Cedar Springs suburb is mostly without hills, and the cars need to be air-conditioned. If you want to use other information about the cars you choose to consider, be sure to be able to defend the information's reliability.

▶ 5. Problem 6 on page 384 asks you to write a report on ways people in your field are going about solving a problem. Now, build on what you did in Exercise 6, and write a report recommending a solution to the problem. The report should be written in a reader-based pattern and should be 750–1,000 words long. Its audience should be educated laypeople.

A ▶ indicates a case study exercise.

16. *Using Research Libraries*

By this point in your college education, you probably have realized the importance of knowing how to use your university's library. For many professionals, needing to know how to research information in a library does not go away upon graduation. Business people, engineers, architects, doctors, and lawyers are just a few of the professionals who find library research a necessary career skill. Here's a simple self-test to allow you to measure your library research skills right now. Try writing out the answers to these questions on a separate sheet of paper:

Library Skills Self-Test

1. Are there basic specialized reference tools in your field, such as style manuals, handbooks, specialized dictionaries, and encyclopedias? If so, name one in each relevant category.
2. Name the three most important current topics in your field. Beside each name write the word used for it in the Library of Congress indexing system.
3. List the three leading periodical publications in your field (these could be either newspapers or journals, and the journals could be professional, trade, or scholarly).
4. Name the most important periodical indexes and abstracting services in your field. Try to name at least two.
5. Name the most important online sources of information in your field.

If you drew a complete blank on two or more of those, your library skills aren't what they should be. This chapter will show you how to do a serious, systematic process of library research, from planning it to writing it up and evaluating it.

> **A Note About Librarians.** Your best friend during any library research project is the librarian. *Ask a librarian* is an option you should freely take at any stage in your research process. You should not avoid, but rather welcome, situations that require you to get a librarian's help. If you can get a librarian involved in your work, even minimally, your research is well begun indeed. When you ask a librarian for help, however, it is important to ask your question carefully. Here are three important guidelines for dealing with librarians:
>
> 1. *Be prepared.* The librarian may never have heard anything about your topic until the moment you come up and ask about it. Be prepared to give the librarian a good short explanation of what you are trying to do. *Remember:* The answer you get can only be as good as the question you ask.
>
> 2. *Be polite.* During term-paper season at a university, each librarian may talk to several hundred people each day. Such situations can make for short tempers, and a little courtesy—just "please" and "thank you"—may well get you a more detailed answer.
>
> 3. *Be persistent.* Frequently a busy librarian will say something like, "It's over on shelf nine," and leave it up to you to find the document in question. If you can't find it in ten minutes of looking, don't hesitate to go back and ask for more help. Be polite, but be persistent.
>
> As a process, library research in many ways parallels other kinds of fact-finding, information-gathering, problem-solving processes (cf.

Figure 16.1   The Process of Library Research

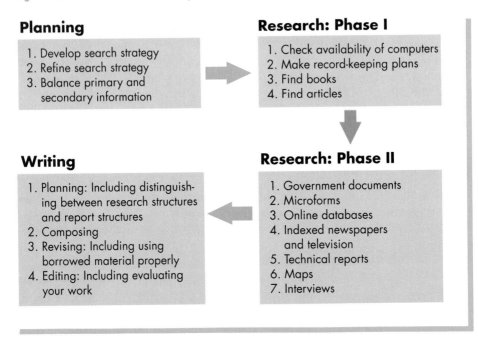

Chapter 14). Ask the right questions, don't focus on a possibly wrong answer too hastily, keep careful records, weigh the value of your findings carefully, and present those findings thoughtfully. As in other such processes, careful planning at the outset reaps large rewards later on, and errors made at the outset tend to be magnified the longer they continue uncorrected. Most important, in library research (as in any research process) you need to be systematic.

This chapter explains the process of library research as something that should take place in a particular, especially systematic way. Figure 16.1 shows the broad outlines of the process.

1. *Planning*

Before you go into a library, work out a search strategy. A typical university library's collection may contain over a million bound books and another one or two million documents of other sorts. If you wait to begin planning your search until you are physically in the presence of that much information, and *then* try to think of where to go and what to look for, you may well not be able to think as clearly as possible about what your plans should be.

1.1　Developing a Search Strategy

A fully developed search strategy will tell you both *where* to look and *what* to look for. You should begin forming a preliminary search strategy as soon as you are presented with a writing task. You need to be able to answer these questions:

1. What is the *topic* you are to write about?
2. What *main point(s)* are you to make about your topic? (In English classes, this is often called your *thesis.*)
3. What are the *important subpoints* you need to cover?
4. What are the *key words* you need to pay attention to? Is each of these the best word for what it signifies?

The *topic* is the subject you want to write about, but at this stage in your work, you often will have a topic (something to write about) without a thesis (something particular to say about it). For example, the student whose research process is profiled in this chapter was a business major who had been a biology major. She originally knew only that she wanted to do her report on "some aspect of DNA research." To take advantage of her current major, however, the slant on the subject she eventually discovered was to focus on the economic implications of such research. Then her main point, her thesis, became to show some of the extent to which that research has become big business and, as she put it, "big bucks as well." After only a little preliminary digging, she decided to include a background on the science behind that research, a thumbnail sketch of the American pharmaceuticals industry, and then closer profiles of two or three of the companies that are "major players in the genetics game." The key words she listed included *DNA, genetics,* and *recombination.*

Some writing tasks allow you to answer the four questions immediately, although those answers will almost certainly change as you proceed. For other tasks, you may not know much for certain beyond the topic until you finish preliminary research. But it's important to keep pushing for those answers, especially for a clear conception of what your one main point, your thesis, is going to be. If you as the researcher and writer don't have the kind of clear sense of your task that allows you to see quite clearly what its beginning, middle, and end will need to look like, how well do you think your reader will understand that you've done?

1.2　Refining the Search Strategy

The flowchart in Figure 16.1 shows a standard or generic search strategy; it tells you in general where to look, but you have to fill in what to look for. One of the best ways to decide what you need to search for is to take a quick look for information on your topic (and your key words) in the various reference tools available for your subject and your field of study. The following reference tools are available for many fields:

Generalized encyclopedias	Specialized dictionaries
Specialized encyclopedias	Directories
Almanacs	Gazetteers
Yearbooks	Atlases
Handbooks	Periodical indexes
Dictionaries	Style manuals

There are many varieties of the reference tools listed here, and there may be kinds of reference tools in your field that are not listed here. Go to your university library's main reference desk and learn what the standard reference tools for your field are. These three books can help you find out what those reference tools are:

Sheehy's *Guide to Reference Books*
Katz's *Magazines for Libraries*
Ulrich's *International Periodicals Directory*

As you continue working on narrowing your topic, defining your thesis, and shaping your search strategy, work through your field's reference tools to see what they say about your topic. From them you can get valuable background, answering such questions as, "What's the current state of knowledge on this subject?" They will also often list the names of authorities on the subject and the sources of more specific information.

Although often a little vague in its outlines, this first stage of your research process is especially important. The more *creative* background digging you do on your topic while you're shaping your search strategy, the better focus your topic will have for the remainder of your project. Through background work with reference tools at this stage, you can avoid making false starts on topics that don't work out and traveling up blind alleys of research. And once you take the trouble to learn what the basic reference tools are in your field, you can use them profitably over and over again.

One other point should be made about this first stage in the process of library research. If you were to ask ten randomly selected students how to start working on a library research paper, seven out of ten would probably say, "Begin with the card catalog." Starting with that step leaves out all of the background information on the topic that reference sources hold. More important, beginning with the card catalog leaves out all of the topic refining and thesis shaping that go on in the writer's mind during the search through basic reference tools. During that reference search, you learn what questions are the right ones to ask within your topic. And you cannot hope to find the right *answers* unless you ask the right *questions*.

Figure 16.2 shows how our DNA research student fleshed out her search strategy based on work with basic reference tools.

1.3 Primary Versus Secondary Information

As you prepare to move out of the planning stage and into library research, it's worth taking a minute to consider the distinction between primary and secondary information (or *sources* of information) and whether your project as planned will have an appropriate mixture of the two. *Primary information* is material you have collected yourself or that you are using directly from its original source. *Secondary information* is material you have received second hand (say, by reading about it). Many writing tasks by their very nature push the writer toward one kind of information and away from another, but the best pieces of writing nearly always have some kind of mix of the two. It may be 80–20, 50–50, or 20–80, but a really good piece of writing will almost never be 100–0 or 0–100.

For example, suppose an accounting major were to be writing about the trend within the Big Six accounting firms in the United States to diversify their operations—specifically to derive less and less of their income from accounting activities and more and more of their income from providing financial counseling services (concerning investment of retirement funds and so on) to businesses. A report like that could be done totally from secondary sources—basically a library research report. But it will almost certainly be a better piece of writing if at least some primary information is mixed in. For this project, that might take the form of interviews with executives from several of the firms involved.

Consider an engineer who works for a company that manufactures the kinds of robots used by law enforcement agencies. The writing task is to write the copy for a promotional brochure that will be mailed to potential customers. Obviously, someone who works where such robots are designed and built could write a detailed piece from first-hand knowledge (primary information), but it would probably be a better piece of writing if it had at least some secondary information mixed in, such as statistics on increasing numbers of crimes of violence and increasing numbers of robots in the field.

Many readers place a higher value on writing that has at least some mix of kinds of information than they do on writing that is all primary or all secondary. So although the distinction isn't always quite as clear-cut as in the examples described here, it's wise for you to pause at some point in your research planning process and consider whether your plans might not be improved by including at least a bit of a mix of kinds of information.

Figure 16.2   A Modified Search Strategy

```
PRELIMINARY SEARCH STRATEGY


Topic: Economics of Recombinant DNA Research
Thesis: Recombinant DNA research is already a multibillion-
        dollar business and is growing rapidly.
Main Points: Explain recombinant DNA research.
             Trace its movement into business.
             Put dollar value on it.
Key Words: Recombinant DNA research, genetic engineering, bio-
           technology.


REFERENCE TOOLS


King's A Dictionary of Genetics
Knight's Dictionary of Genetics
Regar et al., A Glossary of Genetics
Encyclopedia Britannica
Williams and Lunsford's Encyclopedia of Biochemistry
More than 85 journals under "BIOLOGY-Genetics" in Ulrich's
International Periodicals Directory, including:
    Genetic Engineering Letter
    Genetic Engineering News
    Genetic Technology News
    Recombinant DNA
    Recombinant DNA Technical Bulletin
    Biotechnology
Indexes:
    Biological Abstracts
    Index Medicus
    Science Citation Index
```

2. *Research*

If you've done your planning properly, your research itself will be a fairly directed and purposeful activity—a *search* for the best information available. With only a couple of exceptions (noted below), there should be few or no times you find yourself aimlessly wandering from shelf to shelf or item to item (or *browsing*). Although you will need to continue to revise your research plan and may well need the help of librarians along the way, carefully planned library research should not take hours and hours with only meager results. If you're doing library research right, you should feel a positive correspondence between the amount of time and energy you're putting into it and the amount of good information you're getting out of it. If that's not the case, consider sitting down with a reference librarian to go over your plan again.

2.1 Using Computers

Computers play a major role in library research today. Four uses of computers are especially important to consider here: using a laptop for record keeping, using computers for access to the library's collection, the role of computers in producing periodical abstracts and indexes, and other online data bases. As you move into the research phase, you first need to decide how much you intend to use computers in your research; if you can, make a commitment to use computers as much as your resources and your library's facilities allow you to. When you do library research, you're dealing with vast quantities of information, and dealing with vast quantities of information is the computer's best capability.

2.2 Keeping Records

The better records you keep of your research, the easier doing the writing will be, for two reasons. First, good record keeping makes it easier, quicker, and more efficient to retrieve the information you need when you start writing. Second, as Chapter 3 showed, writing tends to work better as you begin the process of actually putting things down on paper sooner. Thus, because doing good record keeping necessarily involves putting at least some of your understanding of your topic on paper as you proceed, that act of record keeping in fact promotes the act of writing. Put more simply, the sooner you start articulating your information and ideas on paper, the better your final written product will be.

Perhaps the best way to keep records of research is to combine your own use of note and bibliography cards with a research notebook (described in Chapter 19, pp. 506–509). Many students find that a *daily log* of their work is the most useful resource when they begin their writing The daily log will help keep you from retracing steps you've already taken but might have forgotten, give you a way to review mentally what you've done as a way to decide what to do next, and provide a fairly unimpeachable record in case anyone ever questions what you actually did.

If you use a laptop or notebook computer to record your daily log while you're in the library, you can combine your note cards, bibliography cards, and log entries in ways that will feed more or less directly into your document's early drafts. An ordinary word processing program will do that work, but there are also many customized programs designed just for library research and note taking. Such programs are often interactive—that is, if you correct an error in a bibliography item in one file, the program will correct it in all other files as well—and are particularly good at helping writers create pages of references and bibliographies.

Many researchers prefer to photocopy articles to use at their leisure rather than take notes from them as they sit in the library. The advantage of this is that you have continued access to the article whenever you need it. But there are two disadvantages: (1) it costs money, sometimes quite a lot of money, and (2) photocopying instead of note taking means you are not encouraged to start coming to your own early understanding of the material, your own conceptual synthesis of how this piece and that piece might fit together in words, the way you might be if you relied a little more on taking notes and a little less on photocopying.

2.3   Finding Books

The traditional way to find books in a library is to use the card catalog, drawers and drawers full of cards referencing and cross-referencing every bound item in a library's collection. Lately those mammoth pieces of furniture often have been replaced by computer work stations, and the card catalog may only be accessed online. That computer system may go under various names, but this chapter will still refer to it as the "card catalog." To give you an idea of the size of the access problem, consider that if you were to put all of the drawers in a typical university's card catalog end to end, you might well have a row over 500 feet long, with over 6 million cards in it. Given that, or a comparable amount of computer memory, how do you find the information you want? That will depend on whether you are doing an *author search*, a *title search*, or (often the most difficult) a *subject search*.

Author Searching. If you know the name of an author who has written on your subject, you can search directly for material by that person.

You specify "author search" on the computer's initial screen or go to the section of the card catalog reserved for author/title cards. You can get names of authors on your topic from textbook bibliographies, from bibliographies in articles you may have seen, or from recommendations from experts on the subject. Many computers also allow you to do author searches if you know only part of the author's last name.

Title Searching. Title searches work much the same way as author searches. Simply specify to the computer that you're doing a title search, or make sure you're in the part of the card catalog that is sorted by title, and the search is straightforward. Because titles tend not to be as unique as author's names, however, title searching can be tricky. One wrong word, especially if it is the first word, can throw you off. Remember that if the title begins with a word like *a, an,* or *the,* you'll probably (but not always) find the title indexed under the first letters of the *next* word.

Subject Searching. Subject searching is the trickiest kind of searching to do, but it may well be the most commonly used. Given a particular subject, there are many different ways you or I might refer to it but probably only one or two ways it might be named in the card catalog. That is, the language we use every day—*natural language*—may have 10 or 15 different ways to refer to something like nuclear energy: atomic power, fission, fusion, and so on. But the language used to index items in the library—*controlled language*—cannot be so redundant. If you look up your topic using the wrong word for it, you may find to your surprise that your university library *seems* to have nothing on your topic. The problem, of course, is that you're looking in the wrong place (or, more properly, looking for the wrong thing). An even worse possible outcome is that you'll find a little information—enough to keep you going in a minimal way but not so little as to alert you that you're making a major tactical error.

To avoid these problems when you're doing subject searching, make sure you're using the right word (frequently called an *index term*). Doing a subject search in a catalog that may well contain over a million items is no time to guess at the term you choose to look up. Instead, use the *Library of Congress Subject Headings List;* bound volumes of it should be right next to the card catalog or computer terminals. Suppose you were doing the research paper on recombinant DNA mentioned earlier: looking in the card catalog under "DNA" would at best lead you to look under "deoxyribonucleic acid," and looking there would lead you to "recombinant DNA," which would lead you to "genetic engineering," which, it turns out, is where most of the appropriate information is indexed. Rather than trying to follow this roundabout sequence, at any point of which you might get

Figure 16.3 General Organization of Library of Congress Subject Headings

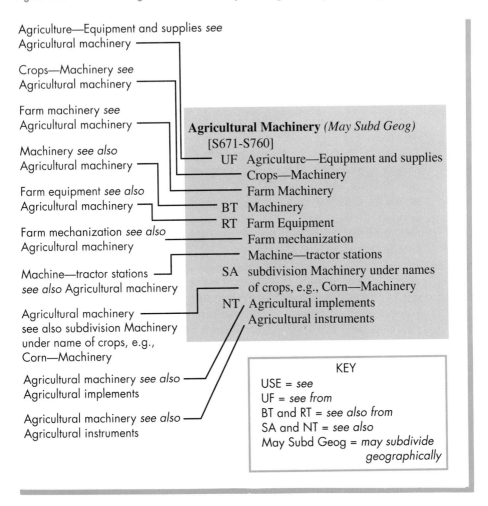

Agriculture—Equipment and supplies *see*
Agricultural machinery

Crops—Machinery *see*
Agricultural machinery

Farm machinery *see*
Agricultural machinery

Machinery *see also*
Agricultural machinery

Farm equipment *see also*
Agricultural machinery

Farm mechanization *see also*
Agricultural machinery

Machine—tractor stations
see also Agricultural machinery

Agricultural machinery
see also subdivision Machinery
under name of crops, e.g.,
Corn—Machinery

Agricultural machinery *see also*
Agricultural implements

Agricultural machinery *see also*
Agricultural instruments

Agricultural Machinery *(May Subd Geog)*
[S671-S760]
UF Agriculture—Equipment and supplies
Crops—Machinery
Farm Machinery
BT Machinery
RT Farm Equipment
Farm mechanization
Machine—tractor stations
SA subdivision Machinery under names
of crops, e.g., Corn—Machinery
NT Agricultural implements
Agricultural instruments

KEY
USE = *see*
UF = *see from*
BT and RT = *see also from*
SA and NT = *see also*
May Subd Geog = *may subdivide
geographically*

lost, give up, or make a mistake, it's much better simply to look up your topic in the *Library of Congress Subject Headings List*. Figure 16.3 shows how the list is organized. Using it this way, you will be able to find exactly the right index term(s) to look up in the card catalog (See Figure 16.4).

Boolean Searches. Some computerized card catalogs allow researchers to do searches that involve combinations of subject terms, generically called *Boolean searches*. For example, you might want to limit your search for information on genetic engineering to sources that deal specifically with business aspects of genetic engineering and limit it

Figure 16.4   One Researcher's Path through the Library of Congress Subject Headings List

1. If you happen to look first under *DNA*, you find the nar-
   rower term (NT) *recombinant DNA*.

2. *Recombinant DNA* can take you to the broader term *genetic
   engineering*.

3. And *genetic engineering* leads you to the two terms *biotechnology* (which would connect you with *biotechnology industries* and thus to *genetic engineering industry*) and *genetic engineering industry*.

For this research project, it turns out that *genetic engineering industry* was the most productive index term. The subject headings entry includes the correct call letters for
items on this subject.

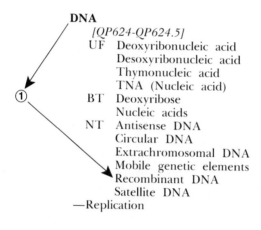

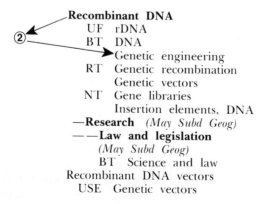

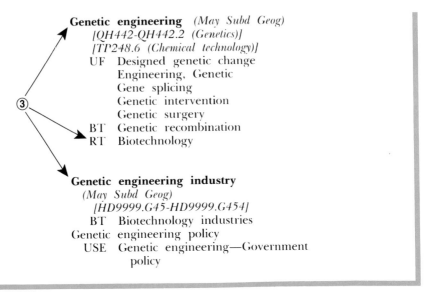

Genetic engineering *(May Subd Geog)*
[QH442-QH442.2 *(Genetics)]*
[TP248.6 *(Chemical technology)]*
UF   Designed genetic change
Engineering, Genetic
Gene splicing
Genetic intervention
Genetic surgery
BT   Genetic recombination
RT   Biotechnology

Genetic engineering industry
(May Subd Geog)
[HD9999.G45-HD9999.G454]
BT   Biotechnology industries
Genetic engineering policy
USE   Genetic engineering—Government
policy

Source: Library of Congress. Reprinted with permission.

still further to sources published since 1985. Then the information you give the computer might look something like this:

genetic engineering // business // 1986–____

While each different computer will require a different code to trigger the Boolean search, that capability is one that is certainly worth using if your library's system supports it. On topics such as genetic engineering, where the first pass through a subject search may turn up hundreds of items, the ability to limit the search one or more ways can prove invaluable.

Catalog Cards or Computer Screens. Some libraries have author/title cards in one place and subject cards in another; other libraries combine the cards. Depending on the kind of search you are doing (author, title, or subject), you may see small variations in the cards (or computer screens) you find. Figure 16.5 shows sample subject, author, and title cards. The main difference between those and what you might get online is that the online version will often tell you whether that book is already checked out, which is valuable information.

Suppose you use the subject headings in the card catalog and compile a list of 30 books that you might look in to find material on your subject. Although it might be best at least to look at all of them, in case you don't have that much time, how do you decide which ones to use and which ones to rank as of secondary importance? From the card catalog you can tell whether a book does or does not include a

Figure 16.5   Subject, Author, and Title Cards

Notice the tracings on the author card showing the subject headings for the title.

Subject
Card

> GENETIC ENGINEERING—SOCIAL
> ASPECTS.
>
> QH
> 442
> .L4   Lear, John.
>   Recombinant DNA : the untold story /
> John Lear. New York : Crown
> Publishers, c1978.
>   280 p., [4] leaves of plates : ill.
> : 24 cm.
>   Includes index.
>
>   TU   09 FEB 79   3608470   TKNUdc   77-29158

Author
Card

> Lear, John.
> QH   Recombinant DNA : the untold story /
> 442   John Lear. New York : Crown
> .L4   Publishers, c1978.
>   280 p., [4] leaves of plates : ill.
> : 24 cm.
>   Includes index.
>
> Tracings———   1. Recombinant DNA.   2. Genetic
>   engineering—Social aspects.
>
>   TU   09 FEB 79   3608470   TKNUdc   77-29158

Title Card

> Recombinant DNA
> QH
> 442   Lear, John.
> .L4   Recombinant DNA : the untold story /
> John Lear. New York : Crown
> Publishers, c1978.
>   280 p., [4] leaves of plates : ill.
> : 24 cm.
>   Includes index.
>
>   TU   09 FEB 79   3608470   TKNUdc   77-29158

Source: Library of Congress. Reprinted with permission.

bibliography (the card will say something like "incl. bib."), and because a book's bibliography usually will lead you to more on the subject, you might want to choose books whose cards show they have bibliographies. If you are looking for ideas for visuals for your report, you might want to choose books with illustrations—listed in the tracings on the card catalog card as "ill." And you will probably want to choose more recent books over older books.

When you find a book on the shelf, take an extra minute to look up and down that shelf for others that might be useful, also. This technique, called *browsing*, can sometimes be very productive. Never rely on it exclusively, but it is a helpful addition to an otherwise carefully directed search.

2.4   Finding Articles

The most up-to-date information on subjects of professional interest is rarely found in books. And in many fields, the books that are available tend to be either textbooks or popularizations. Thus, no library research is complete without a thorough investigation of articles (in periodicals and other such sources) on the subject; in some fields, articles are the only published information you will find. You can find articles by browsing selected periodicals or searching in periodical indexes and abstracts. A thorough researcher will use both techniques; only the most hurried project would rely solely on browsing.

Browsing Selected Periodicals. By skimming recent issues of the three or four professional (or trade or scholarly) journals that are most likely to have articles on your subject, you may well be able to find materials that will move your search ahead quickly. Most university libraries keep the recent issues of journals in a separate place, perhaps called the Current Periodicals Room, and you can go in, take the current issues to a comfortable chair, and scan them. Look for articles on your subject or related to your subject. If you find a few articles that are right, you may have leapfrogged right into the middle of your topic, and you may be able to finish your research much faster. This technique also helps you to refine your own mental picture of your thesis, to have a better idea of what questions you need to ask and what questions you may not be able to answer.

In case you do not know the names of the journals to look in for a particular field of knowledge, the books by Sheehy, Katz, and Ulrich cited earlier (see p. 412) can tell you. Another complete listing of journals organized by field is the *Standard Periodicals Directory*, which also gives readership numbers for many journals. Another way to identify the most important journals in your field of study would be to ask a professor in that field or a reference librarian.

Once again, you would not want browsing through selected periodicals to be the only way you look for articles, but it's a good way to find out quickly whether there is much current periodical information on your subject and to find an occasional up-to-date article to help you with the rest of your research.

By this time you should have refined your idea of your topic down to a fairly specific thesis. From here on, be increasingly specific about exactly what you are looking for. Keep an open mind on your topic initially, but focus more precisely on your specific thesis before too much time passes. Many students find that a week or two of preliminary digging will be enough to make their topic and thesis specific.

Using Periodical Indexes and Abstracts. There are over 70,000 periodicals in the world, and your university library may well have 15,000 to 20,000. How do you find articles on your subject? The most reliable and thorough way is to use periodical indexes. A periodical index organizes the articles in a particular set of journals and during a particular time span into groups by their subjects. You may already be familiar with one periodical index, the *Readers' Guide to Periodical Literature*, from high school or freshman English. The *Readers' Guide* indexes popular journals (listed in the front of each issue), the kind you might subscribe to at home and read for recreation. If you want to know what magazines like *Time, Good Housekeeping, Road and Track,* or *Cosmopolitan* say about your subject, you can find the bibliographical information on their articles in the *Readers' Guide.* Knowing how to use *Readers' Guide* is useful because many other indexes are put out by that same company (Wilson Indexes) and are organized the same way. But if you're in a junior- or senior-level college course, you will probably be expected to use considerably more specialized journals than those in the *Readers' Guide.*

How do you find out what the periodical indexes in your field are? Once again, Sheehy, Ulrich, and Katz will list them. You can also often find the name of the right indexes somewhere on the cover or table-of-contents pages of separate issues of your journals. Or you can ask a reference librarian or a professor in that field. If you want your research to be good, plan on consulting four or five indexes covering at least five years back.

If you have seven or eight indexes to choose from, how do you select which one to use first? This is where *abstracts* come in. A growing number of indexes also contain (in the same volumes or separate volumes) a brief summary of each article cited in the index—an abstract. An index with abstracts can save you a lot of time. If you have bibliographical information on 30 articles, imagine how long it would take you to find and read copies of each one. But you can read 30 abstracts in 30 minutes and quite possibly narrow down to 5 the number of articles you actually need to find.

Most indexes have subject and author volumes; some also have citation volumes. A citation index tells you the names of articles that have referenced a particular name in their bibliography. Thus, if I know that Jane Zimmerman wrote a book on my topic ten years ago, I can look in the appropriate field's citation index to find current articles that include a reference to her book in their bibliographies or notes.

Even with Sheehy's, Ulrich's, and similar sources, it is sometimes difficult to find the indexes for your subject. If you have that problem, ask at your library's reference desk to see a list of the indexes that the library has. The librarian will probably only loan it to you, but you can use it to find other indexes to look in.

Some people find their first experiences with periodical indexes a little troublesome. It can take time to find the indexes, and still more time—and a librarian's help—to learn how to use them. Most indexes are generated by computers that seem to have little regard for your ease of using them. However, as long as you stay in that field, what you have learned will remain valid; you don't have to relearn the indexes with every research project. For example, if you are a mechanical engineering student, once you learn the indexes for the field and how they should be used, you will consult the same indexes for doing research as long as you stay in mechanical engineering. And whatever field you are in, using periodical indexes is probably the only reliable way to find all the articles on your subject.

The following list contains only a small sampling of the diverse assortment of indexes that may be found in most university libraries. A growing number of these indexes can also be accessed online.

A Sampling of Indexes Available in Most University Libraries

Accountant's Index
Agricultural Index
Animal Breeding Abstracts
Applied Science & Technology
   Index
Architectural Index
Biography Index
Biological Abstracts
Biological Index
British Technology Index
Business Periodicals Index
Comprehensive Dissertation
   Abstract Index
Commerce Clearing House
Computer and Control Abstracts
Consumers Index

Current Index to Journals in
   Education
Education Index
Electrical and Electronics Abstracts
Energy Index
Energy Research Abstracts
Engineering Index
Environment Abstracts & Index
Food Science Abstracts
Forestry Abstracts
F&S Index
General Science Index
Geophysical Abstracts
Hospital Literature Index
Humanities Index
Index Medicus

Index to Legal Periodicals
Industrial Arts Index
Journalism Abstracts
Marketing Information Guide
Metals Abstracts
Moody's Indexes
Oceanographical Abstracts
Index to U.S. Government
   Periodicals
Personnel Management Abstracts
Petroleum Abstracts
Physics Abstracts
Psychological Abstracts

Public Affairs Information Service
   (PAIS) Bulletin
Science Citation Index
Social Science Citation Index
Social Sciences Index
Society of Manufacturing
   Engineers Technical Digest
Sociological Abstracts
Standard & Poor's Index
Target Group Index
Wildlife Abstracts
World Textile Abstracts
Zoological Record

You can find the articles you need by browsing, by searching through periodical indexes and abstracts, and by using your library's own list of periodicals. Once you get to the articles you need, write down all the bibliographical information on each one. Remember that an author who has written one article on a subject may well have written others, so you might want to check that name in the author volumes of a couple of indexes. And a journal that publishes two articles on a subject may well have published others, so you may want to skim the contents of other issues.

2.5 Consulting Other Sources

The pages that follow describe seven kinds of other sources for information on your subject. Many other kinds of sources are not discussed here, and one of the measures of your resourcefulness and effort as a researcher will be how many and what kinds of other sources you can find.

Government Documents. By far the biggest "other source" of information on nearly any subject is government documents. The U.S. Government Printing Office distributes over 28 million documents annually to its over 1,000 depository libraries. How do you find out if there are government publications in your field and on your subject? Some university libraries do not index government documents in their card catalogs at all, and others do so only partially. The kinds of periodical indexes discussed earlier in this chapter typically do not include government publications. The best way to find out whether there are government publications about your subject is to ask a librarian who specializes in government documents to help you. Many people who use government documents regularly still rely on librarians' help, and certainly all researchers who are not familiar with how government documents are organized should. The sources discussed here are the simplest ones available for finding your way around government documents, but to use most of them you will still probably need a librarian's help.

> *The Monthly Catalog of U.S. Government Publications.* This may be the best way to find noncongressional publications, such as agency reports, pamphlets, and proceedings.
>
> *The Publications Reference File (PRF)* lists quick access for current government documents still in print.
>
> *The Congressional Information Service (CIS) Index* lists all congressional publications, including hearings, committee reports, and laws, and includes abstracts.
>
> *The Statistical Abstract of the United States* contains most kinds of government statistical information.
>
> *The American Statistics Index* indexes government statistical information and also includes abstracts.
>
> *The Index to U.S. Government Periodicals* will lead you to articles in government periodicals, indexed by both author and subject.
>
> *The Index to Current Urban Documents* and the *Urban Documents Microfiche Collection* give you access to documents issued by cities in the United States and Canada, including annual reports, budgets, environmental impact statements, manuals, questionnaires, and zoning ordinances.

Microforms. In many university libraries there are actually more documents on microforms (microfilm, microfiche, etc.) than there are on paper. As with government documents, some university libraries index many of their microforms, and some index very few. Although the list below will give you an idea of the kinds of documents that may be in your university library, microform collections vary so much from one school to another that you really need to go explore your school's collection for yourself.

> *College Catalogs* for most colleges and universities in the United States are usually kept in microform.
>
> *Special literary collections* are often only available in microform. One example is *Early American Imprints,* including all nonserial American publications before 1819.
>
> *ERIC,* the *Educational Resources Information Center,* indexes many conference papers and otherwise unavailable documents on almost everything concerning any field of education.
>
> *Copies of journals.* Many university libraries keep second copies of their journals on microforms.

Disclosure System. Often you can find the 10-K and annual reports to stockholders for companies traded on the New York and American Stock Exchanges in microform.

The *Visual Search Manufacturer's File* (VSMF) includes manufacturer's catalogs and industry standards for all types of industrial components.

Online Data Bases. Your university library (or your employer) may subscribe to one or more of a growing number of commercially available online data bases. An online data base usually takes the form of a bibliographical resource (such as a periodical index with abstracts). You enter a topic, author, or title, and it searches its records for articles or information on that subject. Depending on the data base, you may be able to get printouts of abstracts sent in addition to bibliographical information. Most data base searching by students should be done only by working in tandem with a librarian. These searches can be expensive. While a quick online search in the ERIC data base may cost only $10, some other data bases—and longer searches—can cost quite a bit more.

Most of these data bases make possible the same kind of Boolean subject search described earlier in this chapter. For example, one researcher wanted to use ERIC online to determine whether anything had been written about the combination of cloze tests (a measurement of readability) and technical writing. ERIC listed over 180 documents on technical writing and over 800 on cloze tests. Searching manually would have required cross-checking the two lists by hand to determine whether any title appeared in both. The computer did the cross-checking in less than a minute, found 18 items the two lists had in common (and that thus discussed both technical writing and cloze tests), and printed out bibliographic information and abstracts. The search was performed for free as a service of the library's reference department. (Again, most such searches cost money.)

Three of the most popular commercial services that provide access to such data bases are ORBIT Search Service, DIALOG Information Services, and BRS Information Services. In addition, the Online Computer Library Center (OCLC) system provides access to nearly 20 million book titles. Here is a short list, organized by field of study, of some of the data bases these services provide access to:

Agriculture—AGRICOLA
Biology—BIOSIS
Business—ABI/INFORM
Chemistry—CA Search
Engineering—COMPENDEX
Environment—ENVIROLINE

Government—NTIS
Mechanical engineering—ISMEC
Physics—SPIN
Pollution—POLLUTION ABSTRACTS
Science—INSPEC
Science and technology—SCISEARCH

Indexed Newspapers and Television News. Nearly all major city newspapers are indexed, and most university libraries have a very good collection of those papers and indexes. The following list is representative:

New York Times	*Washington Post*
Wall Street Journal	*New Orleans Times-Picayune*
London Times	*Chicago Tribune*
Los Angeles Times	*Houston Post*
Pravda	*Atlanta Constitution*

A number of clipping services regularly collect articles in newspapers on predetermined subjects:

The *Newsbank Urban and Public Affairs Library* clips from 190 U.S. newspapers, including one from each state's capital. This allows you to trace different responses to the same news event around the country. Topics include business and economic development, education, health, and transportation.

The *Newsbank Review of the Arts* collects newspaper reviews on books, films, television, art, architecture, and fine arts.

Television news broadcasts represent another source of information for researchers. Increasingly, news broadcasts are indexed, and transcripts are available in university libraries. For example, the *CBS News Index* provides subject-indexed transcripts of all CBS News broadcasts, including morning and evening news and special reports. Check to see which television news transcripts are available in your university library.

Technical Reports. Many libraries contain entire collections of technical reports that are only sporadically included in the card catalog. These may be from institutes affiliated with the university or from nearby research institutes and may concern such subjects as water resources, transportation, or wildlife resources. The major source of technical reports is the *National Technical Information Service* (NTIS), an agency of the U.S. Department of Commerce. There are well over a million titles available in this system, including U.S. government-sponsored research, development, and engineering reports, as well as foreign technical reports. Information about these reports is available through the *Government Reports Announcements & Indexes*, which includes abstracts, and through the NTIS *Weekly Abstract Newsletters*.

Many times you may find that your library does not have the report you need, but ordering NTIS reports is a reasonably efficient process, given enough lead time.

Maps. Again and again, good maps make the difference between nondescript reports and memorable reports. Most universities have excellent map collections, either in the library or in the Geography department. The standard maps to look for are the U.S. Geological Survey topographic maps for the United States. Ask your librarian where to find local and regional maps as well.

The National Oceanic and Atmospheric Administration (NOAA) publishes nautical and aeronautical charts. Other types of maps to look for are geologic maps, soil maps, highway maps, and the Bureau of the Census's GE-50 series, which includes data on per-capita income, employment in different industries, agriculture, and population trends.

You can also find maps in atlases. The following is a partial list:

The Times Atlas of the World

World Atlas of Agriculture

Oxford Economic Atlas of the World

Atlas of World History

Antique Maps of the World

An Atlas of Fantasy

National Atlas of the United States of America

Rand-McNally Commercial Atlas

There are also indexes to maps, such as the *Index to Maps in Books and Periodicals*, the *Index to Printed Maps*, and the *Guide to U.S. Map Resources*.

Interviews. Although they are not part of the information found in libraries, interviews are such an important part of research that it makes sense to discuss them here. Whatever your topic, it is likely that someone at your library is an expert on it. A good interview with such a person can enhance an otherwise run-of-the-mill report. The art of interviewing is too complex to present here, but a few basic guidelines are:

1. Always do an interview by appointment.
2. Before you interview, do some homework:
   - What is the basis of this person's expertise?
   - Find his or her publications or records of achievement, and familiarize yourself with them.
   - Plan your questions before the interview to make the best use of your time and interviewer's time.
3. Send a courtesy note afterward, thanking the person interviewed for his or her time and the information and perhaps offering to send a copy of the finished report.

3. *Writing*

As in Chapter 2, the writing process will be discussed here in terms of four fundamental phases: planning, composing, revising, and editing. The focus here, however, will be on what is unique about each phase when the subject of the writing is the results of library research.

3.1 Distinguishing Between Research Structures and Report Structures

Chapter 2 discusses the planning stage of writing in detail. When you are planning the writing element of a library (or other) research project, the main difference between the process described in Chapter 2 and the process of writing up a research report is that structural planning for the research report needs to be done especially consciously and carefully. Of course, there is a structure ready to hand for writers of such reports—the structure of the research process itself. A writer who unthinkingly uses that research structure (the ordering of a fully developed search strategy) for the structure of the report may be making a fundamental error. It's always tempting to allow the research structure to control the structure of the report you write on the topic—that is, to let the structure of the *input* control the structure of the *output*. But research report writers need to see their role as taking in the research data, making sense of the information in the light of their own best thinking, and then reintegrating the data into a new structure derived from that thinking, a structure selected because it is best for that particular report's purpose and intended audience. Sometimes the structure of the research is the best structure for the report. Sometimes it isn't. The point is that if you don't make a conscious decision about structure, the research structure will insert itself into that void, whether it's in fact the best report structure or not.

At what point in your research process should you start composing? Different writers find different times in the research process the best to begin composing, but in general the sooner you start writing, the better everything will go. Maybe the biggest self-defeating fantasy that research writers are subject to is the one indicated by the writer who says, "In another week I'll have all my research done, and then all I'll have to do is write it up. I know that only leaves a couple of days for the writing, but with all the research done, it should go pretty fast." The person saying it may be a 17-year-old freshman facing his first major research writing experience or a 40-year-old professional facing her twentieth such project; in either case, a statement like that is too often a prelude to disaster. In fact, research is never really over until there's no time left, and writing is something that always takes time—more time than you estimate—to get it done right. The combination of those two characteristics is a powerful force that always works—usually very destructively—against those who would wait until the last minute to start writing.

3.2 Using Borrowed Material Properly

During revising, you need to make sure the use you are making of others' material is correct. Proper use of other people's words and ideas is important both during and after college. The following list gives four guidelines that will help you avoid accidentally using someone else's work without proper credit.

Guidelines for Using Quoted Material

1. Whenever you can, *introduce* the borrowed material in your report with the name of the authority from whom it was taken ("John Smith, project engineer, says that if we . . .").

2. Put *quotation marks* around all quoted material, even in your note cards.

3. When you paraphrase someone else's words, be sure you *really* have put the thought into your *own* words.

4. For everything you borrow, whether it is quoted or paraphrased, *document* its source.

Figure 16.6 presents some short examples to illustrate the four points from the preceding list. Version A is unacceptable because it is not a true paraphrase and because it doesn't credit the source. Although B credits the source, it is still too close to the words of the original. Version C is in the student's own words and credits the source, so it is acceptable. Version D is in the student's own words, credits the source, and *tags* it as well ("As Mobil Oil Corporation points out"), making it the safest use of someone else's material.

In order to make the best use of other people's material in your own writing, you should be aware of three other useful techniques: assimilation, reduction, and insertion. Assimilation and reduction are methods of making long material short enough to use; insertion is a method of introducing quoted material smoothly and gracefully.

Assimilation. When the quoted material you want to use is too long, one way to incorporate it into your own material is to assimilate it. State the core of the borrowed passage in your own words, but present enough of the key phrases or expressions as quotations to give your readers a feeling for the force and flavor of the original. Although the original passage about steam turbines was certainly not

Figure 16.6 An Original Passage and Some Student-Paraphrased Versions of It

The Original (quoted from *Steam Turbines and Their Lubrication*, p. 5):

Because it can readily be built in units of large capacity, and because it is highly efficient and extremely reliable, the steam engine is supreme as the prime mover in the central-station field.

Student Version A (*Unacceptable*):

The steam turbine is supreme as the prime mover in the central-station field because it can readily be built in units of large capacity, is highly efficient, and is extremely reliable.

Student Version B (*Unacceptable*):

The steam turbine is supreme as the prime mover in the central-station field because of its high efficiency, extreme reliability, and because it can readily be built in units of large capacity.[1]

Student Version C (*Acceptable*):

As a central power source the steam turbine is without equal. For power generation its efficiency and reliability, coupled with the ease with which it can be built in units of large capacity, make it the first choice.[1]

Student Version D (*Acceptable*):

As Mobil Oil Corporation points out, the reliability, efficiency, and easily attained size of steam turbine units make them the most popular choice for central power stations.[1]

---

[1] From *Steam Turbines and Their Lubrication* (New York: Mobil Oil Corporation, 1965, 1981), p. 5.

too long to quote in its entirety, we can still use it to illustrate assimilation:

> As Mobil Oil Corporation points out, the steam turbine is "supreme as the prime mover in the central-station field."

Reduction. Another useful way to shorten a quotation down to a usable length is to use ellipsis dots (three spaced periods) to indicate that you have omitted words or phrases. When you omit words in the *middle* of a sentence, use three spaced periods to indicate the omission. When you omit words at the *end* of the sentence, use the ending punctuation plus three spaced periods. This kind of reduction is common in scientific and technical writing. An abuse of it that, although not common, still happens too often occurs when a secondary author uses reduction to change the meaning (by shifting the emphasis) of the primary author's words. For example, it would not be ethical to reduce the original report on steam turbines in this way:

> Because it can readily be built . . . the steam turbine is . . . in the central-station field.

This reduced version makes it appear that Mobil was really saying that ease of construction put steam turbines into the central-station field, which is not at all what the original version said.

Insertion. Insert quoted material into your report so that it does not impair your report's continuity. The worst violation of this rule is the too-common practice of writers' saying what they want to say three times: in their own words, in the quoted material, and then again in their own words. For example:

> Mobil Oil Corporation points out that the steam turbine is the best power source in the central-station power field: "the steam turbine is supreme as the prime mover in the central-station field." Thus we see the supremacy of the steam turbine in the central-station field.

Limitations on Quoting. In some fields, quotation (especially direct quotations) is almost never used in formal reports. Even if quotations are sometimes used in your field, it is likely that there is an unwritten policy dictating that they be used sparingly. The following list provides some good guidelines for when to use quotations in technical material.

Guidelines on When to Use Quotations in Technical Material

- Quote only when your own words will not do the job as well (or better).
- Use only as much material for your quotation as you need—the bare minimum.

- Be sure that, whether you use a direct quotation or not, you credit the source for all borrowed material (unless you are absolutely certain that the material is common knowledge).

Punctuation. One point about quoting material: remember that in most fields—

- Periods and commas go *inside* the quotation marks.
- Colons and semicolons go *outside* the quotation marks.
- Dashes, question marks, and exclamation points go *inside* the quotation marks when they pertain only to the quoted material and *outside* when they apply to the whole sentence.

3.3   Citing Your Sources

As you move from revising to editing, you need to make sure you have properly *cited* all of the borrowed material you use. In the previous section, the superscript number ([1]) in student versions B–D could lead the reader to a number of different kinds of source notes. Those different kinds of source notes all come under the general heading of "reference styles"—methods you use to mark in the text your use of someone else's material and to give your reader the information needed to locate that original material. Reference styles vary from field to field. Although in your previous schooling you may have been required to use MLA, Turabian, or *The Chicago Manual of Style*, by the time you are a junior or senior in college you should be learning your own field's reference style, which may be quite different from those you've used before. Figure 16.7 shows a few examples of how the last line of student version D might look in different reference styles.

Most other forms of referencing are variants of one of the styles shown in Examples A–F.

How to Find Your Field's Reference Style. Most fields of study in college have their own reference styles, established either by the leading journal in that field or by the professional association representing the field. Sometimes a particular academic department will have its own guidelines for reference, and when it comes to theses and dissertations, most universities have schoolwide guidelines. In writing for business and industry, you will often not need references at all. When you do, your employer may well have a specified form. To find out what reference style you should use, either ask a professor in your field (if you're in school) or check the leading journal in your field to find out what it requires. That information is usually printed in small type on a page labeled something like "Notes for Contributors," as Figure 16.8 shows.

Figure 16.7   Examples of Various Reference Styles

Example A: A Footnote. Note appears at the bottom of the page.

```
. . . the most popular choice for central power stations.¹
```

```
¹Steam Turbines and Their Lubrication (New York: Mobil Oil Cor-
poration, 1965, 1981), p. 5.
```

Example B: An Endnote. Note appears with other notes on separate page at end of report.

```
. . . the most popular choice for central power stations.¹
```

```
1. Steam Turbines and Their Lubrication (New York: Mobil Oil
   Corporation, 1965, 1981), p. 5.
```

Example C: One Form of Parenthetical Documentation. Note appears with other notes on separate page at end of report.

```
. . . the most popular choice for central power stations (1)
```

```
1. Steam Turbines and Their Lubrication (New York: Mobil Oil
   Corporation, 1965, 1981), p. 5.
```

Example D: Another Form of Parenthetical Documentation. The first number inside the parentheses—assigned in order of occurrence in the chapter or in the book—indicates the source; the second number gives the page number within the source. By using this system, you can make subsequent references to that document without repeating the entire note: (1:455), etc.

```
. . . the most popular choice for central power stations
(1:5).
```

```
1. Steam Turbines and Their Lubrication (New York: Mobil Oil
   Corporation, 1965, 1981).
```

Example E: Another Parenthetical Form. A short title or author's last name, plus the year of publication, may save most readers from ever checking the note itself, which is an entry in the report's bibliography.

```
. . . . the most popular choice for central power stations
(Steam Turbines, 1981:5).
```

```
Steam Turbines and Their Lubrication (New York: Mobil Oil Cor-
poration, 1965, 1981).
```

Example F: Another Parenthetical Form. This one uses a long entry for first full reference. It makes subsequent references easy, and a ''Notes'' page unnecessary.

(first reference)

```
. . . . the most popular choice for central power stations
(Steam Turbines and Their Lubrication, Mobil Oil Corporation,
1981, p. 5).
```

```
[subsequent references] . . . water contamination which may be
traced to the cooler itself (Steam Turbines, p. 16).
```

To give you an idea of how many reference-style manuals there are, the following list gives some of the more commonly used ones.

Some of the Many Different Style Manuals

The Chicago Manual of Style
Kate Turabian's *A Manual for Writers of Term Papers, Theses, and Dissertations*
The Modern Language Association's *MLA Handbook*
William Campbell, Stephen Ballou, and Carole Slade's *Form and Style* (often called "Campbell/Ballou")
American National Standards for Bibliographic References
Eugene Fleischer's *A Style Manual for Citing Microform and Non-print Media*
The U.S. Government Printing Office's *Style Manual*
The American Psychological Association's *Publications Manual*
The *Council of Biology Editors Style Manual*
The American Chemical Society's *Handbook for Authors*
The American Institute of Physics' *Style Manual*
The American Medical Association's *Stylebook*
Harvard Law Review's *A Uniform System of Citation*
The Royal Society's *General Notes on the Preparation of Scientific Papers*

Figure 16.8 Sample Instructions to Authors on Citation Style

REFERENCES, NOTES, AND BIBLIOGRAPHY

Reference and notes are identified in the text by sequential superscript numerals, except that the original numeral is used when the same page of a previously cited source is cited again. All references and notes are then listed in numerical order at the end of the article, using *Chicago Manual of Style* format:

1. Bergan Evans and Cornelia Evans, *A Dictionary of Contemporary American Usage* (New York: Random House, 1957), pp. 387-88.   *Book*

2. Don Bush, "The Passive Voice Should Be Avoided—Sometimes," *Technical Communication* 28, no.1 (First Quarter 1981): 19-20   *Article in a professional or scholarly journal*

3. "Hurtling Through the Void," *Time*, 20 June 1983, p. 68.   *Article in a popular magazine or newspaper*

4. Paul M. Postal, "On So-called Pronouns in English," *Readings in English Transformational Grammar*, ed. Roderick A. Jacobs and Peter S. Rosenbaum (Waltham, MA: Ginn and Company, 1970), p. 57.   *Article in an anthology or conference proceedings*

5. Evans and Evans, p. 409.   *Shortened form for different page in a previously cited work*

Sources listed as references do not need to be repeated in a bibliography. If, however, you wish to include a bibliography of other related sources, arrange the entries in alphabetical order, again using *Chicago Manual of Style* format:

Fowler, H.W. *A Dictionary of Modern English Usage*. 2nd ed. Revised and edited by Sir Ernest Gowers. New York and Oxford: Oxford University Press, 1965.   *Book*

Held, Julie Stusrud. "Teaching Writers How to Write: What Works?" *Technical Communication* 30, no. 2 (Second Quarter 1983): 17-19.   *Article in a professional or scholarly journal*

Lu, Cary. "Second-generation Microcomputer Report." *High Technology*, June 1983, pp. 28-30.   *Article in a popular magazine or newspaper*

Creager, Cynthia. "Format Design: Help Your Readers Use Your Manuals." In *Proceedings* of the 30th International Technical Communication Conference, pp. W & E 147-150. Washington, D.C.: Society for Technical Communication, 1983.   *Article in an anthology or conference proceedings*

Note: Some regular columns—such as Book Reviews and Recent and Relevant—use other formats that are better suited to their purposes. Contact the column editor (listed inside the front cover of the journal) for guidelines.

Source: Society for Technical Communications. Reprinted with permission.

>   The American Society of Agronomy, Crop Science Society of
>   America, and Soil Science Society of America's *Handbook and
>   Style Manual*
>   The American Journal of Medical Technology's *Handbook for
>   Authors*
>   The American Society for Testing Material's *Manual for Authors
>   of ASTM Papers*
>   The Engineers' Joint Council's *Recommended Practice for Style of
>   References in Engineering Publications*
>   The American Society for Mechanical Engineering *MS-4: An
>   ASME Paper*
>   The American Institute of Industrial Engineers' *The Complete
>   Guide for Writing Technical Articles*
>   A recent reference book, *Business and Technical Writing: An An-
>   notated Bibliography of Books, 1880–1980,* by Gerald Alred,
>   Diana Reep, and Mohan Limaye, also describes a number of
>   typical reference styles.

Publication Style Versus Manuscript-Submission Style. Suppose that in your
attempt to use your field's reference style you take a copy of an article
from your field's leading journal and merely imitate the way it does
references. Will you be right? It depends on whether your reader
wanted publication style or manuscript-submission style. "Publication
style" means the way references finally appear in printed pages of the
journal. "Manuscript-submission style" means the way the reference
appeared in the manuscript the author sent in to the journal. (This
same distinction affects visuals; see Chapter 10, Section 5.2.) The two
styles can be very different. For example, many journals still print
bibliographical references as footnotes (as in Figure 16.7, Example
A), but most such journals require the references in manuscripts submitted for publication to be in end-note form (as in Figure 16.7,
Example B).

Guidelines for References. Whatever reference style you use, there are
at least these three guidelines for using it properly: *clarity, consistency,*
and *common sense.* The main purpose of references (beyond acknowl-
edging that certain material is borrowed) is to enable a curious reader
to find your original source, either to check your accuracy or to learn
more about the subject. If your reference style is clear, a reader can
easily find your original source. Review your own references, and ask
yourself if a reader could trace your borrowed material back to its
source from the information you've given. Once you establish a pattern for your references, follow it consistently, so that your reader
has to figure out the pattern only once. Finally, use common sense

in selecting your reference style; if what you are doing seems strange, it's probably wrong, and you've probably misinterpreted something. If your use of a particular reference scheme employs clarity, consistency, and common sense, it will function adequately for you.

Using Bibliographies. Most reports written in academic settings require some sort of bibliography, a compilation of all the sources consulted in the production of that report. The bibliography may be alphabetized, or items may be numbered in order of appearance in the report. Some bibliographies are ordered chronologically. If a bibliography lists everything the author so much as glanced at, it should be labeled "A Complete Bibliography"; if it lists only the most important sources, it should be labeled "A Selected Bibliography."

Sometimes you may be asked to use shortened forms of bibliographical citation in your own notes and complete forms in your bibliographical entries. A reasonable compromise between no bibliography and a full bibliography is to include in your bibliography only those items that (1) are important for the reader to know you consulted and (2) do not appear in the notes. Such a bibliography should have a title that indicates its unusual nature, such as:

A Selected Bibliography
(Important Sources Not Otherwise Cited)

What kind of bibliography to use, or whether to use one at all, depends on the field you are writing in, and your purpose, subject, and audience. Many kinds of writing for business and industry require neither notes nor bibliography. When referencing is required, a growing trend among journals leads away from bibliographies entirely; most journals can afford only the space for notes.

A Note on "Notes." If you consult the "Guidelines for Contributors" page of the leading journal in your field, it talks about two different kinds of "notes"—literature citations (called "references" here) and notes in which you as author comment on some point. Manuscript style occasionally treats the two kinds of notes differently, with references on a separate page at the end of the article and author's comments at the bottoms of pages. But most journals also discourage all such comments. Most editors probably would delete a note at the bottom of a page, feeling that if your comment is worth making, it should be worked into the text. So when you read the "Guidelines for Contributors," do not be confused by the ambiguous use of the word *notes;* although author's notes may be discouraged, in many fields (and especially in academic settings), bibliographical notes ("references") are expected.

3.4   Evaluating Your Work

After so much attention to the process of library research, you may wonder how your reader, specifically an instructor, evaluates the quality of your research. The following list of questions represents the results of research into how instructors evaluate their students' research:

- Does the report contain enough "hard" information?
- Is the borrowed material properly credited?
- Is there overreliance on any one source?
- Do the sources include the most recent available?
- Are the sources of a variety of kinds?
- Are the sources of the right level?
- Does the report overrely on textbooks?
- Do quotations seem artificially inserted, or are they woven into the fabric of the paper?
- Are all references in the proper form for your field?

Of course one must also ask larger, functional questions in order to evaluate the report's total quality. Those questions include whether the report does the job it needs to do for the audience it is aimed at, and a number of other similar concerns. For a brief checklist of those larger questions, see the introduction to Part Five of this text.

EXERCISES

1. On a separate sheet of paper, write the answers to the following questions:
   - (a)   On what page number of what generalized encyclopedia can you find information on your research topic?
   - (b)   What are the names of the specialized encyclopedias closest to your field?
   - (c)   Under what Library of Congress Subject Headings can information on your topic be found?
   - (d)   What are the names of three to five leading journals in your field?
   - (e)   What periodical indexes are available for your field? Which ones have abstracts?
   - (f)   What is the accepted style manual for your field?
   - (g)   List three other useful reference works specialized for your field.
   - (h)   Create a preliminary bibliography on your research topic. You need to include perhaps two books, five articles, and the names of two local experts you can ask for interviews.

A ➤ indicates a case study exercise.

2. Find a periodical index that is appropriate to your field of study and includes abstracts. Use it to make a priority-ordered list of the five articles you listed in the previous question—the most appropriate, the next most appropriate, and so on.

3. Write a brief (100-word) report on the extent to which your library makes computerized data base searches available to students and on whether such a search would be advisable for your research project. Be sure to consider key factors (beyond availability), such as the appropriateness of your project for such a search and the cost of the search versus its likely benefits.

➤ 4. Choose one of the following topics and prepare a preliminary bibliography for a research report on it. Your audience should be a person interested in possibly investing in that company or industry.

> Johns-Manville and asbestos health hazards
> Apple Computer's management-succession problems
> Lawsuits concerning the Dalkon shield
> Marketing strategies cola companies use to achieve Number One status
> Procter & Gamble's logo problems
> The economics of space exploration

5. Find a copy of the reference style guide for publishing a journal article in your field (see page 434). Make two copies of the "Notes for Contributors" page—one for yourself and one to turn in to your teacher. Note: if the page you find refers you to some multipage document (ASME Report 4, for instance), copy only that page, not the entire document.

PART FIVE

Applications

As the list on the next page shows, there are clearly almost as many kinds of documents as there are writing situations. Faced with this variety, it would be difficult to say any one or two or three kinds are the most important. Thus, Part Five focuses on presenting key elements common to large numbers of such documents—elements that will enable you to respond to each situation appropriately.

Chapter 17 presents a few kinds of documents to add to those discussed previously in this book and introduces the key concept of *discourse communities* as one of the tools for selecting among these many different kinds of documents. Chapter 18 explores one very important kind of document, the proposal, in greater detail. Chapter 19 takes a longer look at one of the most commonly used formats, the long (or major) report. And Chapter 20 explains the basics of making oral presentations and poster presentations.

The Myth of the Perfect Model. Each of the four chapters in Part Five presents patterns and outlines for various kinds of documents. People use patterns like these to assist them in their writing every day. Remember to use all such models and patterns intelligently. In the kinds of complex writing situations that characterize professional life, it doesn't make much sense to use any prefabricated model unchanged from its textbook configuration. To use such a model intelligently, you must consider the important ways in which each specific audience, purpose, and message, as well as your own role as a writer, probably requires alterations in the model. And as your needs and abilities as a writer grow and change, you should find yourself increasingly modifying and combining the patterns this book (or any other textbook) presents. Remember that behind any intelligent use of the

(*continued on p. 444*)

advertisement
analysis of complaint
annual report
answer to a request
article for periodical
brochure
bulletin
business book
catalog
claims adjustment
collection letter
complete periodical
computer documentation
consumer guide
cost estimate
cover (transmittal) letter
credit denial
credit report
design report
employee reprimand
estimate and appraisal
evaluation of facilities
evaluation of personnel
follow-up letter
forecast
grievance

housekeeping memo
house organ
industrial handbook or manual
inspection report
invitation
lab report
letter changing price
letter confirming an order
letter of agreement
major expenditure request
management newsletter
memo for the files
minutes of meeting
monthly or quarterly report
news article
newsletter
periodic report
planning report
policy/procedure bulletin
position paper
preliminary report
press release
procedures
productivity study
progress report
projection
proposal
recommendation
report
request for references
request for information
request to deviate from policy
safety reminder
safety report
summary
survey report
10-K report
thank-you letter
training manual
trip report
trouble/accident report

patterns there will always be principles—such as audience analysis and adaptation—that are more important than the patterns themselves.

Here are four of the most important guidelines you can keep in mind as you work to make intelligent use of the patterns and models this book presents:

1. Always have a specific reader in mind. Be careful not to overestimate that reader's knowledge of (or interest in) your subject.
2. Ensure that your document's purpose is explicitly stated and that every part of the document contributes to fulfilling that purpose.
3. Reread your document with a critical eye before you submit it; clarify any passages in which your reader may misunderstand your meaning. Do not rely on your reader's interest in your subject or goodwill toward you to motivate him or her to fight through needlessly confusing, dense, or dull writing.
4. Ensure that the final version of your document is written in a style and presented in a form that will meet or exceed your reader's expectations.

17. *Varieties of Documents*

For professionals who write as an important part of their jobs, each day brings new problems to solve on paper and new demands on writing skills. For as many situations as there are that can be handled with routine kinds of writing, there are also many situations that require unique kinds of documents whose formats, structures, and approaches are tailored specifically for each particular situation.

The first four parts of this book have demonstrated a number of different kinds of documents: various kinds of business letters, product descriptions, user manuals, abstracts, executive summaries, and so on. This chapter shows you a few of the more common forms that reports and report-type documents may appear in, explains two additional types of documents (the periodic activity report and the lab report), and notes three of the most important factors writers use when deciding what kind of document a particular task requires.

1. *Report Forms in General*

Reports can appear in many different forms. Three of the most frequently used are discussed here: the letter report, the fully compartmentalized report, and the formal report.

1.1   Letter Reports

One of the most convenient ways to present a short report is called the *letter report*, which is used frequently with reports that are only a page or two long. If the entire document (letter plus report) will be more than three or four pages long, you should probably use a different format. (For example, rather than put the report inside the letter, use a cover letter and accompanying report, as in the formal report described later in this chapter.) But if you want a convenient way to put a short report on paper, you may want to consider the form known as the letter report. Figure 17.1 shows the conceptual outline of a letter report.

The letter report contains all of the typical elements of a business letter, but it has a short report sandwiched between the letter's opening and closing paragraphs. The letter should satisfy all of the requirements for an effective letter (covered in Chapter 4), and the report needs to satisfy all the requirements for a good report (covered in the introduction to Part Five).

Strengths and Weaknesses. The letter report form makes your letter and your report inseparable. Its advantage over simply writing a long business letter is that you can change your tone and stance within the document. The letter part can be reasonably personal and sometimes even a little judgmental, while the report remains impersonal and objective. In fairly informal situations, such as when there will probably be only one reader, the letter report offers a convenient format. Figure 17.2 shows such a case; here the letter report is a progress report, but contained within it is the early draft of a technical report.

The letter report's advantage over the letter-plus-report is also its weakness. If the report will need to be passed on through an organization, the letter report format may not be right. You may say things in the letter parts of the report that would be inappropriate for readers other than the one to whom the letter is addressed. This can also apply if the report will be kept on file; you may not want the letter's life to be that long.

1.2   Fully Compartmentalized Reports

A report form that is growing in popularity for all kinds of reports is the fully compartmentalized report (also called the programmed report or the modular report). Its main distinctive feature is that there is a systematic relationship between the length of the report sections, the use of visuals, and the places where page breaks occur (places where the reader has to turn the page). Figure 17.3 shows the conceptual design of a fully compartmentalized report.

Figure 17.1 The Conceptual Outline of a Letter Report

```
                                            Inside heading

    Inside address


    Subject line


    Salutation

        Introductory paragraph of letter——may be written in a
    less formal style than that used in the report proper.

    First heading of report
        First paragraph of report

    Report continues one or two pages, with more headings.
                                .
                                .
                                .

                    [Continuation of page format]
    Last paragraph of report.

                    *   *   *   *   *   *


    Concluding paragraph of letter. Again, the tone may be
    somewhat less formal than in the report proper.

                                            Signature block
```

Figure 17.2   **A Typical Letter Report**

Dear Fred:

Here are the preliminary findings for the report on personnel requirements, 1995–2000. Take a look at the areas I've covered and tell me
if I've left out anything major that you think we'll need.

Background

Table S.1 shows the employment of technical personnel in the energy program at the present time. Engineers make up about 60% of
the total professional employment but only about 30% of the total
doctoral employment. In Table S.2 these data are broken down by
budget category.

[Tables omitted]

These data on current employment were used to model staffing
patterns . . . .

[Two pages omitted]

Conclusion

This report suggests that we foresee several kinds of personnel problems. Spot shortages of highly specialized, experienced workers in
fields such as cryogenics and superconducting magnets are expected. Temporary shortages caused by sharp increases and de-
creases in the program's required staffing will probably occur. General shortages in certain disciplines, such as nuclear engineering
and computer science, are almost certain. And the university programs in energy technology may require additional support.

For a long-range activity like this one, these can be viewed as ordinary operational problems that can be expected to be adequately
managed, given the energy program's already demonstrated sensitivity to personnel and education issues.

*   *   *   *   *   *

That about wraps it up. Anything else you want covered in the
final report, let me know in the next week or so. Otherwise I'll proceed to write it up as outlined here.

Sincerely,

David

Figure 17.3 Conceptual Design of a Fully Compartmentalized Report

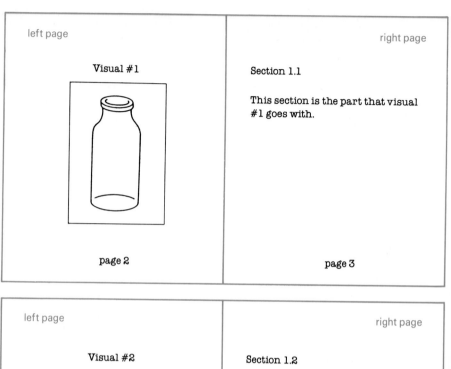

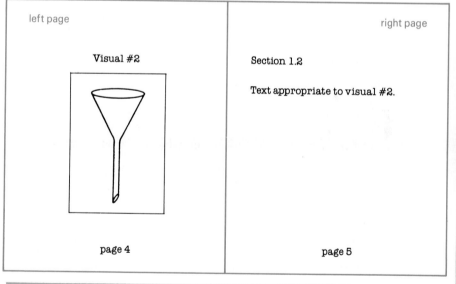

This pattern goes through the entire document. That is, each new section of text starts a new page, and most, if not all, of those sections have visuals of some sort associated with them. Thus, the reader is always looking at the appropriate visual at the same time as he or she is reading the matching text. When a page must be replaced (as in the looseleaf binding that is commonly used in repair manuals), there is usually room for the text on that page to be longer or shorter without materially affecting the rest of the report. With this form you lose the economy of having every page filled with text; instead, you have lots of short pages (pages with several inches of extra white space). What you gain is a one-to-one match of visuals and text, a match achieved effortlessly, and the ability to change sections, to let them grow (or shorten) without having to change the rest of the text.

1.3   Formal Reports

Another alternative to a letter report is the formal report; what you may have known up to now as the "term paper" format is a recognizable derivative of the format widely known as a formal report. Chapter 19 discusses this format and all its parts in complete detail, but you'll understand more about the varieties of reports presented here if you know a little about the format of a formal report. Here is an outline of the parts of a very complete formal report:

Letter of transmittal
Distinctive cover
Executive summary (or abstract)
Title page
Distribution list (optional)
Contents page
List of figures and illustrations (may be two separate lists)
Introduction
Body of the report
Conclusion
References
Appendixes (if needed)

Of course, not all formal reports have all these parts, and a few have more.

2.   *Additional Types of Reports*

There is no comprehensive and authoritative list of all the types of reports, or of the standard types of reports, nor is there agreement among authorities about what should be the "basic" kinds. The types presented in this book are among the ones that occur most often, but do not be surprised if someone someday asks you to write a type of report not covered here. The only thing authorities agree on is that, however the names may change and come into and out of vogue, the only really new form of report is the proposal. Because proposal writing is so important, it is given its own chapter (Chapter 18). Here are descriptions and examples of two additional types of reports.

2.1 Periodic Activity Reports

Periodic activity reports (also called progress reports) give employers, customers, supervisors, or investors necessary information about how work on a particular project has gone. Such reports can be turned in at intervals determined by time (such as weekly or monthly) or by task (completion of planning, completion of preliminary site work, and so forth).

Organization. Your reader's purposes should help shape the organization of your progress report. Three elements usually appear in progress reports:

1. An explanation of the relationship between the part of the project this report covers and parts earlier reports covered.
2. An explanation of the project's recent developments.
3. At least some mention of the project's next phases.

These three elements—the look backwards, at the present, and to the future—are present one way or another in most progress reports. But they appear in so many different ways that to be more specific about them would be misleading. Some reports carry all three, some only two, and some only one.

Functions. Besides keeping others informed of your progress, progress reports force you, the writer, to focus systematically on how your project is going. Progress reports can also form the first draft of your final report. In many cases progress reports are the only contacts between the person doing the work and the one for whom the work is being done. In view of all these functions, the length and organization of your progress reports are especially significant.

Length. Most readers expect progress reports to be short. But you can't decide what "short" means without some fairly precise knowledge of your reader's purpose in requiring the report. Is the report totally routine, done solely for record-keeping purposes? Or is the report significant in that it will contribute to future decision making, maybe even decisions about whether to continue to support the project? To know what "short" means, you have to know what your reader expects in the report and how your reader will use it. Only then can you tell what level of detail to use, what to include, and what to exclude. As in other kinds of writing for business and industry, concern for your reader's purpose should shape the way you write progress reports. Figure 17.4 shows a typical progress report, written regarding the student report at the end of Chapter 19.

2.2   Lab Reports

This section describes elements common to reports of routine laboratory testing. When your goal is to report on original research, you would normally use a much fuller format, such as the one presented in Chapter 19, "Written Reports."

Pattern for Lab Reports. Many times routine laboratory work is written up on preprinted forms. There's not much to say about using such forms, except to remind you that when you need to go into more detail than the form allows, you can nearly always attach a supplemental letter or short report to the standard form. When a preprinted form is not used, you are usually given a set of headings, such as the ones in the following list, to structure your report around. You can cover other areas as well, but you'll usually be expected to include at least these:

> Purpose
> Results
> Discussion
> Procedures

or

> Purpose
> Procedures
> Results
> Discussion

Figure 17.4   A Typical Progress Report

This report presents my progress on the English 461 major report, "Economic Implications of Recombinant DNA Research,"
from the time of our last conference (Jan. 20, 1993) to this
date (Feb. 15, 1993). At that time, we agreed that the report
would have two major sections: the first on current research
and recombinant DNA technology, and the second on the economic implications of that research and technology. The research for those two sections is well under way; a major
change in the report's focus, if you approve, will be to add
a new section at the front of the report that presents the
basics about recombinant DNA.

Research is well under way on the section on current research and technology regarding recombinant DNA. Attached to
this report is a list of books and articles that have been
consulted so far, as well as a second list of sources yet to
be read. The material on economic implications presents no
problems at all; it is quite accessible and easy to read. The
material on current developments in research and technology,
on the other hand, is somewhat harder to find and much harder
to understand. Even for a person with a college-level background in biology and a special interest in genetics, the typical article in, for example, Biotechnology is nearly impossible to understand, much less interpret.

The extremely technical and difficult nature of literature
on current research and technology in recombinant DNA has two
implications for this project: (1) it will not be possible to
give quite so full a presentation of the current research and
technology as originally conceived, and (2) I propose the addition of a new first section to the report, one that presents the basic elements of DNA recombination. Thus, the report will have three sections, proceeding from the basics of
recombinant DNA to a section on the current state of research
and technology (slightly shorter than originally proposed but
easier to understand because of the presence of the new Part
One), to a section on the economics of recombinant DNA. For
an executive/layperson reader, this three-part structure will
make the report much easier to understand.

The remaining work to be done mostly involves the exploration of other sources (government documents and interviews,
especially) and the actual writing of the report. With your
approval of the changes described here, writing can begin almost at once. Current plans are to finish writing by Feb. 23,
keyboard on the 24th and the 25th, and have ample time left
over for careful proofreading before the due date (March 1).

Purpose of the Test. Why is this work being done? To determine the compression strength of a particular sample, to determine the cause of failure of a certain structural element, to isolate and identify a particular enzyme? Be explicit in your lab report about why the test is being done. That way, when your reader sees your report, he or she will not think you did more than you actually did or mistakenly fault you for doing less than you actually did.

Most readers will check to see whether your purpose and results are on the same scale. For example, if you set out to determine why a certain sample of steel pipe has stress fractures, you can legitimately conclude something like, "The stress fractures are caused by prolonged exposure to (a particular chemical) under (a particular pressure)." You *cannot* conclude, however, that is the reason a certain pipeline exploded. Your lab report might well be a part of that larger investigation into the pipeline's explosion, which would then become a long report of the sort shown in Chapter 19, but you still have to keep the scope of your results comparable to the scope of your purposes.

Results of the Test. Many readers like to see *results* right after *purpose* and will use what you say there to decide whether to read your procedure. Others will feel the order of purpose-procedure-results (which mirrors the steps the experimenter goes through) should be kept intact. You, as the writer, have the choice of using a reader- or writer-based structure. You should make the decision about which structure to use based on the nature of your subject, your purpose, and especially your audience and their purpose in reading. An expert may be as interested in the procedure as in the results, so in writing for an expert audience you would put procedure *first*. An executive/layperson might be interested in procedures only if something in the results piques that interest, so in writing for that audience, you would keep procedure *last*. When in doubt, you're probably better off to stay with the more traditional pattern of purpose-procedure-results.

Be careful how you phrase results. Let the reader know explicitly when you are merely reporting the test's results and when you are drawing conclusions from those results.

Discussion of the Results. If you want to add more information to the report that does not spring directly from the test itself, you can add "Discussion of the Results" as an optional section. Here you would discuss any anomalies or unexpected results. You might, for example, mention how the results fulfill (or fail to fulfill) the test's purpose or recommend whether and which further tests should be performed. Here you can also enlarge on the implications of the test results.

Procedures. Under "Procedures" you might write a narrative description of the tests that were run and under exactly what conditions. Use visuals to help explain what was done, but remember that you have to explain in words all of the content of the visuals. You cannot just say, "I placed the sample on the plate as shown in Figure 3"; you must first explain how the sample was placed on the plate, and then refer the reader to Figure 3.

General instructions for writing up procedures in laboratory reports are the same as those for processes, described in Chapter 13. Be specific, and be sure you don't lose your reader in the process. You probably know all of the steps in the process before you begin the report, but your professional reader probably doesn't. Those steps should be forecast at the beginning of the procedures section. Break the test into steps, forecast them, and use subheadings to help your reader follow the steps. Most important, be specific: *exactly* what materials were used, *exactly* what machine, *exactly* how long, and so forth.

Style in Laboratory Reports. Most writing works better if it uses present tense and active voice and includes personal pronouns. That is, in most situations most readers would rather read, "Next I place the sample on the plate" or "Next I placed the sample on the plate" than "Next the sample was placed on the plate." Although present tense, active voice, and personal pronouns can make laboratory reports much more readable, the tradition requiring past tense, passive voice, and omission of personal pronouns is still quite strong. It can be persuasively argued, on historical evidence, that past-passive-impersonal is *not* the essence of scientific style (*clarity* is); on psychological evidence, that past-passive is harder to read and increases reader errors; and, on philosophical evidence, that impersonal style falsifies the experiment's reporting (because the experiment was conducted by a human agent, the report should show it). The discussion of objectivity and advocacy in Chapter 3 is also relevant here. As the example of objective style on page 70 shows, such writing does not need to be past-passive-impersonal. Nevertheless, the majority of readers of lab reports seem to want past-passive-impersonal, forcing the majority of writers to deliver it. The person writing a lab report is usually not in a position to select the style it is to be written in. As a writer, be aware of the problem, and do your best to give your reader the best possible report within the bounds of what your reader will accept.

Figure 17.5 shows a fairly typical laboratory report. But remember that, other than preprinted forms, there is no one standard format for laboratory reports. Generally, each laboratory has its own form for such reports.

Figure 17.5   **A Typical Laboratory Report**

CONTROLS FOR ENZYMATIC AND NONENZYMATIC BROWNING

In food systems several types of browning reactions occur
which may be favorable depending on the food product being
produced. Browning reactions are enzymatic or nonenzymatic
in nature. Enzymatic browning occurs in some fruits and
vegetables when they are cut or damaged. Nonenzymatic
browning includes carbonyl—amine browning, caramelization,
and ascorbic acid browning. In this experiment the factors
that inhibit or control these reactions were demonstrated
in apples, chocolate cakes, and caramels (Campbell et al.,
1979).

Method
Slices of apples were treated with different solutions to
illustrate the inhibition of enzymatic browning. The treatments used are listed in Table 1. Untreated cut apples
served as the control. The samples were evaluated one hour
after treatment.

Table 1. Treatments used on apples to retard enzymatic
browning.

Treatment

A. No treatment[a]

B. Water

C. 50% sucrose solution[b]

D. Ascorbic citric acid solution

E. Lemon juice

F. 0.05% KH sulfite solution

---

[a]The control

[b]50g sugar dissolved in 50ml water

Five chocolate cake batters were prepared to show the
effects of pH on carbonyl—amine browning and the crumb
color of the baked product. The proportions of the leavening agents, baking soda, baking powder, and buttermilk
were modified in each batter to yield a different pH con-
centration. Two of the cake batters were neutral, one was
acidic, one was basic, and the final one was extremely al-
kaline. The formulas for the cake batters appear in Table
2. The procedure for mixing was that of Campbell et al.

(1979). The pH of the cake batters was determined with pH
paper. The color of the batter was also noted. The cakes
were evaluated on the basis of crumb color, grain, volume,
and flavor.

Table 2. Formulas for cake batters with different pH.

| Ingredient | Formulas | | | | |
	Acidic	Neutral[a]	Neutral[b]	Basic	Alkaline
Cake flour, g	100	100	100	100	100
Sodium bicarb, g	—	— —	0.7	1.3	5.0
Baking powder, g	5.0	5.0	2.2	—	—
Salt, g	1	1	1	1	1
Cocoa, g	18	18	18	18	18
Sugar, g	150	150	150	150	150
Shortening, g	66	66	66	66	66
Eggs, g	100	100	100	100	100
Milk, g	—	68	—	68	68
Buttermilk, g	68	—	68	—	—
Vanilla, ml	2	2	2	2	2

[a]Neutral with 5.0g baking powder and no baking soda.
[b]Neutral with 2.2g baking powder and 0.7g baking soda.

Caramels were made using two different sugars, sucrose
and glucose, to demonstrate carbonyl–amine browning. The
basic formula for caramels appears in Table 3. The formula
containing sucrose was heated to a boiling point of 120°C,
whereas the formula with glucose was heated to 138°C.

Table 3. Basic formula for caramels.

Ingredient	Quantity
Sucrose[a]	200g
Margarine	14g
Milk	127ml
Half and Half	110ml

[a]Repeat substituting glucose for sucrose.

Results
All the solutions used to treat the apples were effective
in stopping enzymatic browning. The apples treated with
lemon juice were slightly yellow while all the others re-
mained white (excluding the control).

The evaluation of the cake batters and cakes appears in
Table 4. The pH readings for the supposedly neutral batter
containing 0.7g baking soda and 2.2g baking powder and for
the batter with 1.3g baking soda are both too acidic. The
inaccuracy in pH may be attributed to human error when taking the pH. The crumb color of the cakes turned darker as
the pH increased.

The caramels containing sucrose were lighter in color
and hard in nature. The ones produced with glucose were
darker in color with a more desirable texture.

Table 4. Evaluation of chocolate cakes based on batter color,
batter pH, crumb color, grain, volume, and flavor.

| Characteristic | Formulas | | | | |
	Acidic	Neutral[a]	Neutral[b]	Basic	Alkaline
batter color	light	medium	medium	medium	darker
pH	5.5	6.4	5.5	5.5	8.0
crumb color	light	light	darker	darker	darkest
grain	poor	excellent	very good	good	poor
volume (cm)	5.5	6.5	7.0	5.0	5.5
flavor	good	very good	excellent	good	good

[a]Neutral with 5.0g baking powder and no baking soda.
[b]Neutral with 2.2g baking powder and 0.7g baking soda.

Discussion
Enzymatic browning is inhibited by inactivating the poly-
phenoloxidase (the enzyme responsible for the reaction),
altering the substrate, or eliminating oxygen. Polyphenoloxidases, present naturally in fruits and vegetables,
transform the polyphenolic compounds (in fruits) in the
presence of oxygen to quinones. The quinones polymerize,
and thus the fruit turns brown (Campbell et al., 1979). In
this experiment the water and the 50% sucrose solution
treatments prevented browning by eliminating the fruit's

contact with oxygen. Ascorbic citric acid and lemon juice
retarded browning by producing a pH medium (pH 4 or less)
in which polyphenoloxidases do not function. The final
treatment of 0.5% kH sulfite solution stops the browning
process by repressing the enzyme system (Campbell et al.,
1979).

The darkening that occurred in the cakes is primarily
due to the increase in pH by manipulation of the leavening
agents. Basic pH mediums increase crust browning as carbonyl—amine browning is accelerated (Campbell et al.,
1979), and they alter pigments found in cocoa. Thus, a
darkened and red crumb color results (Lee, 1975).

Carbonyl—amine browning results from the condensation
of a carbonyl group of a reducing sugar and an available
amino group, followed by a series of reactions that ultimately produce brown pigments known as melanoidins. High
temperatures, high pH's, and low moisture levels accelerate the reaction. Sucrose was ineffective in producing satisfactory caramels since it is not a reducing sugar and
does not participate in carbonyl—amine browning. This accounts for the light product produced. Glucose, on the
other hand, is a reducing sugar, participates in carbonyl-
amine browning, and consequently darker caramels resulted.
The hard consistency of the sucrose product (caramels) is
due to excessive crystal formation since sucrose is unable
to invert (Campbell et al., 1979).

Conclusion

The results of this study have shown that enzymatic and
nonenzymatic browning can be inhibited or controlled in order to produce a desirable food product. Conditions that
affect browning include pH, the availability of oxygen,
and inactivation of enzyme systems.

References

Campbell, A.M., Penfield, M.P., and Griswold, R.M., 1979.
"The Experimental Study of Food." Houghton Mifflin Co.,
Boston. Lee, F.A., 1975. "Basic Food Chemistry," p. 331.
Avi Publishing Co., Inc.

3. Creating Formats

There are conceivably as many kinds of documents as there are recurring situations in business and industry. Many times you may have to construct the format from scratch, with no routine or model to work from. The next three sections offer guidelines for situations in which you have to create your own formats. The first section explains the concept of *discourse communities* as used in the field of rhetoric today, a concept that can be especially useful when a writer is trying to determine the particular look or sound a document needs to have. The second section discusses one-of-a-kind documents, and the third deals with creating new formats for documents that will become routine.

3.1 Reading the Community of Discourse

When writers try to make decisions about various aspects of documents they're working on, often a key factor in the decision is an audience-based one. Whether you're trying to decide about using first-person pronouns (still frowned on in some circles), about integrating visuals and text or keeping them separate, about using an executive summary or just an abstract, or any of a number of different factors, over and over again some facet of your audience is a critical consideration in the decision making. Treating all audiences the same is clearly unacceptable; on the other hand, treating every audience and every communication situation as completely unique may involve more decision making than you want to do every time you write something. There is an alternative between these two positions.

Many times it is possible to put various audiences into groups defined by the communication channels and messages they share. For example, aeronautical engineers and scientists who are interested in new uses of ceramic materials (such as in thin-wall engine castings) tend to constitute an identifiable communication group, or *discourse community,* defined by the journals, newsletters, conventions, books, and networks of professional friends and acquaintances they share. Countless discourse communities exist within professional life and public life as well. Stamp collectors, Honda Gold Wing owners, and organic gardeners are all members of their own various discourse communities. In professional life, the members of each discourse community—defined and united primarily by communication—tend to expect the professional documents they read to look and sound a certain way. Once you figure out how a piece of writing needs to be done for a particular community, you can take it as a starting point in designing another document for other members of the same group on the assumption that they will want the same kind of look and sound that the first one did. Once you've figured out how to do a report for, say, one U.S. senator, you may well have a good beginning understanding of how to do a report for another one. Once you've figured out how to do a brochure for one group of state-employed social workers, you may well have a good beginning understanding of how to do a brochure for another such group. And once you've figured out how to do a proposal for one National Cancer Foundation review board, you probably have a good pattern for how to do the next one.

Thus, when it comes to creating either tailor-made documents or new routine formats for documents, looking at the likes and dislikes, customary document forms and practices, of the discourse community involved is a good place to start in making decisions. Even if you haven't worked for this exact audience before, so long as you've worked with members of that community you should have a good idea of the kind of writing you need to be doing.

3.2 Creating Tailor-Made Documents

When you need to write a document without a model, you can easily construct an outline. To do so, combine three lists:

1. The list of topics you want to cover.
2. The list of questions your reader will want answered.
3. The list of topics that need to be included for record-keeping purposes.

Combine items from the three lists when you can, and then organize the resulting list into a coherent structure, one that makes sense from the reader's point of view.

Consider the following example: Suppose you're the Minnesota/ Wisconsin regional manager for XYZ Kitchen Implements, assigned to write a report for the national office of a situation that occurred in your region. Jones Cafeteria in Duluth, a large and successful cafeteria in a major suburban shopping center, uses six of your machines, worth more than $150,000, including a food slicer. On January 7 they reported to your office in Minneapolis that their slicer had stopped working, and you dispatched a service representative, who reported by telephone that a new motor was needed. He told Jones Cafeteria it would take three to five days to get a new motor from Chicago, and they agreed to slice by hand for that time, a process that uses more food (the slices aren't nearly as thin) and takes much more time. Then the serviceman went on vacation, you were out for three days with the flu, and Chicago somehow lost the order. Nine days later you got a furious call from Jones Cafeteria, realized what had happened, and frantically called Chicago. They said they could have a new motor in Duluth in one day, so you got another one-day extension from Jones Cafeteria. The motor didn't show up. So twelve days after they reported the breakdown, you personally delivered to Jones Cafeteria a new slicer to use free of charge while you personally see to getting theirs fixed. They were happy to have a functioning slicer again but still understandably angry about the poor service they received.

Now you have to write a report to the marketing vice-president in the national office about the whole situation. Here are the three lists you come up with:

Topics You Want to Cover
What happened—background
What you did
What went wrong
How you dealt with it
How to prevent future occurrences

Topics the Reader Wants
What happened—background
What went wrong
Analysis of why the problem occurred
How it was handled
How to prevent future occurrences

Topics for Record Keeping
What went wrong
How it was handled
How to prevent future occurrences

Here is the composite list of topics, which will become the outline of the actual report:

The Background
The Problem
   What Went Wrong
   How It Was Handled
   Analysis of Why the Problem Occurred
The Future
   How to Prevent Future Occurrences

In a similar way, you can make a structure for anything you need to write, even when you cannot find a model.

3.3   Creating Routine Formats

You may be called on to create a format to be used regularly for a certain kind of document in your organization. Usually this will mean someone is looking for a one- or two-page form to use every time a particular situation comes up. The process you should follow is an extension of the one described in the preceding section. Begin by making three lists (topics you want to cover, questions your reader will want answered, and topics you need to include for record-keeping purposes). Combine and organize those three lists into one form, just as was done in the preceding example. But the next step is different. Through your company's records, trace the accounts of previous similar situations, and try writing them up on the draft of your form. You can also test your form by sharing it with others who have dealt with such situations in the past. A third alternative is to monitor the first uses of the newly designed routine form, to continue to refine it through its first year of use. Your goal should be a form that is short, easy to use, and designed to prevent its users from making mistakes in its use. To accomplish the goals, design the form carefully and test it thoroughly.

EXERCISES

1. Write a progress report describing your progress on the major report you are writing for this class.

2. Write a short (200–250 words) progress report to your academic adviser detailing your progress toward a degree.

3. Find a lab report you wrote for another class, and rewrite it for the students and teacher in this class to read. Pay particular attention to the ways this class's setting, purpose, and audience require changes in your original report. Be sure to consult with your instructor about the desired length.

4. Write a short (4–5 pages) report on one of the topics listed in the Exercises at the end of Chapter 16. The audience is laypeople.

➤ 5. You are a student intern in a local telephone company office. Your first task is to design a form to use in handling customer complaints. The office already has such a form, but your supervisor doesn't want you to see it "because it's so bad you'll do better just to start from scratch."

A ➤ indicates a case study exercise.

➤ 6. Consider again the Jones Cafeteria situation described on page 462. Put yourself in the place of the regional manager for XYZ Kitchen Implements, and design and write a report to the home office summarizing what happened and recommending action to prevent similar problems in the future. Invent details such as names and dates if you need to, but try to stay within the situation as described in the text. *Note:* You may want to come up with a different pattern for your report than the one suggested in the text. Whatever pattern you choose, be prepared to explain your choice in class.

➤ 7. You are doing volunteer work for a local organization dedicated to convincing the public that threats to public health and safety from high-technology industry, from incinerators, from hazardous waste dumping, and so on need to be dealt with by improved communication on both sides rather than polarization. Your task is to take the relevant material from Figure 19.9, a student report entitled, "A Consumer's Guide to Risk Communication," and plan a brochure to be given to consumers. The purpose of the brochure will be to help consumers do a better job of evaluating the risk communications they encounter. You need to produce a mock-up of what you want the brochure to look like and to use the information in the report to give you the material to write the body copy.

18. *Proposals*

What mechanism connects people who can do a job with the people who need the job done? In business, in science, or in engineering, your way of communicating with the people you want to work for will usually be a *proposal.* For example, if your company does business with the federal government, you may begin projects by responding to a request for proposals

(RFP) in the *Federal Register* or *Commerce Business Daily*. If you are an architect preparing a design for a new city hall, a researcher requesting funding for a new project, a production or maintenance engineer suggesting a way to deal with an engineering problem, or a student suggesting a term paper project to a teacher, the proposal is probably the required format.

There are many different varieties of proposals. They may vary in length from short (a page or less) to long (400–500 pages for a proposal for a major government contract). And proposals can address any kind of audience: an engineer proposes a new maintenance procedure to management, a manufacturing company proposes a new product to its customers (who may be private individuals or the staff of a government agency), or a research scientist proposes federally funded research in a document that will be reviewed by grant administrators and other research scientists.

Proposals may also be *solicited* (in response to a request) or *unsolicited*. As in unsolicited letters of request (such as job applications), the unsolicited proposal usually emphasizes reader benefits from the start much more strongly than does the solicited proposal. And as in unsolicited letters of request, unsolicited proposals have a much lower rate of return.

1. *Basic Elements of Proposals*

Despite all the different varieties of proposals, there is a set of elements common to nearly all of them. When you write a basic proposal, you should choose its elements from the list in Figure 18.1. Of course, depending on your topic, purpose, and audience, you may omit some elements or change their order.

Before going into a more detailed discussion of the typical parts of proposals, it is useful for you to see a sample. Figure 18.2 shows a student's proposal for her major report project in a typical advanced technical writing course. As you can see, this student's proposal follows very closely the outline suggested in Figure 18.1. In many cases the writer will expand some areas of the outline and condense others, and, of course, proposals written by practicing professionals are usually considerably more complicated. But the parts, with minor changes, are approximately the same.

1.1 Introduction

The opening sections of a proposal typically explain the motivation for the project and give a brief overview of it. Write these opening sections as though your audience will make up its mind on the basis of them alone. Although we all hope that every line of any proposal we write will be read and considered carefully, the opening section establishes in the reader a tendency to approve or disapprove the rest. Its importance cannot be overstated.

Figure 18.1   Elements of Proposals
Depending on whether your proposal will be long or short, you would go into each section in more or less detail. The order of elements may change from that presented here.

> **The Introduction:** Play your three strongest cards.
>> *Need:* What is the problem? Explain why the project needs to be undertaken.
>>
>> *Goals:* What is the solution? What, specifically, do you want to accomplish?
>>
>> *Benefits:* What is to be gained? Remember to take the reader's point of view.
>
> **The Body of the Proposal:** Tell your entire story.
>> *Detailed description:* Explain fully what you intend to do and how you intend to do it. In a five-page proposal, this might be three or four pages.
>>
>> *Timetable for completion:* You can use increments here based on project tasks ("site preparation," "preliminary excavation," and so on) or time periods ("Week 1," "Week 2," and so on).
>>
>> *Facilities and equipment:* Explain what facilities and equipment you will use, with specific attention to what is already on hand and what needs to be acquired.
>>
>> *Budget:* In most cases there will be a prescribed form for the budget. Whatever form you use, check your figures and your math very carefully.
>>
>> *Personnel:* Explain who will do this work. What are these people's qualifications, relevant experience, and references? If you need to include résumés, they should be in an appendix (appropriately cross-referenced in this section).
>
> **The Conclusion:** Typically, the conclusion will be a restatement of the proposal's needs and benefits. Reemphasize these points:
>
> Why should this be done?
> Why are you and your firm the ones to do this work?
> What will this project's benefits be?
> What are the project's longer-term or larger implications?

Figure 18.2   A Student's Proposal for a Major Report

Proposal for a Major Report

In recent years, people have become very concerned with
avoiding hazards. There is an increased expectation now
that the public will be informed of known or suspected
health hazards, such as radiation or chemical exposure.
For this reason, government and industry are now more
likely to communicate with the public concerning risks to
public health. Risk communication is increasing, and a recent study by the National Research Council has identified
a need for the recipients of risk messages to become better informed about risk communication. In response to your
request for a major report, I propose to write a report
that provides important information that we, as the public, need to know about risk communication.

The goal of this report will be to orient the consumer
to the uses and abuses of risk communication, to enhance
the consumer's ability to participate in the risk communication process. After defining and explaining risk communication, the report will focus on ways the consumer can
evaluate both risk messages and risk messengers. It will
also advocate active participation of the interested public in risk communication. The report will increase your
awareness of risk communication as an interactive process,
and provide both reasons for becoming an active participant and methods for doing so.

Detailed Description
    Content. Following an introduction to risk communica-
tion, the major report will explain the common cognitive
liabilities that impair our understanding of the primary
measures of risk. Next, a section on evaluating the purpose of the risk message will provide ways to identify misleading or missing information. Following this, the report

1

will provide information on recognizing credibility or advocacy problems on the part of the risk messenger. A final
section will briefly address incentives for participation
in the risk communication process, ways to participate effectively, and methods of obtaining more information about
the risk.

    Sources. Improving Risk Communication, written by a com-
mittee for the National Research Council, makes a recommendation that a "consumer's guide" to risk communication
should be written. According to Dr. Milton Russell, who
served on the NRC committee, the proposed guide is under
consideration by several research groups, but has not yet
been written. Because Improving Risk Communication contains many recommendations for the recipient of risk communication, I plan to use this 330-page book as the main
source for recommendations about consumer evaluation of
risk communication. I have gathered general information
about risk communication from four other book sources, one
government publication, and seven periodical sources.
These sources will provide the information and concrete examples necessary to explain such an abstract subject. They
will also provide some of the visuals for the report. One
remaining challenge will be to develop additional visuals,
using either a conceptual approach, or illustrations from
additional sources. Two additional periodicals that would
contain such illustrations are Risk Analysis and Science,
Technology, and Human Values.

Personnel

    As a Liberal Arts student I have successfully written
analytical research papers, and therefore I am qualified
to do the research necessary to complete the major report.
I have also had recent experience in writing a long analytical research report. For a political science class last
semester, I analyzed the conflict with Iraq, integrating
information from 30 sources, and received an "A."

2

Facilities and equipment

This report will require library research. In doing the preliminary research for this report, I have used Hodges Library and also the Oak Ridge Public Library, and have determined that adequate resources are available on this topic. I have already obtained twelve sources and have a list of five additional articles on risk communication (in technical journals) that are available at Hodges Library.

I will have adequate equipment to produce this report, because I have recently purchased a computer, word processing software, and a printer.

Specific plans and schedules

Jan. 10   Review technical communications periodicals.

Jan. 15   Attend meeting of the STC re: risk communication. Locate and photocopy periodical information and obtain a book source on risk communication. Begin research.

Jan. 22   Interview Amy McCabe (at your suggestion), and obtain two additional book sources and one EPA publication.

Jan. 29   Prepare a draft of a report defining and describing risk communication. Obtain a peer review of the report and make revisions.

Feb. 5   Submit definition report for editorial review.

Feb. 12   Revise definition and description report, and continue research on risk communication as a process.

Feb. 19   Prepare a draft of a report on risk communication as a process, and obtain a peer review. Revise the report and submit it for editorial review.

Feb. 26   Revise process description and obtain major book source to develop focus for the major report.

Mar. 5   Continue notetaking on sources and begin sorting notes by topic.

3

Mar. 12   Take new notes on prior sources, focusing on providing a guide to risk communication.

Mar. 18   Sort all research notes by topic, make a tentative outline for a consumer's guide to risk communication. Write a draft of that report.

Mar. 26   Obtain additional source material concerning research on risk communication, and take notes. Interview one of the authors of <u>Improving Risk Communication</u> to find out the status of other possible sources. Revise the macro-structure of "consumer's guide."

Mar. 29   Meeting to get preliminary approval for major report topic. Draft proposal following approval of topic.

Apr. 2   Review previously used visuals and develop new visuals to illustrate the "Consumer's Guide" portion of the major report. Develop cover or title page art. Begin adapting information from definition and process reports for inclusion in the "Consumer's Guide" report.

Apr. 9   Draft letter of transmittal, abstract, list of illustrations, introduction and conclusion. Adapt working outline of report to create table of contents. Integrate appropriate information from definition and process papers into the major report, and revise macrostructure. Begin revisions on microstructure.

Apr. 16   Complete revisions and edit the report. Obtain photocopy of proper style sheet and edit references accordingly. Get peer review and revise. Obtain brown envelope and assemble notes and copies. Edit front matter. Label proposal as an appendix. Obtain an editorial review of the entire report. Create title page.

Apr. 23   Make final corrections, print final copy of report and assemble in folder, with illustrations.

Apr. 29   Turn in completed report and research materials.

4

Conclusion

As a technical communicator, your knowledge of risk communication is likely to be knowledge from a writer's perspective. The major report I am proposing will be speaking
to the recipient of risk communication, because in the future you are increasingly likely to encounter risk communication from a recipient's perspective. Your approval of
this topic will give you exposure to some interesting information on cognitive liabilities and risk. It will also
increase your awareness of risk communications as a dialogue and provide practical methods for participating in
risk communications. If you have any questions about this
proposal, I will be happy to answer them.

5

Need. If your proposal is short, all of the opening sections (need,
goals, benefits) might well combine to form an introduction to the
proposal. If the proposal is long, in addition to these opening sections
being separate, the proposal may well have a separate executive summary of the sort described in Chapter 7.

The "need" section explains what the occasion for the proposal is.
Is it in response to a request, provoked by a problem, or made possible by the availability of new techniques and procedures? What brings
this proposal up, to this audience, now? (Rhetoricians call this element
exigency.) This section also makes explicit any earlier connection you
might have had with the people or agency you are writing to. Have
you worked for them before, talked with them about the project, read
a brochure they sent out describing the problem?

Goals. The second part of the proposal, still often within the introductory sections and thus still relatively short, consists of a *brief* overview
of your project, expressed in terms of its goals. The way you explain
the goals should not only convince the audience to accept your proposal but also show them that you understand the project's requirements (how long the project is to last, other limitations, criteria, and
so on). Get to the essence of the problem, to show your reader you
have carefully thought through exactly what should be done. Don't
limit yourself to echoing the words of the announcement you are

answering; don't run the risk of suggesting that all you have done is read the announcement. Show quite clearly that you have understood the problem behind the announcement and have devised a solution.

Benefits. The last of the proposal's opening sections should list the benefits to your reader from accepting (agreeing with, funding) your proposal. What, specifically, will result? You may want to use a short-term, long-term approach here.

These three introductory sections should be short, perhaps three paragraphs in a three- to five-page document, or a page or two in a ten- to twenty-page proposal. Many times these sections will be done as lists, with each item introduced by a number if the ordering of items is significant, or some typographic convention such as a bullet (•) if the ordering is not significant. Figure 18.3 shows the introduction section of a student's term paper proposal.

1.2   Body

The body of your proposal should go into more detail than the introduction, and it will thus be longer. If the opening section has gained your reader's interest, then this section will have a chance to do some convincing. The detailed description, timetable, explanation of facilities and equipment, budget, and list of personnel constitute the body of your proposal. Make these sections complete and specific. Although you may have used some generalizations in the introduction, use specifics in the body.

Detailed Description. In the detailed description, explain the project's *scope*, in terms of both its time and its tasks. When would it start and finish? What event will mark the project's beginning and ending? Explain what *methods* you will be using. Will they be traditional or innovative? Another important part of many proposals is an account of how you propose to solve the problems you encounter within the project. What preliminary problems do you anticipate, and how do you intend to solve them? How have you broken down the separate *tasks* within your project, and how do you expect to accomplish each?

For many research proposals, this opening section also includes a review of (or at least reference to) earlier studies or approaches. This may be a review of the background of the problem, or of previous approaches to solving the problem, or both. The detailed description should also include a schedule for your preliminary and progress reports (and an indication of what their distribution will be). Longer proposals may well include a significant section here exploring the

Figure 18.3   An Example of a Proposal's Introduction

PROPOSAL FOR TERM PAPER

Computers play a major role in today's technologically advanced society. Use of computers for design, manufacturing, distribution, and management continues to grow at an
amazing rate. As more people's jobs involve using computers for several to many hours each day, the subject of the
efficiency of the computer system's interface with the
user becomes more and more important. In response to your
request for a term paper, I propose to write mine on the
human-factors approach to designing computer systems.

The goal of this project will be to identify the ways
in which the new field of human-factors engineering is being used to improve the quality and efficiency of interactions between people and computers. After summarizing the
growing importance of human-factors engineering on such areas as computer languages, programs, and systems, the report will focus in on the physical relationship between
the video-display terminal (VDT) and the operator, with
specific attention to avoiding possible health hazards for
operators and promoting ease of use.

The report's benefits include increasing your own awareness and knowledge of an area that affects all of us who
are involved with computers on a daily basis. For you, the
report will bring you up to date on the most recent research about what constitutes the most efficient configuration of operator, keyboard, screen, lighting, and seating.
For me, the report lets me increase my knowledge in a
field that may well be an important part of my career.

problem this project is supposed to solve (see Chapter 14) and recommending the proposal's course of action as the solution (see Chapter 15).

Timetable. You also need to include a detailed time schedule for the project. Its increments can be defined in terms of *project tasks* and *milestones* (such as "land acquisition," "site preparation," "preliminary excavation," and so on) or in terms of time periods ("Week 1," "Week 2," and so on).

Facilities and Equipment. Discuss in this section the facilities and equipment that will be used in your project. What facilities and equipment are available, and what remains to be acquired? Will the equipment have to be leased, bought, or designed from scratch? Who will pay? And who will get the equipment at the end of the project?

Budget. As you might expect, the budget is an important part of your proposal. Depending on how long, detailed, or expensive your proposal is, your budget section may well need a short summary in words in addition to a detailed tabular presentation. You may also want to break the budget down in several ways—certainly by year but possibly also by source of funds (such as requested funds, matching funds, etc.). Whatever other parts of your proposal may or may not be examined closely, you know your budget will be, so make it a careful and visually attractive presentation. Even if all you are doing is one simple table for the budget, remember to make sure that each horizontal row is accurately totaled in the right-hand column, that each column is accurately totaled in the bottom row, and that the bottom row of totals and the right-hand column of totals add up to the same amount. Showing the readers of your proposal that you are careful and thorough may make the difference between having your proposal accepted and having it rejected.

Personnel. The final part of your proposal's body should focus on the people who will be working on your project. Who are your personnel, and what are their qualifications? What previous experience on similar work do the people in your group have? What are the references for the people on your team? What is the match-up between your people's qualifications and the approach you will take?

Be careful not to load up the proposal's body with résumés. A few very short biosketches of key personnel may be acceptable in the body; anything longer should be in an appendix.

Figure 18.4   The Conclusion of a Student's Term Paper Proposal

> •
>
> •
>
> •
>
> CONCLUSION
> Your approval of this report will mean we can both learn
> more about a topic that will affect the lives of every
> worker in the United States in the decades to come. I will
> gain valuable experience in writing the kind of report my
> future jobs will require, and you will be able to evaluate
> my skills in writing on a significant topic at some
> length. The project awaits only your final approval to
> proceed.

1.3   Conclusion

At the end of your proposal, you can take one more opportunity to convince your reader that you are the right person for that job. A restatement of needs and benefits usually will function as your proposal's conclusion. It is one last chance to convince your reader that you have anticipated and can deal with foreseeable problems, to restate the merits of your project (or your approach), and to urge the reader to take the action you're requesting. Figure 18.4 shows the conclusion to the student proposal whose introduction you saw in Figure 18.3.

Not every proposal will have all the parts mentioned here; many big proposals will have more. Every time you write another proposal, you should expect to make slightly different decisions about which elements to include or to exclude and the order in which to put your elements.

2.   *Writing Major Proposals*

While students frequently do their proposals as solo efforts, in professional life proposal writing is nearly always a group effort. (Chapter 11 discusses writing as part of a group in more detail.) Another important difference between student and professional writing is the way the writing of major proposals is especially driven by criteria established externally That is, most requests for proposals (RFPs) establish a fairly detailed list of points that the subsequent proposal *must* deal with if it is to be considered at all.

For example, one government agency recently specified these criteria for a particular round of proposals:

- Must deal with a population base of at least 5 million people.
- Cannot request more than $500,000 per year, for maximum of five years.
- Cannot be more than 100 pages, excluding appendixes.
- Cannot be permanently bound.
- Must be double spaced.
- Must demonstrate uniqueness of project in that area.
- Must differentiate strictly between goals and objectives.
- Must provide matching funds from other sources of at least 33 percent of the total budget.
- Must be submitted in five copies to the specified office no later than twelve o'clock noon on the specified date.
- Must conform to the following outline:

  Title page
  Title of project
        Principal investigators' names and signatures
        Names of sponsoring organization(s)
        Date of submission
  One-page summary
  Goals
  Objectives
  Plan of work
  Personnel
  Budget

In fact, from the earliest planning meeting, the team responsible for writing and preparing the proposal identified a number of other criteria that they also set for themselves:

- Must demonstrate no government funding in the past has ever been spent in this region for this kind of use.
- Must show that this organization already has a demonstrated track record of service and accomplishment in this kind of work.
- Must not only tie each goal to a specific set of objectives but explain how the meeting of those objectives will be measured.
- Must demonstrate that this region is especially in need of these services.
- Must emphasize the benefits, both short and long term, to the people of this region that will come from the funding of this proposal.

Because so many such concrete and detailed demands are made of proposals and because so many proposals involve more than just one person in the research, writing, and production of the documents, the way you go about working on a major proposal needs to be a little different from the way you would go about most other writing projects or even most group writing projects. With so many detailed criteria to meet and so many people involved in the project, it is easy for criteria to fall through the cracks or for whole sections not to do what they should. The way to avoid having these kinds of problems, problems that effectively kill too many proposals, is to create and use a *proposal compliance matrix.*

2.1   The Proposal Compliance Matrix

The proposal compliance matrix is a tabular presentation (usually done as a spreadsheet though sometimes as a wall chart) that shows who on the proposal development team is responsible for each part of the proposal and for each point of compliance, when that part or point is to be provided, and where in the finished proposal that part or point will appear. There are a number of possible layouts for proposal compliance matrixes; Figure 18.5 shows the matrix (somewhat abbreviated) for the government proposal described above.

Depending on the length and complexity of the proposal and the size of the development team, a matrix can become very intricate. But no matter how involved the matrix becomes, it's vital for it to reflect in precise detail the aims of the project and its development team's assignments. Notice two of the items in Figure 18.5, "Our demonstrated track record" and "Explain how objectives are measured." How well—precisely and fully—do those capsule statements reflect the real goals of the team? The first ("Our demonstrated track record") omits the key phrase "in this kind of work," and the second ("Explain how objectives are measured") omits the vitally important element of "tie each goal to a specific set of objectives." These two elements are important and, depending on the quantity and quality of communication within the proposal development team, may well get lost in the shuffle if the matrix is left as Figure 18.5 shows it. The first ("in this kind of work") can probably be simply added to the boxed capsule statement. The second ("tie each goal . . .") probably needs to be a separate item on the matrix because of its importance.

2.2   A More Complex Proposal

The proposal structure this sample matrix was designed for is quite simple:

> Title page
>> Title of project
>> Principal investigators' names and signatures
>> Names of sponsoring organization(s)
>> Date of submission

Figure 18.5   A Proposal Compliance Matrix (Somewhat Abbreviated)
Checkmarks show work has begun; dates show completion.

Compliance point	Handled by	Researching	Writing	Reviewing	Completed	Section and Page Nos.	Checked
Title page and signatures	Bob						
Summary	Bob	Completed 1/15	Completed 1/20	✓			
Goals	Sandra	Completed 1/15	Completed 1/20	✓			
Objectives	Judy	✓					
Plan of work	Sandra	✓					
Budget	Phil	✓					
No government funding spent here in the past	Sandra						
Our demonstrated track record	Sandra						
Explain how objectives are measured	Judy						
Demonstrate this region's need	Sandra	✓					
Short-term benefits	Phil	✓					
Long-term benefits	Bob						

One-page summary
Goals
Objectives
Plan of work
Personnel
Budget

And the team working on the proposal (Bob, Sandra, Judy, and Phil) is relatively small as such projects go. Imagine how complicated the proposal writing process could become, and how much more important a carefully designed matrix would be, if the team comprised 25 people and the proposal outline looked like this:

Title page
      Title
      Submitted by
      Submitted to
      Project period
      Funds requested
      Project director

Proposal Abstract

   I.   Introduction
       A.   Background of the Organization
          1.   History and location
          2.   Purpose of organization
       B.   Identification of the Problem
          1.   How the problem relates to the organization's purpose
          2.   Documentation of the problem's significance
  II.   Statement of Need
       A.   Summary of proposed action
       B.   Listing of objectives
 III. Plan of Action
       A.   What you plan to do
       B.   How you plan to do it
       C.   Personnel
       D.   Facilities and Equipment
  IV. Evaluation—How Will These Factors Be Measured?
       A.   Accomplishment of objectives
       B.   Monitoring of projected actions
       C.   Attribution of causality [how do you know it was *your* activities that solved the problem?]
       D.   Further actions [depending on the outcome of your work, what further activities might be justified?]

E.   Reporting [to whom and on what schedule will
interim progress reports and final reports be
sent?]
V.   Plan for Future Actions
VI.   Budget
A.   Direct costs
B.   Indirect costs
C.   Matching funds
D.   In-kind contributions
E.   Fringe benefits

It's always tempting in a complex situation to assign each section
to a different person (or group of people) and thus simplify
the compliance chart considerably. The problem with that plan is
the same kind of problem as the ones pointed out on page 478: the
various sections of the proposal and the points those sections make
need to be as seamless a fabric as possible. If (as is likely to happen
with too fragmented a development effort) the organization's history
and purpose don't seem to match the project's purpose, the plan of
action doesn't seem to match the described need very closely, the
evaluation methods don't seem to match the proposed objectives
tightly, the personnel qualifications don't seem to fit with the proposed task, and so on, the proposal stands a good chance of failing.
Using the fully developed, detailed, and specific compliance matrix
is the best way to ensure that a large team produces a major proposal
that reads like the tightly woven, carefully thought-out document you
want it to be.

3.   *Proposals as Problem-Solving and Recommendation Reports*

Comparing proposals to problem-solving/recommendation reports
may help you to understand the internal logic of proposals better.
Although not all proposals are done in response to problems, viewing
proposals that way for a few minutes may help you to appreciate what
proposals are all about. One clear comparison between proposals and
problem-solving/recommendation reports involves the introduction's
three sections—need, goals, and benefits. In the "need" section, explain what the problem is: Has water quality in Town Lake deterio-
rated to the point that fish kills are becoming commonplace? Has new
technology made it possible to retrieve and repair malfunctioning
orbiting satellites? Has the increase in suspended particulate matter
in the atmosphere made studies of particulate deposition on tree
leaves in the mountains necessary? The problem should be something
that the reader will either recognize immediately or that he or she

can be educated to appreciate relatively quickly. Making it that way is a matter of careful writing on your part.

If the "need" section explains *what* the problem is, then the "goals" section discusses *how* to solve the problem: Do you propose a study of upstream sources of pollution, with an eye toward increased regulation of their discharges into the lake? Do you propose a space shuttle mission to retrieve and repair hitherto-inoperative satellites? Do you propose establishing measurement techniques and base information for levels of particulate matter deposition on tree leaves at various elevations and locations in the mountains? Your reader will look particularly carefully at the match (or lack of match) between what you propose to do and what the reader has requested be done. If the government has requested proposals for a new plane to replace the T-28 basic jet trainer, you cannot submit a proposal for both a new plane and a new fleet of aircraft carriers from which to launch them.

The final section of your introduction, "benefits," tells the reader how your solution of the problem will help him or her. The introduction already has sections on what the problem is and how it should be solved; this section is on *why* it should be solved. Remember to look at the benefits from the reader's point of view (just as you did in the "reader benefits" part of Chapter 4 on business letters). There may well also be benefits for you in the project, and in some cases you may make them explicit, but they must take a definite back seat to the benefits for the reader.

If the first three sections of the proposal (need, goals, and benefits) explain *what* the problem is, *how* it is to be solved, and *why* it should be solved, then what does the rest of the proposal do? In terms of our problem-solving example, the rest of the proposal presents specifics to support the generalizations made in the introduction. Write the first three sections so well and so strongly (though not at such length) that by the time the reader has finished the introduction, he or she is already looking favorably on your request. If the problem-solving essence of the introduction has been done properly, the remainder of the report simply fills in the details.

Many recommended proposal outlines, such as the major proposal outline on pages 480–481, provide a section or sections for both the analysis of the problem (there, it is Section I.B and Section II) and the recommendation (Section III in the major proposal outline). It's important to remember that these two elements in your report—the problem analysis and the recommendation—need to be tightly connected in terms of their logic. As Section 2.1 of Chapter 15 explains, a fully developed three-step logic is required in such situations. That is, you don't want to say in one place,

Town Lake is polluted primarily by untreated sewage

and then in another place,

> We recommend solving the pollution of Town Lake by building additional overflow diversion channels and settling basins between the sewage treatment plants and the lake

without making the *connection* between those two elements clear:

> Town Lake is polluted primarily by untreated sewage that overflows from treatment plants and sewers during heavy rains; we do not need to spend the major funds required to expand the everyday capacity of the sewage treatment plants because 95 percent of the time they in fact have excess capacity. What is needed is a way to handle the overflow caused by occasional (three or four times a year) heavy rains. The relatively inexpensive tactic of building additional overflow diversion channels and settling basins between the treatment plants and the lake will handle all such situations very well.

It's easy for proposal writers to deceive themselves into thinking, "Well, if the people who read this proposal just think about the problem for a minute or two they'll see for themselves how our analysis of the problem connects with the solutions we propose." That is too often wishful thinking. For the key points in your proposal, you need to make the connections between the evidence you supply and the conclusions you draw from it quite explicit. Usually that means using the kind of three-step logic demonstrated here and explained in Chapter 15.

4. Two Common Problems with Proposals

One of the most common problems with proposals has already been discussed in this chapter: the longer the proposal and the more people working on it, the greater is the tendency for important parts and points not to fit closely together *in the reader's mind*. Researchers and writers who live with their subjects for months or years often carry their work's important elements of coherence and cohesion in their minds very strongly, but those same people often fail to make those elements explicit on paper. There are three ways to deal with this problem; doing a thorough job of audience analysis (see Chapters 1–3), using a three-step logic (see Chapter 15), and employing a detailed and specific proposal compliance matrix (described earlier in this chapter). Two other major problems occur often enough in proposals to be important to describe here: writer-based proposals (versus reader-based proposals) and "blue-sky" proposals (versus substantive proposals).

4.1   Writer- Versus Reader-Based Proposals

When a proposal's recommended outline begins with something like this

I. Introduction
   A. Background of the Organization
      1. History and location
      2. Purpose of organization
   B. Identification of the Problem
      1. How the problem relates to the organization's purpose
      2. Documentation of the problem's significance

it almost invites writers to give in to the natural tendency to produce writer-based, catalogical writing.* For example, an engineer who may be struggling to get started writing will find it quite easy to retrieve a written history of the company from another document, merge that with a statement of the company's purpose from another document, and then derive the "identification of the problem" sections from the company's preliminary problem-solving reports on the technical question at hand. This common technique produces a proposal whose very important opening sections are a kind of a collage of pages and paragraphs taken from other sources and roughly stuck together. The need for the proposal to be as *seamless* as possible has already been violated. Beyond that, in doing what may be easiest (and hence best in the short term) for the writer, such a writer is doing what probably is hardest and worst for the readers: *all that readers need or want to know about the history of the organization is the part of the history that relates to this project.* That's also all the readers want to know about the purpose of the organization. And the readers only want to know as much about the identification of the problem—especially the evolution of your organization's understanding of it—as is necessary to make your solution make sense.

While proposals often contain significant sections that trace a company's history and that detail how a particular technical problem was analyzed and solved, there is an important limiting factor on the length and the kinds and amount of detail in such sections. Everything you say about those topics needs to connect directly to this proposal and this project. Otherwise, your reader will be strongly tempted to move on to the next proposal in the stack (remember

---

* These terms have been explained in previous chapters: *catalogical* and *analytical* in Chapters 11 and 14 and *writer* versus *reader based* in Chapters 12 and 15. Additionally, the closely related concept of *system-* versus *task-oriented structures* is explained in Chapter 13, and *research* versus *report structures* are explained in Chapter 16.

there are *always* other proposals in the stack) to see if it might be just a little bit more to the point.

4.2 "Blue-Sky" Versus Substantive Proposals

In some senses proposals are like job application letters; the writer is trying to sell something. What would you, as a prospective employer, make of this kind of job-application letter:

```
Dear Prospective Employer:

I expect to graduate soon from State University with a de-
gree in general studies. I would be interested in knowing
if you have any employment possibilities for me.


Sincerely,
```

Obviously, few people would send out such a letter, and fewer still would consider responding positively to it. What's wrong with it? It's almost totally lacking in substance. Professionals call such a letter a *blue-sky* application—it's made up out of thin air and nothing more. The same distinction—between being blue-sky and being substantive—applies to proposals. Unfortunately, the blue-sky proposal is all too common. Unlike the blue-sky application letter, the blue-sky proposal typically has many pages, pretty pictures, flashy printing, and expensive binding, but it still has almost nothing to say in any kind of substantive sense.

Consider again the sample student proposal shown in Figure 18.2. Compare it with one that does not forecast the eventual report in detail, does not list potential sources of information, offers no personnel qualifications, fails to show the availability of appropriate resources to carry through the proposed work (here, the library facilities), and gives a schedule that looks like this:

Week One—Do library research.
Week Two—Begin rough draft.
Week Three—Revise rough draft.
Week Four—Edit rough draft.
Week Five—Turn in report.

By this comparison, the sample proposal at the beginning of this chapter is thus said to be *substantive* and the one described here *blue-sky*.

People who regularly read and evaluate proposals (and job applications) are especially sensitive to this distinction. Obviously there is a continuum between substantive and blue-sky proposals, and no professional would turn in a proposal that was totally devoid of substantive content. One cannot build a bridge out of the breeze, and few professionals will try to fake their way through the rigorous kind of scrutiny most proposals receive. The more common manifestation of this problem is proposals that *fail to give the appearance* of substantive content because the writers omit important details—details they may in fact have at hand but fail to include. The erroneous assumption seems to exist in the minds of some proposal writers that the readers will "read into" the proposal occasional missing bits of data, will see between the lines things like narrowly defined objectives when in fact the proposal's stated objectives are quite broad and very fuzzy, and will generally make better sense out of the document than it in fact makes on its own. Nothing could be further from the case. Proposal readers are usually skeptical, critical readers. Where holes exist, such readers are far more likely to shine a spotlight on them than gloss over them; where details are fuzzy, such readers are far more likely to see them as *really* fuzzy; and where material doesn't make sense, such readers are more likely to go to the next proposal in the stack.

There are two ways to avoid running into this problem:

1. No matter how good a writer you are or how sophisticated a desktop publishing setup you have access to, do not expect to get a proposal accepted unless you have a well-developed idea at the heart of it.
2. If you do have a good idea at the heart of your proposal, make sure you have the evidence—hard data, expert testimony, solid arguments, and so on—*in the proposal* to support that idea. Do not rely on your reader's goodwill or kind and gentle heart; proposal readers are tougher than that.

5. Guidelines and Checklists for Proposals

Often the RFP itself will give important guidelines for proposals written in response to it. Figure 18.6 is the RFP to which the sample student proposal in Figure 18.2 responded; notice the guidelines it contains.

In a similar way, many RFPs in professional life contain within them various guidelines for proposal writers. Often the RFP attracts letters of interest from prospective proposal writers, who receive in response full guidelines for that particular proposal from the agency

Figure 18.6   RFP for a Student Project

REQUEST FOR PROPOSALS FOR RESEARCH PAPERS

Proposals are being solicited requesting approval of topics for research reports on a question of significance in
the proposer's field of study. Such reports are an important way people assess professional knowledge and technical communication skills, and they also constitute an important learning mechanism for writers as well as readers.
If your proposal is accepted, the <u>deliverable</u> will be a report written according to the specifications detailed in
"Major Report Guidelines."
 Deadline for submission is Nov. 1, 1992. Proposals are
to be between 500 and 750 words long, double spaced, with
a cover sheet, introduction, detailed description (including a tentative outline of the proposed report), conclusion, and at least one visual. Your proposal should identify (at a minimum) the question(s) your report will seek
to answer, the purpose of the report, the tentative content of the report, and its audience. (For further information on the content of your proposal, see the relevant
chapter in *Effective Professional and Technical Writing*.)
 Proposals will be evaluated on both the quality of writing they display (it is seen as an important indicator of
the quality of the eventual major report) and their substantive content. Proposals that give the impression of
lacking substance––that sound like important elements are
made up out of blue sky––will be returned for revision and
resubmission.

or company that initiated the RFP. Figure 18.7 shows two fairly typical RFPs.

5.1   National Science Foundation Checklist

Many large organizations that regularly receive hundreds of proposals publish general guidelines for proposal writers. Here is a fairly

Figure 18.7 **Two Sample RFPs.**

Regional Contracting Department, Naval Supply Center, Oakland, CA 94625-5000

66—LABORATORY AND OFFICE SYSTEM FURNITURE Sol N00228-92-R-3041. Due 030992. Contact Bid Room 510/302-4238. Requirements contract to design, manufacture, install quality and functional interior furnishings for the laboratory, clinic and administrative offices, (i.e., Pharmacy, Laboratory, Radiology. Emergency, ICU, etc), at Naval Hospital, Oakland, CA. The contract will be one base year with four (4) option years. (017).

70 General Purpose ADP Equipment Software, Supplies and Support Equipment, incl Leasing

Computer Sciences Corp., 16511 Space Center Blvd., Houston, TX., 77058, NASA, Johnson Space Center, Prime Contract No. NAS 9-17920

70—SOFTWARE MAINT FOR THE AMDAHL 5890-600E S/N 10832 Sol 92-11. Due 013192. Contact Carey Joan White/Purchasing Agent, by FAX (713)280-3853. No phone calls please. Computer Sciences Corporation proposes to place a contract against GSA ADP Schedule Contract No. GS00K91AGS5895-PS01 with Grade 500 Oracle Parkway M.D. 15-10 Redwood Shores, CA 94065, for the purchase of proprietary software licensed on the Amdahl 5890-600E consisting of the following product numbers: 1, SQL*REPORTWRITER 2, SQL*FORMS 3, SQL*PLUS 4, PRO*PLI 5, RDBMS 6, CASE*DICTIONARY 7, SQL*QMX 8, 3270/NET. This acquisition is supported by an approved Justification For Other Than Full and Open Competition for specific make and model. Vendors who can furnish the required equipment are invited to submit a written substantive statement clearly stating their ability to fill this requirement. Oral communications are not acceptable. Substantiveness would be indicated by a statement of exactly what equipment is being offered, firm pricing, and delivery schedule. (Delivery must be received at the above address 30 days after receipt of the contract). FOB point, installation and delivery charges, if any, and other information which shows a bonafide ability to meet this specific requirement. Vendors responding to this notice with GSA Schedule Contract, include contract number, and expiration date. No solicitation will be issued, nor will a contract award be made on the basis of responses received to this notice, since this notice cannot be considered a formal solicitation document. Written responses must be submitted within 10 days of this notice. If no written responses are received an order shall be placed in accordance with the terms of the above referenced schedule contract. When a response is received from a schedule Vendor for an item(s) that meet the requirement at the lowest overall cost, an order will be placed against the ADP Schedule Contract. When a response is received from a non Schedule vendor that meets the requirement and an analysis indicates that a competitive acquisition would be more advantageous to the Computer Sciences Corp., a formal solicitation will be issued. The above is in accordance with FIRMR 201-32-206. All sources may submit an offer and it will be considered by this Corporation. (016)

Source: Commerce Business Daily, Issue No. PSA-0517, January 24, 1992.

representative checklist for proposal submission, taken from the application guide for National Science Foundation (NSF) grants for research and education in science and engineering. The guidelines presented here have been slightly edited for this book's audience.

____ NSF cover form
____ NSF cover sheet
____ Appropriate boxes on cover sheet checked
____ All required signatures
____ Full text certification
____ Human subjects certification, if required
____ Special provisions for research in Greenland or Antarctica
____ Animal care and use statement, if required
____ Table of contents
____ Project summary
____ Results from prior NSF support
____ Detailed description of proposed activity
____ Bibliography of pertinent literature
____ Statement of the impact of the proposed research
____ Eligibility or other special program-required statement
____ Vitae of all senior personnel
____ List of up to 5 publications most closely related to the proposed work and up to 5 others, including those in press, for each investigator
____ NSF budget form

___ Brief description and justification of major equipment requested

___ Current and pending support

___ Description of available facilities and major items of equipment to be used

___ Required number of copies of the proposal, including the original signed copy

___ Residual funds statement

___ Proposal packages properly addressed

___ If proposal length exceeds 15 pages, justification has been discussed with and approved by appropriate program officer.

___ Additional documentation (letters of commitment, eligibility statements, etc.), as required

5.2 Commercially Prepared Questions and Checklists

In a similar way, there are commercially available lists of questions and checklists for preparers of proposals to use. Here is a list of seven questions from *The Winning Proposal: How to Write It,* Herman Hultz and Terry Schmidt (New York: McGraw-Hill, 1981). These are questions that preparers of proposals need to make sure to answer explicitly and convincingly:

1. Do you fully understand the project's problems and needs?

2. Are you expert enough at whatever skills and technologies are needed to furnish the planning and performance that will satisfy the project's needs and solve the project's problems?

3. Will you provide fully qualified staff people to do the work?

4. Have you done such work successfully before?

5. What, specifically, do you promise to deliver?

6. Do you have a track record of success to prove that you can and will deliver?

7. Can you prove your abilities at all of the above?

Another, more complicated, checklist is offered in Norman J. Kirtiz's "Guide to Proposal Writing," from the Grantsmanship Center in Los Angeles. The checklist is abbreviated here:

Summary

• Appears at beginning of proposal
• Is brief
• Is clear
• Is interesting

Introduction

- Clearly establishes who is applying
- Provides evidence of applicant's accomplishments
- Leads logically to the problem statement
- Is as brief as possible
- Is interesting
- Is free of jargon

Problem Statement or Needs Assessment

- Is stated in terms of client's needs and problems, not the applicant's
- Makes no unsupported assumptions
- Makes a compelling case
- Is as brief as possible

Objectives

- At least one objective for each problem or need
- Objectives are outcomes, not methods
- Time required for each objective to be accomplished is specified
- Objectives are measurable

Methods

- Flows naturally from problems and objectives
- Clearly describes and states reasons for activities
- Presents reasonable scope of activities

Evaluation

- Presents plan for evaluating and plan for modifying evaluation methods if necessary
- Clearly states criteria of success
- Explains any test instruments, questionnaires, or methods of data analysis to be used
- Describes any evaluation reports to be produced

Further or Other Necessary Funding

- Presents specific plan for obtaining any necessary future funding
- Has minimal reliance on future grant support

Budget

- Tells same story as proposal narrative
- Is detailed in all aspects
- Includes all items
- Includes indirect costs where appropriate
- Is sufficient to perform the tasks described

6.   *Examples of Proposals*

This section contains three sample proposals to consider in addition
to the sample student proposal in Figure 18.2: a sample proposal
written within a company (as an *internal proposal*, it's simpler than
those described in the rest of this chapter), a proposal written by a
freelancer for an engineering firm, and part of a professional proposal.

6.1   An Internal Proposal

TO: J.T. Lomax

FROM: Susan Jones

SUBJECT: Discussion of 8/14/92 (recarpeting of P-3)

DATE: 8/18/92

The carpeting in P-3 should be replaced as soon as possible. Its poor
condition creates several safety hazards in this heavily traveled area,
and its continued deterioration (despite repeated requests to have it
replaced) adversely affects the morale of all the employees who
work in that room. Two alternatives exist for replacing the carpet:
new carpet or new tile. Given that the tile under the current carpet
is already broken beyond repair and would have to be replaced, a
procedure more expensive than putting down new carpet, I recommend we replace the old carpet with new carpet.

The current torn carpet presents two kinds of safety hazards. In
many places the carpet is pieced together, with the seams joining
the pieces directly under desk chairs. Several of those seams have
opened up, creating a situation in which chair legs (or casters) and
employees' feet can catch in the carpet. A fall in this room, with its
many desks, tables, and dividers, could lead to serious injury. The
second source of hazards is the many large wrinkles in the carpet.
Again the potential exists for falls that could injure employees.
These conditions become worse each day.

The tile under the current carpet is already broken and patched,
so we cannot simply remove the carpet. Between new tile and new
carpet, the advantage is clearly with new carpet. It is quicker and
cheaper to install, and it offers other advantages (such as sound ab-
sorption) not offered by tile.

According to Maintenance, recarpeting will cost $18.75 per square yard (installed), for a total of $1518.75. Because the carpeting can be done in one day, we can do it on a Saturday and thus cause a minimal disruption of the work in this room. Retiling, on the other hand, will cost $5500 for labor and materials alone. It would require three days to do, resulting in a total cost (including materials, labor, and lost time for the work that should go on in that room and cannot be moved) of almost $7500.

6.2   A Freelancer's Proposal

Proposal for Preparing a Style Guide for XYZ Engineering

Based on our discussion last Thursday, May 20, I would like to propose the preparation of a report style guide to be used specifically by XYZ Engineering in the preparation of their reports for commercial clients. This style guide will promote more efficient production of reports, reduce the cost of those reports, and improve their quality.

Brief Overview

Technical reports prepared by XYZ Engineering are written by nuclear engineers as the result of their analyses of nuclear power plants and chemical refineries. Customers often judge the engineering accuracy of the report's results and recommendations by the writing quality of the report. Sloppiness, inconsistencies in format, stylistic flaws, and grammatical errors detract from the report and lower the firm's credibility in the customer's eyes. The elimination of such problems will be the major goal of this report guide. Secondary goals include standardizing the process of document production and establishing a clear system of document review for quality control.

I plan to use three major sources for this guide. Many of XYZ's engineers are graduates of State University and are familiar with the *University Thesis and Dissertation Manual*. To the extent that XYZ's current reports have a model, it is that document, which I will use as a basis. Supplemental information will come primarily from two sources. Because many of our reports are done for the National Laboratory, I will consult the Laboratory's *Technical Reports Preparation Manual* when necessary. Additional stylistic information will be taken from *The Chicago Manual of Style*, perhaps the most widely used style manual.

The style guide will have three parts:

- A full description of the document production and review process, by task and time.

- Guidelines for the treatment (including format and placement) of each component of a report, from the title page to the list of references.

- A handbook with rules for consistent punctuation, abbreviation, capitalization, pagination, typing of equations, enumeration, and referencing in reports from XYZ Engineering.

Timetable

This style guide should take fifteen weeks to complete, followed by a two-month trial period, and two weeks of revisions. If I begin work on it June 1, you should have a trial document on or about September 15, and a finished document in use December 1.

Personnel and Facilities

Because I worked as Office Manager for XYZ Engineering for two years before returning to school to work on my Master's degree, I am already familiar with the nature of these reports. I understand that as part of this agreement I am to have after-hours and weekend access to XYZ's computer workroom and all paper supplies needed for this project. I have enclosed a copy of my résumé for your review.

Both the State University *Manual* and the Laboratory *Manual* are currently undergoing revision. I have contacted the head of each committee to be assured of current information concerning any important changes in those documents.

Fee

As we discussed, my fee for this work will be $2500.00, payable upon your acceptance of the finished, trial-tested document.

Conclusion

A report-writing style guide for XYZ Engineering will make your reports more efficient and less expensive to produce while improving their quality. If you have any questions about this proposal I will be happy to answer them. I look forward to providing this service for XYZ Engineering.

6.3   Part of a Major Proposal

The next few pages contain several parts of *The Metro-Area Cancer Information Service,* a professional proposal written from a hospital to a funding organization. Because the entire proposal is over 300 pages (of which all but the first 99 are in appendixes), contained in a three-ring binder with a four-inch spine, it's not possible here to give you more than a sense of the tone and style of it. Here's the outline of the proposal:

> Cover Page
> One-Page Summary
> Introduction
> > Background of organization
> > Aim of organization
> Identification of Problem
> > Relationship to organization's purpose
> > Significance of problem
> Statement of Need
> > Goals
> > Objectives
> Program of Work
> > Overall plan
> > Methods
> > Personnel
> > Facilities
> Evaluation
> Future Directions
> Budget

Following is part of the introduction, part of the identification of the problem, part of the goals and objectives, and part of the program of work.

INTRODUCTION

Today five million Americans who have at one time or another been diagnosed with cancer are still alive, and the number who are considered cured grows every day. Although cancer will still kill nearly five hundred thousand Americans this year, the number who are cancer survivors proves that early diagnosis and prompt treatment are keys to lowering the death rate and increasing the survival rate. And the key to early diagnosis and treatment is education.

For ten years, the Central County Cancer Information Service (CCCIS) has played a key role in public and professional education promoting the early diagnosis and treatment of cancer. In those ten years:

* Telephone specialists have responded to over 5,000 calls about cancer and cancer-related problems.

* More than 100 cancer-related educational programs have been presented to a variety of targeted audiences.

[Two pages omitted]

IDENTIFICATION OF THE PROBLEM

When people know more about cancer, when they receive early diagnosis and prompt treatment, cancer deaths are prevented. The key elements in accomplishing this prevention are education and communication. As the public increases its knowledge about healthier lifestyles, cancer screening, up-to-date methods of treatment, and other available support services, the chances that greater numbers of those diagnosed with cancer will also be survivors also increases.

For the over one million people in the five-county metropolitan area with Center County at its heart, the CCCIS is the only major vehicle for cancer information. Due to budgetary limitations, the CCCIS has been able to serve only Center County residents. With the additional funding from the American Cancer Institute (ACI) requested through this proposal, the CCCIS will be able to double the size of its population service base, bringing to an additional 500,000 people the important information and services that are essential to increasing the cancer survivor rate.

To this point, no ACI-funded offices exist in the five-county area. Yet in counties such as Harper, West, and Pike, with a total population of over 250,000, the people have higher than average risk factors, incidence, and mortality for certain types of cancer. Each of these counties has large numbers of people who are poor, very rural, 65 years of age or older, and unemployed; all of these are factors that increase cancer risk.

[Three pages omitted]

Goals

The subject of this proposal is the creation of the Metro Area Cancer
Information Service, based on the established and successful CCCIS.
The improvements in cancer education and communication that
will follow from creation of the Metro Area Cancer Information Service will significantly contribute toward reducing the cancer mortality rate and increasing the cancer survivor rate through the five-
county area. Three additional goals have been identified:

[Two pages omitted]

Program of Work

To achieve the goals identified above, seven key tasks have been
identified:

Task One:	Expand Service Area
Task Two:	Expand Telephone Service
Task Three:	Expand and Increase Information Activities
Task Four:	Increase Staff Training
Task Five:	Increase Quality Assurance
Task Six:	Create Liaison with Local Organizations in Expanded Area
Task Seven:	Expand Facilities and Equipment

Task 1: Expand Service Area
The proposed service area would include Center, Harper, West,
Pike, and Valley counties with a resulting population base of approximately one million. Of that population, approximately 500,000 are
urban, living in Center County, primarily in the city of Center City.
Another 250,000 are suburban, living primarily in Valley County
and in the parts of Harper, West, and Pike counties immediately adjacent to Center County. The final 250,000 are rural, primarily in the
relatively isolated areas of Harper, West, and Pike counties. Currently the only Cancer Information Service in the region is the
CCCIS, with services limited to residents of Center County.

Expansion to serve the five-county area is a natural step for the
CCCIS. The new counties are physically adjacent to Center County
as Figure 3 [not reproduced here] shows, and joined by a network of
roads and other transportation facilities, making travel for outreach
programs, etc., relatively easy. Telephone calls within the five-
county area incur no extra charges for callers. Figure 4 [not repro-

duced here] shows the population distribution in this geographical area.

Three of the new counties in the expanded service area present particular obstacles to health educators seeking to provide information about improving lifestyles to minimize cancer risk factors, implementing prevention programs, and instituting programs for early detection.

[Remaining 20 pages omitted]

EXERCISES

1. Write a proposal for a research-and-report-writing project for this class. Begin by defining a technical subject and explaining the important processes that take place within it or that it is part of. Try to isolate a problem area within the subject—one your research will investigate and your eventual report will describe in depth. Your work may not result in a solution to the problem, but you can certainly look at available (or imaginable) solutions and evaluate them. You can put the report in the communication situation of this class—written by a student, for classmates and a teacher— or you can invent your own simulated real-world context: for example, an accountant providing basic background information on a problem in retirement funds investment planning for a corporate client or an engineer writing a basic report on a technical problem for management.

2. With two or three of your classmates, write a proposal to set up a workshop on effective writing for local high school students. The purpose of the workshop—to be held on two consecutive Saturday mornings, for two hours each morning—is to give high school students who intend to attend college extra training in the kinds of writing they will have to do there. The audience for the proposal should be the high school principal.

3. Pick a university procedure that you think isn't handled as well as it should be—enrollment, student parking, allocation of tickets for athletic events, student job placement services, and so on—and write a 300–500 word proposal to the appropriate campus authority recommending a specific change or changes in the procedure.

4. From examination of recent copies of *Commerce Business Daily* or the *Federal Register*, find RFPs that might be of interest to the kind of company you expect to work for upon graduation. Write a short analysis of the sample.

A ▶ indicates a case study exercise.

How are the RFPs alike? How are they different? How do the things they want match or differ within the sample? How do the kinds of proposals they seem to want match or differ from the patterns presented in this chapter?

➤ 5. The short proposal in Section 6.1 is written from the section head of a company, Susan Jones, to her superior, the division head, J. T. Lomax. As an in-house communication, it is in the form of a memo. Put yourself in the position of being a summer intern in Jones's office. She has brought this memo, labeled DRAFT, to you so that you can rewrite it to reflect more clearly the key elements of proposals she understands you learned in the writing class you recently took. Rewrite the memo to improve its effectiveness.

➤ 6. As part of the new-employee training program on your new job, you have been asked to evaluate different brands of notebook computers and make a recommendation (in the form of a proposal) to the company's executive vice-president, Lindsay Hollingsworth, for the purchase of one model for all of the company's management-level employees who want one. The criteria you are given are very loose: "Get the best you can for $1,500 per computer. Do some checking around, and make the best recommendation you can, but I have to have it no later than 8:00 Monday morning."

19. *Written Reports*

1. *The Importance of Long Reports*

Writing a long report as part of your professional employment is the major test of any professional's writing skills. Whether 20 or 200 pages long, the major report gives you a chance to take a significant topic and give it a thorough treatment. Success can mean much more than personal satisfaction for a job well done, and failure can mean much more than simply having to do the report over.

   Writing an effective major report can require you to employ any and all of the skills and knowledge presented in earlier chapters, but most of all it requires your determination to produce a document that presents its subject clearly for its specific reader(s). With skill and determination you can write a report that will fulfill its purpose and make you proud of being its author. (Of course, in professional life, *group* authorship is common; that is covered in Chapter 11.)

1.1 Types of Long Reports

Any of the types of documents discussed in earlier chapters of this textbook can appear as a long report. Long reports include proposals, feasibility studies, problem-solving reports, progress reports, lab reports, meeting reports, trouble or accident reports, and most other typical kinds of reports. But many long reports cannot easily be classified as one "kind" or another: the report is a unique structure of thought, tailor-made to present a specific subject to a specific audience. Particular sections within such a report may be readily identifiable as this "kind" or that "kind," but the document itself, taken as a whole, is unique.

1.2 Parts of Long Reports

Most long reports have these basic parts:
> Distinctive cover
> Letter of transmittal*
> Title page
> Abstract or executive summary
> Contents page
> List of figures and illustrations
> Glossary*
> Introduction
> Body
> Conclusion
> Appendixes
> References*

These individual parts can be grouped together into three sections: *front matter* (everything up to and including the glossary), the *report proper* (the introduction, body, and conclusion), and *back matter* (the appendixes and references). It's not unusual for the front matter and

---

*These items are less frequent than the others.

back matter to include more pages than the rest of the report, especially when the report proper merely lays out a problem and presents conclusions or recommendations, and the appendixes contain all of the technical data (discussed further in Section 2.5, "Typical Structural Patterns," later in this chapter).

1.3   Physical Characteristics of Long Reports

Besides a distinctive cover, often with artwork on it, long reports usually share a number of other features: good paper quality, laser or laser-quality print, 1- to 1 1/2-inch margins on every page, extensive use of headings and subheadings, and full and frequent use of visuals. Although student reports are always double spaced, professional reports are occasionally single spaced (or space and a half), with double spacing between paragraphs. Usually reports are bound, especially if they are to go outside the company, and at least several copies are printed. Of course, these elements also depend on whether the report is a Class A, B, or C document (see Chapter 11, Section 1.2, for more on this).

1.4   Assumptions About Long Reports

Perhaps because of tradition, people assume several important qualities about long reports: some are obvious, such as that the content will justify the length, or that the report's structure will make sense to its reader. But the biggest assumption is less than obvious: that the report is assumed to be *self-contained.*

"Self-contained," as used in this context, means a number of different things. It means that anything your reader needs to know about the circumstances of the report's composition in order to read and understand the report should be in the report. For example, if you decided to focus only on data up to 1990, perhaps because more recent data are not available, you should state that explicitly somewhere in the report, probably in the letter of transmittal. Or if you decided to put all of the technical data into appendixes at the end of the report, resulting in a report with a very short body but long appendixes, you should explain that in the report, probably in the introduction. Do not force your reader to try to figure out what the report's structure is or why.

"Self-contained" also means you cannot assume that the reader you have in mind for the report is the *only* reader who will see it; your report's circulation and lifespan may be much greater than you expected when you wrote it. Do what you can to ensure that the report explains itself to *whoever* reads your report, *whenever* they read it.

2. *Techniques for Producing Long Reports*

In terms of producing the document, the biggest difference between the long, or major, report and other reports you may have written is length. Many writers find that producing a report with 15 or 20 (or more) pages of report proper—not counting front and back matter—presents not just a different *degree* of difficulty but also a different *kind* of difficulty than writing shorter reports. Because of its length, producing a long report requires a special emphasis on planning.

People approach planning the process of producing a long report in a variety of ways. For group projects, something like the compliance matrix (shown in Chapter 18) is a good idea. Some writers take voluminous notes, make extensive outlines, write many rough drafts, and plan their projects extensively. Others write as they research and plan as they write, counting on doing substantial revising later to draw the document's scattered pieces together. Neither process is ideal. The techniques involved in producing a long report include planning, researching, and writing.

2.1 Planning Long Reports

Planning for producing long reports falls into two distinct areas: Defining the Purpose and Scope, and Using the Calendar.

Defining the Purpose and Scope. You and the person for whom you are writing the report must agree in detail on what your report will and won't do, what it will and will not discuss. For example, reports written by juniors and seniors in an advanced report-writing class may have any of several different purposes: you may want to present the results of original research on a subject, or you may want to use the library to define the current state of knowledge on a subject. You may want your paper to solve a problem, to lead to a recommendation, to point to a prediction, or merely to present the facts. You and your reader should specifically discuss which of those purposes your paper is to fulfill—whether it is to be persuasive or merely expository.

Just as you and your reader must clearly understand the purpose of your report, so you must also agree on its scope. What years are you covering? what countries? How do you define your topic's boundaries, and what will be your criteria for including one fact and excluding others? Are your resources to be limited somehow—say to those in your university's library—or are you allowed to find information wherever you can? To what extent are you encouraged, or expected, to find information outside the library?

Purpose and scope are overlapping aspects of reports, and they both must be clearly understood in order for you to write a good report. One of the best ways to ensure this understanding is to plan at least three meetings with the person you're writing for:

1. Determine the topic initially, to find what "ballpark" you'll be working in.
2. Narrow the topic's scope—what you'll include and exclude—by defining the paper's purpose and approach to the topic in some detail.
3. Discuss your progress and test whether reader and writer are still in agreement on purpose and scope.

The First Meeting. Suppose your instructor requires a major report with 15 to 20 pages of body on a topic relating to your major field of study. In your first conference, you explain that although you're a business major, for your first three years at college you majored in biology, and you'd like to do a paper in that field—specifically, "something with recombinant DNA research." Your instructor further stipulates that the paper have something to do with business (the college you're enrolled in and the area you'll be writing in when you leave college). So the two of you agree on the topic "Economic Implications of Recombinant DNA Research." At that point, with a partial idea of purpose and scope, you have a general area within which to work, and you and your reader have agreed on enough for you to begin collecting information on the subject.

The Second Meeting. In the second conference, you work out with your reader more precisely the purpose and scope of your report. In English classes this step is often called "determining a thesis." To the extent that "thesis" only means a point you're trying to prove, "determining a thesis" is a misleading name for this step. During this second conference, you work out the details of purpose and scope. For example, on the topic "Economic Implications of Recombinant DNA Research," you may decide that you will first explain the biology of recombinant DNA, then follow a couple of cases of university researchers forming private companies to develop their research, and then, for the paper's last ten pages or so, trace the financial growth of several companies involved in capitalizing financially on recombinant DNA research.

During this second meeting, you discuss with your instructor the paper's purpose. The two of you agree that the paper should do more than just trace the subject's history; because you're working in a past/present/future structure, anyway, one logical conclusion would be to predict the financial future for such companies. Your instructor points out that there are several other issues involved: the relationship between big business and university research; the financial

soundness of a firm that may consist of only six people, some laboratory equipment, and a patent, but which begins trading on the stock
market with $6 million worth of shares; and so forth. The two of you
agree that both of these would be legitimate topics to include in the
paper but that to orient the report toward making a judgmental statement about either topic would require totally rethinking the entire
paper. At this point, by the end of the second conference, as a result
of negotiations between you and the person you're writing the report
for, you have the paper's purpose and scope fully defined.

The Third Meeting. In the third of this series of three conferences,
you report to your instructor that you have finished the research for
the discussion of the biology of recombinant DNA, that you've found
plenty of information on the current status of companies involved in
marketing the results of DNA research. But you think that the last
section is a little thin—too thin to support any prediction about the
future. Your instructor is pleased with your progress and suggests
that you select the company for which the most information is available and develop your prediction based on that company's situation.
That way, even if you cannot find enough information on all of the
companies, you can still construct a reasonable estimate of the future
based on one company's performance.

Less dialogue between you and your reader risks your writing a
report that doesn't satisfy your reader's need. More dialogue can be
helpful but isn't necessary. However many times you meet with your
reader, discuss the subjects of purpose and scope in at least as much
detail as we have gone into here.

Using the Calendar. Because of the major report's length, you need to
plan carefully how to use your time. The best way to do that is to
establish the set of tasks that producing the report requires and to
plot those tasks out on your calendar. The list of tasks will be different
for each writer and each project, but the following list can be taken
as representative. It allows the writer nine weeks to produce the paper
on "Economic Implications of Recombinant DNA Research" discussed here.

Schedule for Producing a Typical Long Report

Week 1: Topic-clearance conference. Collect preliminary bibliography.

Week 2: Continue preliminary bibliography and ensure that available sources can supply enough information.

Week 3: Second conference, to establish purpose and scope. Do research for Part I, "Biology of Recombinant DNA."

Week 4: Begin writing first draft of Part I and research Part II, "University Research Meets Big Business."

Week 5: Write the first draft of Part II and research Part III, "The New Companies."

Week 6: Progress report conference. Research Genentech.

Week 7: Write Genentech section and conclusion.

Week 8: Revise all sections and produce visuals.

Week 9: Print and proofread final manuscript, copy it, and turn it in.

Notice that this task-based calendar breaks the larger project down into four parts, and you research one part while writing the previous part in rough-draft form. This method spreads the actual writing of the report out over almost the entire nine-week span and is usually more effective than the alternative of leaving all the writing for the last week. If you leave all the writing for the last week, you run a number of needless risks: if you catch the flu that week, or another instructor surprises you with a major exam, or if you find you haven't done enough research, or if the words just won't come, you don't have the time to do a good job of writing the report. Subdividing the long paper into several smaller papers means that whatever problems you encounter in any one week, you still have the majority of the paper—the result of all the other weeks' work—well in hand.

These two techniques—dividing the paper into sections and writing the sections separately (not waiting until the last week to write)—will enable you to handle any long report more efficiently and with less stress. The sooner you begin putting words on paper for your report, in however rough a form your first draft is, the better your final report is likely to be.

2.2 Researching Long Reports

The research that goes into long reports can be of any variety, including physical research (such as laboratory tests), original investigation (surveying the territory or talking to the people yourself), and library research. The techniques of good physical research and original investigation are too varied to detail here, and the process of library research is thoroughly presented in Chapter 16. But to do *any* kind of research well you must do two things: keep records and be thorough.

Keeping Records. The best way to keep records, regardless of the kind of research you do, is to keep a notebook. This technique applies to library research as well as to laboratory research. The notebook

Figure 19.1   A Typical Research Notebook Entry

> February 26, 1992
>
> Went to the Main Library at 10:00 to take another look at the September, 1981 <u>Scientific American</u>, looking especially for the article on microbial production of pharmaceuticals. Went to the microfilm/microfiche collection, found the right film, and copied all of the pages of that article (even at 10 cents a page, it's worth it to have my own copy). It looks like big chunks of the article will be useful in my major report—have to be careful about quoting, paraphrasing, references, etc.

should contain dated and timed entries summarizing each block of time you spend on research (see Figure 19.1). Writers with access to laptop computers find them invaluable for this kind of record keeping.

Although the entries certainly need not be grammatically correct, they should be written clearly. Each entry should summarize what you did in that time: for example, if on the date specified you found five journal articles, took notes from two and copied one, the notebook entry should list the names and other bibliographic information on all five and in one or two sentences summarize each of the three you found usable.

The research notebook does not replace other methods of keeping notes such as bibliography cards and note cards (see Figures 19.2 and

Figure 19.2   Sample Bibliography Card

Scientific American, September 1981, Volume 245, Number 3 pp. 65-75, "Industrial Microbiology," by Arnold L. Demain and Nadine A. Solomon
T1. S5 Main Library Basement
This whole issue contains articles that may be useful. Note especially the lead article, "Industrial Microbiology," which surveys the whole field. Pay attention also to "The Microbiological Production of Pharmaceuticals," which should be right on target.

19.3). But those cards do not help you control the process of your research, so that you know what you have done and what you have not yet done, the way the research notebook does. The research notebook also can serve as a way of guaranteeing the authenticity of your work.

Being Thorough. The second key ingredient to good research of any variety is thoroughness:

1. Follow up all of the leads your research uncovers, even when you suspect that what you find may challenge your understanding or working hypothesis about your subject.

2. Accept the appearance of phenomena you haven't predicted, follow up on them, and analyze them.

3. Try to use the best-quality materials and information, not simply that which is easiest to come by.

When you think about how thorough your research needs to be, remember that thoroughness of research is one very important mea-

Figure 19.3   Two Sides of a Typical Note Card

"Industrial Microbiology," by Arnold L. Demain
and Nadine A. Solomon. Pages 66-75 in Scientific
American, September 1981, Volume 245, Number 3.
p. 67   Started with fermentation (yeast to make
        alcohol) (Babylonians before 6000 B.C.)
p 68   Pasteur's role
pp. 69-70 General discussion of nature of
        microorganisms that make the
        process possible

p. 70   Four products: microbial cells, large molecules
        (i.e., enzymes) that they synthesize, primary
        metabolic products, secondary metabolic products

N.B.: This article is mostly useful for general
background, definitions, etc. The issue
contains another article, "The Microbiological
Production of Pharmaceuticals," that is more
to the point.

sure of the quality of any kind of research project, especially in academic settings. Professionals read the abstract of a report and then immediately check the thoroughness of its bibliography or references section. A thorough bibliography

- cites the standard works on the subject.
- cites the most recent works on the subject.
- cites a variety of different kinds of sources.
- cites works that cover a good span of time.
- cites works in proper form.
- cites works on the appropriate levels.
- does not overrely on any one or two works.

Many times people make snap judgments—rightly or wrongly— about the quality of your research on the basis of such a quick examination of the bibliography, an element of the report that the writer may have seen as relatively less important than many others. It's a good idea, therefore, not only to be thorough in your research but also to ensure that your bibliography (as well as the other aspects of your finished report) clearly reflects that thoroughness.

2.3   Writing Long Reports

Writing reports of 20 pages or more requires different kinds of skills than writing papers 3 to 5 pages long. Breaking your long report down into several major parts will help you to deal with its length. Also, remember to separate the writing you to do *understand* your subject from the writing you do to *explain* the subject, and be sure to adjust the report's structure to help your *reader* deal with the report's length.

2.4   Exploratory and Presentational Writing

Keeping a research notebook and writing the entries in it only for yourself is a different kind of writing from the kind appropriate for your report. The notebook sample in Figure 19.4 shows the kind of *exploratory* writing characteristic of a person who is still thinking through a subject. The writing is writer centered, characteristic in tone and structure of a person who doesn't yet know where the topic is going.

Readers of reports expect writing different from exploratory writing. The differences in tone and structure on both the sentence and paragraph level can be seen by comparing the exploratory, writer-centered writing in Figure 19.4 with the presentational, reader-centered writing in Figure 19.5. That is, exploratory writing *sounds like* someone is still working through a subject for the first time, while

Figure 19.4 Writer-Based Prose
This is how one passage looked in its first draft, which came right out of the writer's research notes.

Industries want to keep up with what is going on in recombinant DNA research; it has industrial, commercial, and scientific applications, both positive and negative. Positive applications include pharmaceutical, agricultural, and chemical ones. Negative applications include biological warfare, accidents, or use by terrorists.

presentational writing *sounds like* someone has mastered the subject and is presenting it in the best possible way to a particular reader.

The kind of thinking that goes into doing research for a major report can cause problems in the report's finished version in yet another way. A common error that inexperienced writers make is to allow the sequence in which they *learn* the data to dictate the sequence in which they *present* the data. (See Chapter 16, Section 3.1, and Chapter 18, Section 4.1 for more on this topic.) The order that your subject has when it first becomes clear in your mind is probably not the best order for you to use to make that subject clear to your reader. The structure of your long report should be the one that is most appropriate for that topic and your expected audience; that structure may not resemble at all the order of your research.

Allow some time at the end of your research to consider the way your paper represents your subject. Is the structure you have been using so far really the best way to present that topic to that reader in order to fulfill that purpose? Most writers become committed to a certain structure rather early in the research process, and it's tempting to become unthinkingly committed to that structure for the duration of the project. It may well be that the structure you became committed to early was the best way for you to understand the subject and around which to organize subsequent research and writing, but

Figure 19.5   Reader-Based Prose
This shows the same passage as in Figure 19.4 revised once to make it fit the particular report's purpose, message, audience, and situation. The style is still rough, but it's clearly becoming more presentational.

```
It is natural for industry to follow closely the achieve-
ments required in recombinant DNA research. As with nu-
clear science, recombinant DNA procedures have industrial,
commercial, and scientific applications. As with nuclear
science, scientific knowledge about recombinant DNA can
apply positively or negatively. Positive applications in-
clude developments in the pharmaceutical, agricultural,
and chemical industries. The negative applications include
biological warfare or use by terrorists.
```

it may well not be the best structure in which to present your report to your reader. Student writers are often afraid even to question their structure, fearing that the answer to the question may mean that they may have to write the whole paper over again with a different structure. But professional writers know that doesn't necessarily mean rewriting the whole paper, though it may mean reorganizing already-written sections and perhaps writing a new introduction and a few new internal summaries. That kind of reorganization takes relatively little time, and it can be well worth the effort to restructure the paper so that it is right for the reader, the subject, and the paper's purpose.

2.5   Typical Structural Patterns

Certain structural patterns occur over and over again in long reports. Some reports have only one of the structures listed on pages 37–38; other reports will use a combination of two or more.

There are many other types of structures. The subjects of other parts of this text could be structures, also: proposal, recommendation, problem solving, and various definition techniques. Deciding which organization is right for your subject depends on the topic, your purpose, and the nature and purpose of your audience.

2.6   Audience-Centered Structural Adaptations

Adapting the structure of a long report to a particular audience makes a big difference in the audience's response to the report. Your report should unfold and open itself up to its reader; the reader should not have to struggle to understand what you have written. If your audience is composed of one kind of person, that structural adaptation is fairly straightforward. If the audience is more complex, such as executive/layperson, adapting the structure to it is still fairly simple. If, however, the audience is both complex and multiple—for example, composed of some executives, some experts, and some technicians—adapting the structure can be more of a challenge. There are three ways to accomplish such a structural adaptation to a complex and multiple audience:

1. Write for the lowest common denominator.
2. Write several different versions.
3. Compartmentalize the report.

Writing for the Lowest Common Denominator. Writing for the lowest common denominator is generally the least satisfactory technique. By lowering the level of the concepts you use, your choice of vocabulary, and the kind and amount of detail you use to the level of the least-interested and least-informed audience, you run the risk of all your other audiences' losing interest.

Writing Multiple Versions. Writing several different versions of your report is probably the best solution to the problem of a complex and multiple audience—if you can successfully direct a different version of the report to each audience. Unfortunately, this solution is not always possible.

Compartmentalizing the Report. A good solution to the problem of a complex, multiple audience is to compartmentalize your report—that is, to put each part in a clearly labeled section and to organize the sections so that each reader can easily find the section or sections most appropriate to that reader. (You may recall "the compartmentalized report" as a type of report discussed in Chapter 17. The kind of compartmentalization discussed here is a less fully developed application of the same principle.) Thus, for example, in a problem-solving report, the first major section may describe the solution, for executives who are interested only in that. The next section may describe the implementation of the solution, for the technicians who will have to apply it on the practical level. Final sections on the method of solving the problem and on technical data or calculations are available for experts who may want to know the "how" and "why" as much as the "what." Thus, you have produced, not a one-level report, but a two- (or more)-level report (see Chapter 11, Section 1.2).

Any well-written report will be compartmentalized to some extent, but when faced with a complex, multiple audience, you will find that thorough compartmentalization is a good practice. In such a situation, the letter of transmittal and/or the introduction to the report should make the structure clear to all readers.

3. Basic Elements of Reports

Certain elements are common to nearly all major reports. Those elements can be conveniently discussed as front matter, the report proper, and back matter. (Refer to the student reports at the end of this chapter for examples of the elements in the following discussion, as used in a Class C report.)

3.1 Front Matter

Included in front matter are the cover (with art and label), the letter of transmittal, the abstract, the table of contents, the list of figures and illustrations, and (optionally) the glossary.

Cover (with Art and Label). Reports usually have a distinctive cover, or binder, often of a sort specified by the company that the writer works for. Cover art is becoming more and more common as a way of making the report distinctive. Usually the cover also has a label giving the report's title, the date, the author's name, and the name of the person or company for whom the report was written. Page i of the sample student report on DNA at the end of this chapter shows a typical cover, with art and label.

Letter of Transmittal. Most major reports contain at their beginning a letter from the report's author to the report's reader, called a letter of transmittal. It officially submits the report to the reader. This letter should be treated not as a totally routine or trivial piece of communication but as the writer's first chance to talk to his or her reader, as an introduction to the whole report that goes in front of the report's formal Introduction. In the letter of transmittal, the writer typically mentions a number of relevant details concerning the report: explanations of its structure, details about what it includes and what it excludes, and often a brief summary of the report. If, as the result of ongoing conversations between reader and writer, there are special characteristics of the report's content and structure (such as a certain year's data omitted or engineering specifications moved into appendixes), that should be explained in the letter of transmittal. The attitude the writer establishes toward the reader (goodwill, positive emphasis, and "you" attitude) is especially important in the letter of transmittal. Page ii of the same sample student report (Fig. 19.8) shows a letter of transmittal.

Title Page. Like a book, the major report carries a title page, with the report's title, the author's name, the name of the person, agency, or corporation the report is written for, and the date. Often the title page carries a list of the people or offices the report is being sent to, called the report's *distribution*. (The title page of the report on DNA is not shown, because it is almost identical to the cover.)

Abstract or Summary. The first page after the title page usually will be an abstract or summary of the report. The abstract or summary should be typed triple-spaced for easy reading, and should be no longer than a page. Page 520 shows a sample abstract.

Table of Contents. The report's table of contents should list the report's sections, down to the level of at least B- or C-level headings. The important thing here is to organize the contents page(s) visually so that the reader sees the report's structure at a glance. To do this, use vertical and horizontal spacing on the page, different kinds of typeface for emphasis, and the general ordering of elements to organize the table of contents. Page 521 shows a sample table of contents.

List of Illustrations. Most reports have a list of illustrations separate from the contents page. Page 521 shows one such list.

Glossary. If your report is being written for a layperson, it's often wise to include a glossary. If you can define eight to ten words at the front of the report and make the reader's understanding of the whole document easier, taking a page to do it can be a good idea. This usually would only be done for lay readers, however. (The glossary is not shown.)

All the pages of the report's front matter should be counted, and all but the letter of transmittal and title page usually carry lowercase roman numerals (i, ii, iii, iv, etc.). Use arabic numbers for the body of the report (beginning with 1 for the Introduction) for every page, right on through whatever endnotes, bibliography, or appendixes you use.

3.2 The Report Proper

The report proper usually has at least three parts: the introduction, the body, and the conclusion.

The Introduction. This section should be on a separate page or pages, with its own A-level heading. The *cpo* model presented in Chapter 8 is a good model for this introduction. Its length should vary, depending on the length of the report. For a 20-page report, an introduction of one or two pages is appropriate. The Introductions of the sample student reports at the end of this chapter are good examples; notice their extended use of the *cpo* structure.

The Body. This section usually begins on a new page following the Introduction, beginning with the report's first heading (for example, "Background" or "Statement of the Problem"). The report should use headings and subheadings throughout and make extensive use of visuals. The sample student reports (pp. 518–561) provide examples.

The Conclusion. This section of your report, like the Introduction, should be clearly labeled as a separate section, beginning on a new page. Conclusions to reports, however, take many forms and different names, from "Conclusion" to "Recommendations," to "Feasibility," to any number of other possibilities. Whatever title you use, ensure that your reader will be able to find your report's "bottom line," and that it is clearly labeled and marked as a separate part of the report's structure. The sample student reports provide examples (pp. 525 and 558).

3.3   Back Matter

Any number of different kinds of elements can appear after the Conclusion of a major report. If the report is written in an academic setting, the writer will usually include some form of endnotes and/or a bibliography. The way these two parts of the report's documentation are done varies widely from one field to another; the Appendix discusses different documentation styles in detail. The sample student report provides examples of one popular form of documentation (page 560).

You may elect to include an appendix or appendixes at the end of the report, containing such elements as specifications, alternate solutions, technical data, drawings, personal recommendations by the author, or suggestions for the future. The assumption about such appendixes is that the reader can read them if he or she is interested but that the appendixes are not essential to the meaning of the report. The choice of what to put in an appendix versus what to include in the report proper is one of the writer's main opportunities to adapt the report's structure to its specific reader.

4.   Evaluating Reports

Learn how to evaluate the quality of the reports you write, so that you *know* the report is good *before* you turn it in. The guidelines shown in the following list should help you evaluate the reports you write.

Part One: Purpose and Effectiveness

- Does the document fulfill its purpose effectively by providing material that can be easily assimilated by the intended reader?
- Are the central ideas or the purpose clearly stated, and does the content carry out the stated purpose?

Part Two: Writing and Editing

- Is the text easy to read and understand?
- Does the text use appropriate grammar, syntax, spelling, punctuation, and capitalization?
- Are nomenclature, abbreviation, and capitalization consistent?
- Is there a variety of sentence structure and originality of expression?
- Are the words well chosen and used in their correct meanings, without continuous repetition of the same words and phrases (except where required)? Has jargon been held to a minimum?
- Is the pattern of organization logical and easily recognized from headings and other organizational dividers?
- Are text headings, figure captions, and table titles well written?

Part Three: Graphics, Layout, and Production

- Do the illustrations contribute to the usefulness of the document and provide adequate detail for the purpose of the document?
- Are the illustrations legible and neatly constructed?
- Are the photographs in sharp focus, do they have a good range of tones, and are they properly cropped?
- Does the artwork show imagination and creativity where possible?
- Is the type easy to read?
- Do the size, shape, and binding of the document fit its intended purpose?
- Is the overall design layout effective and well executed?
- Is the printing of good quality?

5. Sample Long Reports

Here you will find key elements of one student report ("The Economic Implications of Recombinant DNA Research"), all of another one ("Risk Communication"), and key parts of a Class B professional report ("The Electric Car: Is It Still the Vehicle of the Future?"). For samples of a Class A report, see the excerpts from "Steam Turbines" in Chapter 11.

Figure 19.6   Main Elements from a Student Paper, a Class C Report

THE ECONOMIC IMPLICATIONS

OF

RECOMBINANT DNA RESEARCH

By

Terrell Alverson

For

Dr. Michael L. Keene

English 461

March 1, 1984

1600 West Sylvan Boulevard
Knoxville, TN 37919
March 1, 1984

Dr. Michael L. Keene
Department of English
The University of Tennessee
Knoxville, TN 37916

Dear Dr. Keene:

As you requested, I am submitting the following report on the
economic implications of recombinant DNA research. This area
of biotechnology is receiving a great deal of attention from
scientists, industrialists, and Wall Street analysts. Investment opportunities are rapidly increasing in the growing area
of genetic engineering.

This report is organized as you suggested in our planning
conferences, with an opening section on the basics of recombinant DNA, a second section on research and recombinant DNA
technology, and a third section on its economics. The report
has three appendixes, providing the results of the Asilomar
Conference, the NIH guidelines on DNA research, and a listing
of DNA-related patents.

This report provides useful information for anyone interested in recombinant DNA research or for anyone curious about
the links between big science and big business. Thank you for
all the help you have given me in preparing this paper. If
you have any questions regarding this report, please call me
at 555-1250.

Sincerely,

Terrell Alverson

Terrell Alverson
Business student

ABSTRACT

One of the most widely reported but least understood areas in which high technology is affecting the marketplace is recombinant DNA research. Gene splicing, in its simplest terms, is easy to understand, but its possible dangers have alarmed the scientific community. These risks can take many forms, but so can the benefits of the many possible applications of recombinant DNA technology. As a result of the opportunities for profit, biotechnology has become big business.

TABLE OF CONTENTS

iv

[The List of Illustrations would normally begin on a new page.]

LIST OF ILLUSTRATIONS

v

INTRODUCTION

The acknowledged landmark date for biotechnology is June 16,
1980, the date the U.S. Supreme Court struck down the Patent
and Trademark Office's stipulation that there was a patent
law distinction between living and nonliving matter. One of
the most important effects of this decision was its psychological effect on the business community, especially on potential investors in research and development. This decision, a
result of <u>Diamond</u> v. <u>Chakrabarty</u> opened the door for biotechnology to become a billion-dollar industry. Stock market analysts project that by 1995 total sales in the genetic engineering market may reach $13 billion.

The general public will be affected by these developments
in many ways. As recombinant DNA technology is refined and
turned to large-scale production techniques, new medicines,
foods, and chemical products will be increasingly apparent.
The general public also has the opportunity to participate in
this phenomenon by becoming investors in the rapidly growing
genetic-engineering industry. Genentech initiated this investment opportunity by being the first genetic engineering company to make its stock available to the public, and most
other companies in the field have now also gone public.

The purpose of this report is to inform the general public
and potential investors about recombinant DNA and its economic potential. The report is broken down into three sections. The first section explains the basic process of DNA recombination. The second section details some of the more
recent technological applications of recombinant DNA technology. The third section analyzes the relationship between big
business and the big science of biotechnology. Three appendixes offer further details on specific points mentioned in
the report: the results of the Asilomar conference, the NIH
guidelines on recombinant DNA research, and a list of DNA-
related patents and their holders. Based on this report, readers should have the knowledge to allow them to understand a
subject that will soon shape the lives of all of us.

1. SCIENTIFIC BACKGROUND OF RECOMBINANT DNA

In 1972 Jackson, Symons, and Berg described the biochemical method for cutting DNA from two different organisms and recombining the fragments to produce biologically functional DNA molecules (Johnson and Burnett, 1978). This method, known as gene splicing or DNA recombination, depends on using certain enzymes to cut and splice the DNA at specific points in the molecule's structure. DNA is the genetic coding for all living organisms. Recombination of DNA gives us the ability to take genes from one organism and splice them into the gene set of another organism.

1.1 The DNA Molecule

The DNA molecule consists of two nucleotide chains which twist around each other, forming a double helix (Figure 1). DNA's chains contain alternating sugars and phosphates. The four chemical bases, Adenine (A), Thymine (T), Cystosine (C), and Guanine (G), bond to the sugars in the nucleotide chain. The bases in one strand then match with the base partner in the other strand. Adenine pairs with Thymine, and Guanine with Cystosine. When one of the code words (A–T, C–G) matches incorrectly, genetic diseases such as sickle–cell anemia occur.

Recombination of DNA depends on restriction enzymes and vector systems (bacterial plasmids). Restriction enzymes break or cut up DNA at specific sites along the base sequence. These restriction enzymes leave sticky ends and unmatched bases on the broken DNA molecule.

1.2 The Procedure for Gene Transfer

The following steps explain the procedure for transferring genes to a foreign cell using plasmids:

1. Plasmids are isolated from the bacterial cell.
2. The restriction enzymes break the plasmids leaving sticky ends and unmatched bases on the DNA molecule.
3. The desired foreign gene is chemically sequenced using methods that determine the order of the base pairs.

2

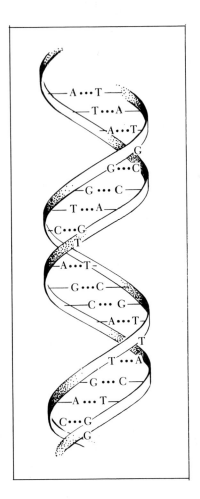

Figure 1. Diagram of the DNA double helix showing the base pairing.

3

Source: Gary E. Maciel, Daniel D. Traficante, and David Lavallee, *Chemistry* (Lexington, Mass.: D. C. Heath and Co., 1978), p. 379.

[The remainder of the body of the report occupies pp. 10–19.]

CONCLUSION

Although some of the initial euphoria has dwindled, enthusiasm still prevails in the genetic engineering field. Even
though stock prices have settled from the original first-day
excitement, projected sales are still into the billions by
the mid-1980s. Those sales rely on the markets created by
products such as antiviral or anticancer interferon.

Because the NIH heavily funds medical research and the
pharmaceutical industry already has large-scale facilities,
pharmaceuticals will be the first genetically engineered products available. Research and development in the agricultural
and chemical fields are taking place with more results daily.
Through the benefits of genetic engineering, our generation
may eventually have the ability to end famine and disease.
Even cancer may become a disease of the past.

BIBLIOGRAPHY

Agricultural Microbiology. Winton J. Brill in Scientific American, Volume 245, No. 3, pp. 199–212; September 1981.

Asilomar Conference on Recombinant DNA Molecules. Berg, et al., in Science, Volume 188, pp. 991–994; American Association for the Advancement of Science, 1975.

Bringing Biotechnology to Market. Steven J. Hochhauser in High Technology, Vol. 3, No. 2, pp. 55–60; February 1983.

Cell Biology: Structure, Biochemistry, and Function. Phillip Sheeler and Donald E. Bianchi, pp. 134, 443; John Wiley and Sons, 1980.

Cloning Gold Rush Turns Basic Biology Into Big Business. Science, vol. 208, pp. 678–681; May 16, 1980.

Dangers of Legislative and Regulatory Approaches Concerning the Hypothetical Risks of the Recombinant DNA Technique. W. Szybalski in Genetic Engineering, pp. 253–276; Elsevier/North Holland Press, 1978.

Genentech: Is Its Glamor Gone? Marjorie Sun in Science, vol. 211, p. 262; January 16, 1981.

Genetic Engineering: A Key to Innovation in Industrial R&D. J. Boldingh in Genetic Engineering, pp. 203–208; Elsevier/North Holland Press, 1978.

Genetic Engineering: Building New Profits. Yale L. Meltzer in Chemical Marketing Reporter, pp. 34–41; April 6, 1981.

21

Source: Used with the permission of Terrell Alverson.

Figure 19.7   A Student Paper, A Class C Report

A Consumer's Guide
to
Risk Communication

April 28, 1991

GLENDA TOLLIVER

727 Robertsville Rd.
Oak Ridge, TN 37830
April 28, 1991

Dr. Michael L. Keene
Department of English
The University of Tennessee
Knoxville, TN 37916

Dear Mike,

As you requested, here is the report on risk communication. This report provides important information for recipients of risk communication in the form of a consumer's guide. The National Research Council has identified a need for this type of guide. Informing the public about risk communication has become necessary because of new developments in risk communication.

This report contains an opening section defining and describing risk communication. It then explains two new developments in risk communication and their implications for the consumer. The report then provides methods that the consumer can use to evaluate the content and purposes of risk communication. In a final section, the report advocates active participation in risk communication. An appendix to the report provides greater detail about the process of risk communication. The references are done according to The Chicago Manual of Style, Style B.

As we discussed in the planning conference, one section of the report relies mainly on information provided by the National Research Council, in Improving Risk Communication. This is the only available source of information on consumer evaluation of risk messages.

The information in this report will be useful for anyone who wants to be well informed about health risks. Thank you for your help in planning this project. If you have any question regarding this report, please call me at 555-0044.

Sincerely,

Glenda Tolliver

Glenda Tolliver

EXECUTIVE SUMMARY

Many organizations today are communicating with the public about health risks. Because risk communication is more common now, it will be helpful to understand the practice of risk communication. But there are also other reasons for having a guide to risk communication. Some organizations are beginning to see risk communication as an interactive process. There will be greater opportunities for the public to participate in risk communication, and people need to understand these opportunities. This consumer's guide provides reasons, methods, and goals for participating. Another reason for a guide to risk communication is the problem of persuasive risk communication. It has become common for presentations about risks to aim at persuading the audience rather than informing them about risks. This guide to risk communication will help you to be able to serve your own interests in finding out about health risks.

TABLE OF CONTENTS

List of Illustrations

INTRODUCTION

In recent years, people have become very concerned with avoiding hazards. There is an increased expectation now that the public will be informed of known or suspected health hazards, such as radiation or chemical exposure. For this reason, government and industry are now more likely to communicate with the public concerning risks to public health. Risk communication is increasing, and a recent study by the National Research Council has identified a need for the recipients of risk messages to become better informed about risk communication (1:177). The following report provides important information that we, as the public, need to know about risk communication.

The goal of this report will be to orient you, the consumer, to the uses and abuses of risk communication, to enhance your ability to participate in the risk communication process. After defining risk communication and explaining its context, the report will describe two new concepts of risk communication and their implications for the consumer. The report will then describe ways the consumer can evaluate risk messages. The report will increase your awareness of risk communication as an interactive process, and provide both reasons for becoming an active participant and methods for doing so.

1. THE CONCEPT OF RISK COMMUNICATION

The words Bhopal, Chernobyl, and radon remind us that people have become very concerned in recent years with avoiding hazards. Government and industry frequently discover possible hazards to the public and have responsibilities to inform those who may be affected. The process of communicating with the public about hazards is known as risk communication. In order to understand that process better, you will need to know something about views of risk, obligations to inform,

1

and efforts to communicate. Because we are all a part of the
"public," we need to know what to expect of industry and government when they attempt to communicate with us about possible dangers.

1.1 Views of Risk

Risk may be thought of as a danger of harm (2:6). But experts view risk as a mathematical quantity. To experts, dangers have two important features: probability ("How likely is
it to happen?") and magnitude ("How bad would it be?"). Technical experts say that risk is a combined measure of these
two questions: that probability $\times$ magnitude = risk (3:137).
But most people do not view risk as a measure; rather, they
see it as a threat which may arise from a particular source.

1.2 Sources of Risk

Some risk arises from nature, such as the risk of flood,
lightning, or earthquake. Some risk arises from society, such
as the risk of murder. Other risks come along with technology, whether that technology is spear thrower, automobile, or
space shuttle (2:3). As you can see, risk has always been
with us, and we ignore or accept many of the risks around us.
To a great extent, this is because the exposure to risk is
balanced by some benefit. For example, we accept the risk of
automobile accident because of the benefits of mobility. In
the past, the benefits of new technology have clearly outweighed the risks, and people have accepted those risks. But
today, the complexity of technology and the scientific uncertainty about risks and benefits are changing the acceptance
of risks. Today, people are likely to label some of the risks
of technology as "unacceptable," or at least questionable.
For example, people question the hazards of radiation exposure from nuclear power plants, the risks of pesticide resi-
dues on produce, and the hazards posed by toxic wastes.

1.3 Obligations to Inform

Both government and industry have obligations to inform
the public of risks. These obligations may be legal, moral,

2

or ethical (2:4). For example, a professional code of conduct may require that known risks be disclosed. Freedom of information legislation compels the federal government and industry to "communicate risks on which they have information to those who might suffer the consequences" (2:4). And there is a duty in a democracy for the government to educate the public about risks, to promote informed decision making (4:39). The obligations to inform are multiple, and the process of informing requires an organized communications effort.

1.4 Efforts to Communicate

To communicate is to "impart information through a one-way process, or through a two-way process where messages are exchanged" (2:8). Communicating risk is complicated by the fact that simply giving an audience a message may not inform the audience of the risk (2:9). For example, a few years ago, the EPA tried to communicate with the public about the risks of the pesticide EDB in grain products. Complicated scientific terms and ideas were used, and when the presentation was over, most laypeople were still unable to answer the question that was uppermost in their minds (1:192), namely, "Should I eat the bread?"

1.5 Purpose of Risk Communication

Government, industry, and even interest groups may seek to inform the public about risks. These attempts to inform are generally more successful than in the past because risk communication has improved. But risk communication does not always seek to inform, and it may be used to persuade. Covello, a risk communications scholar, defines risk communication as "purposefully" conveying information to "interested parties" about:

(1) levels of risk to health (see Fig. 1),

(2) the significance of such risks (see Fig. 2), or

(3) efforts to manage such risks through decisions, actions or policies (see Fig. 3) (5:172).

3

Figure 1 Levels of Risk. Some risk communications reveal levels of risk.

Table II. Benzene "Exposure Budget": Major Sources of Benzene Exposures and Risks

Activity	Intake (μg/day)	Pop. at risk ($\times 10^6$)	Total risk (%)
Smoking	1800	53	50
Unknown personal	150	240	20
Ambient	120	240	20
Passive smoking	50	190	5
Occupational	10000	0.25	1
Filling gas tank	10	100	<1

Source: Lance Wallace, Risk Analysis 10, no. 1, p. 61.

4

Figure 2   The Significance of a Risk. Some risk communications focus on the health consequences of a risk.

Table VI.   Distribution of Exposures and Leukemia Risk Due to Benzene in the Indoor Air Environment

Exposures to Benzene Median: 10 $\mu g/m^3$

Assessment of health impact of total population exposure

Total population base: 230 million; TLV: 30 mg/m^3
Leukemia risk: 0.1/million people/year/$\mu g/m^3$ exposure
Annual background rate: 6 per 100,000 for a total of 13,800
Threshold level: 0 $\mu g/m^3$

Population (%)		Exposure, $\mu g/m^3$				Excess due to	
Cum.	Fract	Night	Day	Outdoor	Breath	Outdoor	Total
10	10	1.0	2	0.3	0.5	0	2
25	15	4.7	4	1.2	3.5	0	6
50	25	13.0	8	4.9	9.0	3	18
75	25	25.0	14	11.0	18.0	6	32
90	15	42.0	27	16.0	33.0	6	38
95	5	61.0	49	21.0	48.0	2	22
99	4	210.0	299	32.0	80.0	3	110
100	1	350.0	770	40.0	105.0	1	71
Median expos.:		3.5 TLV/ 10,000			Total	21	299

According to this estimate this exposure causes   2% of incidence

Source: Ibid., p. 55.

Figure 3 Efforts to Manage a Risk. Some risk communications stress the management of risks through decisions, actions, or plans.

Can I reduce exposure to benzene? Yes, some, by doing as follows:
- Work with local facilities to reduce benzene emissions.
- Stand so you don't smell gasoline when filling your car. Encourage gas stations to install the devices that allow you to start filling the pump and then walk away.
- Use paints, glues, and cleaning products with the windows open to provide fresh air.
- Use protective equipment. Always use impermeable gloves to avoid exposure to skin. Use a cartridge respirator if you are a frequent user of paints, glues, etc.
- If skin contact occurs, wash thoroughly.
- Choose paints and glues without benzene if possible.
- Read product labels and follow instructions for safe use.

Source: Chemical Fact Sheet (U.S. Public Health Service).

6

The fact that risk communication is "purposeful" is significant for the consumer. Industries and government agencies now
employ professional communicators to design and present messages about risks. This gives the source organization of the
message a new level of effectiveness, whether in informing or
in persuading. Correspondingly, as a part of the public, you
need to have a new level of effectiveness in understanding
risk communication. You now understand that risks are communicated to the public for varying purposes. It will also be useful to understand why the process of risk communication has
developed.

2. THE CONTEXT OF RISK COMMUNICATION

About twenty years ago, industries and federal agencies began trying to respond to public concerns about health risks.
Unfortunately, little was known about how to communicate with
the public about risks. For example, in 1977, the U.S. Geological Survey was given a new responsibility of issuing warnings whenever their research revealed a possible natural hazard. This federal agency then had a professional staff of
scientists, but few professional communicators. The following
year, geologists studied an area near Kodiak, Alaska, and concluded that a gigantic mudslide could occur. They predicted
that so much mud falling into Kodiak harbor could create a
tidal wave capable of destroying the low-lying town that surrounded the harbor. The USGS concluded that they had a duty
to inform, and so they announced their prediction to the national news media, without talking to anyone in Kodiak. One
result of the announcement was that economic investment in Kodiak dried up, and businesses were devastated by risk communication rather than by a tidal wave (see Fig. 4). But there
was no federal disaster relief to help the local people overcome the economic disaster. More problems ensued because the
USGS had warned too quickly and had no evidence to back up
their prediction. Furthermore, the research to establish the

7

probability and magnitude of the risk would cost $500,000, and the USGS had no funding to do such research. The town of Kodiak was by then too poor to do the research, either. There was a long period of intense public hostility toward the USGS, and lasting suspicion of "expert" opinion, because the tidal wave never materialized (2:284–286).

Poor risk communication can be disastrous. And disaster in risk communication, although not usually as severe as in the Kodiak debacle, was very common during early attempts at risk communication. Risk communication as an organized process began to develop as professional communicators were called upon to help communicate risks, and research was done to determine how risks could be communicated better.

2.1 The Development of an Organized Process

There were two different reasons that risk communication began to develop as an organized process, the need to communicate better, and the motive of making technology acceptable.

2.1.1 The Need to Communicate Better

Risk communication was developing as a process, in part because of the need to communicate better, as in the Kodiak case. But for many organizations and experts there was an underlying motive for informing the public about risks.

2.1.2 The Motive of Making Technology Acceptable

Many organizations and experts hoped that the risks of new technology would become acceptable, if only the risks were understood. As a government spokesman put it—"if they [the public] had the same knowledge we have, they would see things our way" (6:127). For example, they believed that if people could understand the smallness of the risks of living near a nuclear waste repository, people would accept that risk. So, much of the improvement in risk communication had an ulterior motivation. But whatever the motive, many improvements were made in communicating risk.

8

Figure 4 The "Tidal Wave" That Hit Kodiak. Poor risk communication can be disastrous.

2.2 The Improvements in Risk Communication

Communications professionals improved risk communication by meeting the needs of the audience. They realized that laypeople need explanations of technical terms and ideas about risk (4:38). Communicators also knew that laypeople need risk messages that address their particular concerns (2:167). And they realized that an audience needs to trust the source organization that issues the message before they will believe the risk message (4:39). Because of the new emphasis on the needs of the audience, risk communicators carefully began to design and present risk messages that could be understood and believed by a particular audience (see Fig. 5). Also, a great deal of research was done to improve risk messages further. (More information about the process of risk communication is available in Appendix A [omitted].)

3. TWO NEW VIEWS OF RISK COMMUNICATION

Even with elaborate new preparations of the risk message, technical experts and risk communicators eventually began to feel frustrated. Audiences still were not interpreting risks to their health in the same way as the experts interpreted the same risks. For example, scientists might evaluate the risk of runoff pollution from a hazardous waste storage facility and find the risk to be very small. But the public might still demand that precautions be taken to prevent runoff. These disagreements between expert opinion about risk and laypeople's perception of risk have led to two new concepts of risk communication: a technocratic view and a democratic view (7:294).

3.1 The Technocratic View

It will be useful for you, the recipient of risk messages, to be aware of the technocratic view of risk communication.

10

Figure 5 Old and New Patterns of Risk Communication. Risk communication originally consisted of experts revealing their findings to the public:

Communications channel

A new, organized process of risk communication includes pro-fessional communicators who design and present messages that are tailored to the needs of a particular audience:

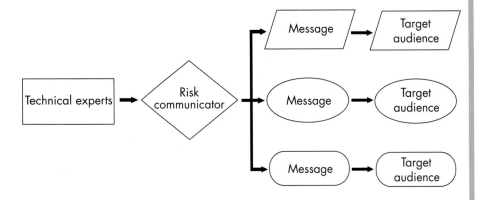

Communications channel

11

This view developed as experts began to give up on informing the public about risks (3:145). It became common for risk communications research to aim at "proving" that the public is irrational in its perceptions of risk, and thus cannot evaluate risks (6:126). These studies might show, for example, that people who take action because there is a hazardous waste site near their home usually have taken no action to determine radon levels in their homes. Since the risk posed by radon is much greater than the risk posed by the hazardous waste site, such a study calls into question the ability of people to evaluate risks rationally (7:296). These studies have been used by industry and by experts to justify a new type of risk communication characterized by a one-way persuasive flow of information (3:145) (see Fig. 6).

The idea behind this new, technocratic view of risk communication (see Fig. 7) is that because the public cannot evaluate risks, the public needs to be calmed with reassuring risk messages. These messages are designed to persuade the public that the official view of the risk is the correct view. The major problem with this idea of risk communication is that it subverts the informative purpose of communicating about risks. The proliferation of persuasive risk communication is significant for consumers. It has become increasingly important for people to be able to evaluate the content and purpose of risk messages and the sources of risk messages.

3.2 The Democratic View

Another, competing concept of risk communication, the democratic view, is also significant for consumers. This concept, like the technocratic concept, is supported by risk communications research findings. In both concepts, it is agreed that there are marked differences in the ways experts and laypeople evaluate risks. It is agreed that an expert's idea of risk is impersonal and narrowly defined (8). For the expert, as you will recall, risk is often a number that indicates the probability and magnitude of a risk, for example, 7.9×10^{-6}

Figure 6 Persuasive Risk Communication. Risk communication that is designed to persuade is characterized by a one—way flow of information from the communicator to an inactive audience.

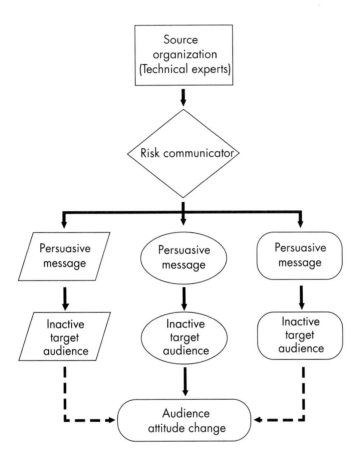

Figure 7 The Technocratic View. Technocratic risk communication is characterized by the following statements:

*"The lay public is irrational in considering risks."**

*"Let the experts decide!"***

*"The public needs to be calmed and reassured—not informed—by risk communication."****

* (6:126)
** (1:94)
*** (3:140)

14

The layperson, on the other hand, perceives risk subjectively and personally, incorporating his or her social, moral, and ethical concerns in evaluating risks (9:12). For example, when a high-level nuclear waste repository was proposed for Nevada, residents rejected the idea, even though experts felt certain that it could be operated safely. Research revealed that the public was primarily concerned with the uncertain impact of the facility on future generations (10:469).

The democratic concept of risk communication differs from the technocratic concept by viewing laypeople's perceptions about risk as legitimate. The democratic concept is supported by recent risk communications research, which has demonstrated that there is an "underlying richness and sophistication" in the views of non-experts (7:294). Advocates of a democratic concept of risk communication point out that the non-expert public evaluates risk "based on its social and political consequences" (7:296) (see Fig. 8). A democratic concept of risk communication calls for a two-way flow of information, designed to inform the public about risks for the purpose of allowing the public to serve their own interests in risk decisions (see Fig. 9). This pattern of risk communication incorporates feedback from the audience to improve decision making about risks. In accepting feedback from the public, an organization recognizes that an informed layperson's judgment about a risk includes criteria that experts ignore. In this view, public judgments about risk are "not inferior," and are necessary for a balanced evaluation of risks (7:296).

Because a democratic concept of risk communication is gaining recognition, it is likely that there will be greater compromises between the technocratic approach to risk communication and the democratic approach. One likely outcome, predicted by the National Research Council, is that two-way interactions about risks will become more common (1:177). This will afford you a much greater opportunity actually to participate in risk communication (1:177).

Figure 8 The Democratic View. Democratic risk communication is characterized by the following statements:

*"The laypublic are not fools."**

*"Laypeople add values to the process of decision making about risks."***

*"The knowledge of the public complements the knowledge of technical experts."****

* (7:294)
** (3:141)
*** (6:128)

16

Figure 9 A New, Two-Way Pattern of Risk Communication. This pattern of risk communication incorporates feedback from the interested public to improve decision making about risk.

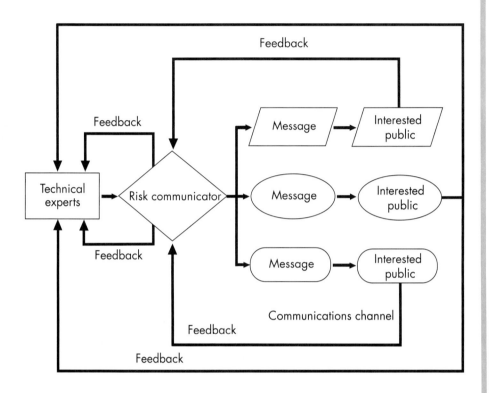

4. PARTICIPATION IN RISK COMMUNICATION

Experts on risk communication are predicting that risk communication will become more and more an interactive process
in which audience feedback is expected. Also, the practice of
persuasive risk communication is becoming common. This means
that if you, the consumer, want to serve your own interests
in decisions about risks, you must learn to ask the right
questions. To participate effectively as recipients of risk
communication, people must be able to (1) evaluate risk messages, (2) evaluate source organizations, (3) recognize the
value of participation, and (4) enhance their participation
skills.

4.1 Evaluating the Risk Message

In order to become informed about a health risk, you will
need to look critically at the purpose and content of the
risk message.

It is very important to evaluate the purpose behind the
risk message. Is the message intended simply to inform, or
does it attempt to influence your behavior or persuade you?
This distinction is important because "people's insensitivity
to the importance of how risk issues are presented exposes
them to manipulation." People can become concerned, for example, about "doubling their risk" without realizing that this
means a risk of .0000002 instead of .0000001 (11:357). As a
recipient of risk messages, you need to be aware that "a
broad spectrum of techniques" may be used in risk messages to
influence the audience (1:82).

In risk messages, one way that a source organization may
try to persuade an audience is through highlighting selected
facts. Highlighting may be accomplished through the source organization's choice of what to present and what to omit. As a
recipient of a risk message, how can you know whether facts
may have been omitted from that message? Checking the message
for completeness, using the following questions, may help
you.

18

Has the risk message discussed:

(1) The nature of the risk, for example, a pesticide which may cause up to 100 new cancers this year,

(2) the benefits associated with the risk, for example, that use of the pesticide will lower food production costs and improve public nutrition,

(3) the alternatives to the risk, for example, a discussion of the health risks of other pesticides which might be used instead, and the costs of using no pesticides,

(4) the uncertainties in knowledge about the risk, for example, that there may be other health risks posed by the pesticide, but the scientific studies are incomplete, and

(5) the management of the risk, for example, who will decide whether the pesticide is to be used, and on what basis (1:174)?

Information about risks can also be highlighted in other ways. Certain portions of the message may be emphasized with visuals, color, or larger print (1:82). Paying closer attention to the portions of the message which are not emphasized by typography or graphics will help you to evaluate how well the evidence supports the source organization's interpretation.

The way the facts are presented can affect how you interpret those facts. "A medical treatment that is reported as 60% successful, for example, is usually preferred to a treatment that has a 40% failure rate, although the two are objectively equal" (3:139). The same information about a risk can be presented in different ways to make it sound less frightening, or more credible (1:130). For example, the Soviet government analyzed the probability of health risks from the Chernobyl disaster. They reported that they expected "less than .05% cancer fatalities." This sounds much less alarming than reporting the actual figure, which "works out to be 35,000 to 45,000 premature deaths" (1:122).

Information can also be framed using risk comparisons. These comparisons are used to relate a poorly understood risk to other risks with which the audience is familiar (see Fig. 10). "But risk comparisons can also be used to influence or even mislead" (1:84). For example, a risk comparison may imply that if a person is already taking the larger of two risks (e.g., eating peanut butter often), she should be willing to take the smaller risk as well (e.g., living near a nuclear power plant) (12:375).

A risk message may influence some members of the audience by supplying the opinions about the risk of a respected person or group (1:84). For example, researchers have found that more people are willing to eat irradiated food if they are told that astronauts eat this type of food (13:499). Risk messages may also appeal to the emotions, such as fear, community spirit, or parental concern (1:85). For example, when the National Resources Defense Council communicated with the public about the risks of the chemical alar, their focus was on the risk to children.

4.2 Evaluating the Content

It is important for you to remember that scientific information is not necessarily neutral information--that "expert" opinion has become a commodity (6:125). This means that "technical information is often skewed, both in the questions that are asked and the results that are obtained" (14:743). First, you can ask for the credentials of the staff members who have evaluated the risk. Next, you can ask, "How was this information gathered?" and "How was it interpreted?" For example, in Los Angeles, city officials wanted a garbage incinerator and paid scientists to test the safety of incineration. The experts performed a test to determine the toxicity of the ash residue and submitted their results to health officials. These officials pronounced the ash "nonhazardous." However, when this result was communicated to the public, it occurred to someone to ask exactly what had been burned to produce the ash for the test. As it turned out, there was no documentation of what had been burned, or whether it contained any metals, even though "the central purpose of the test was to de-

20

Figure 10 Risk Comparisons. Risk "ladders," such as the one below, are sometimes used to compare different risks. This ladder
compares the annual mortality rate from various risks, per one
million people. Source: James K. Hammitt, Risk Analysis 10, no.
3, p. 369.

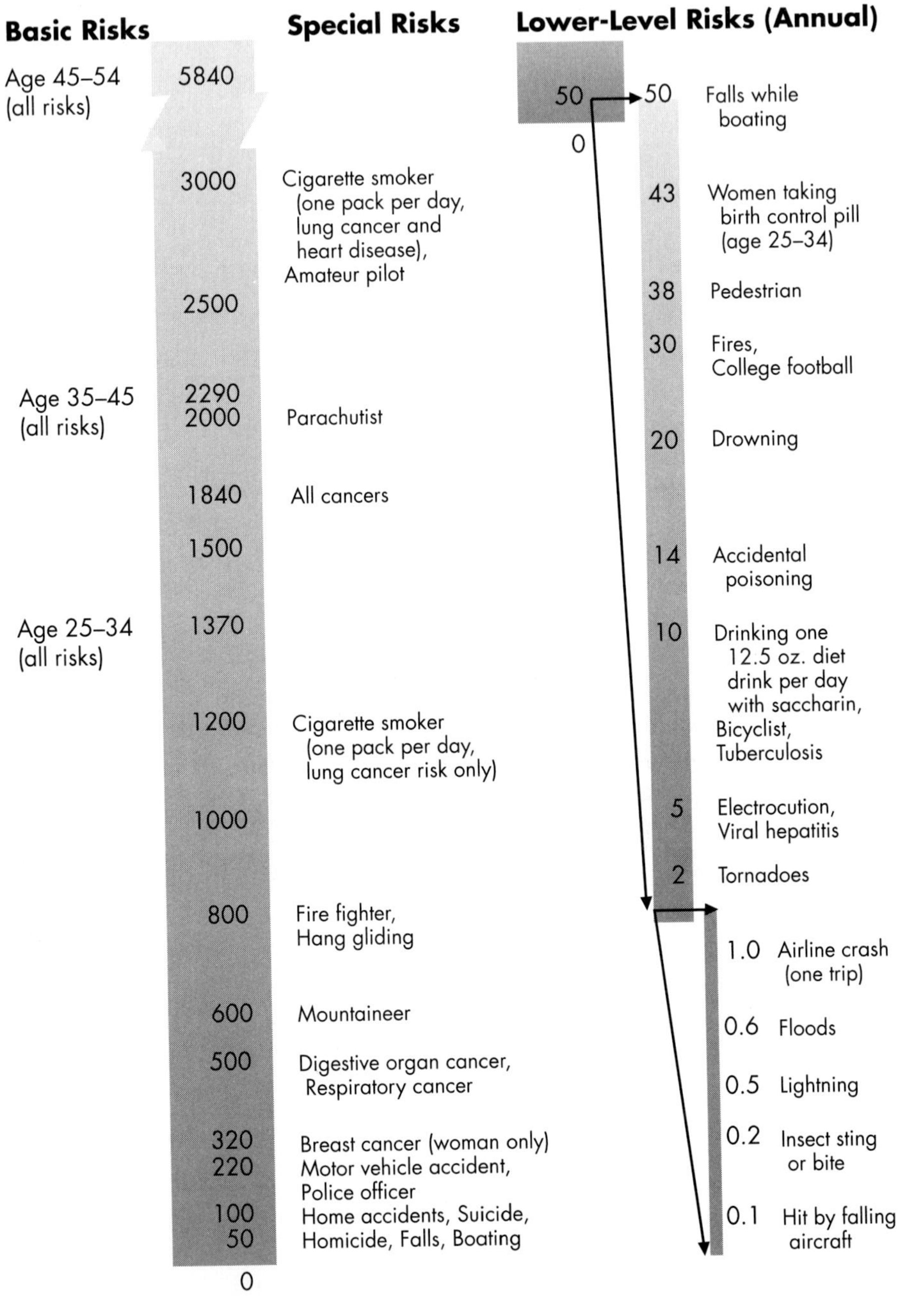

Basic Risks

Age 45–54 (all risks)	5840
	3000
	2500
Age 35–45 (all risks)	2290
	2000
	1840
	1500
Age 25–34 (all risks)	1370
	1200
	1000
	800
	600
	500
	320
	220
	100
	50
	0

Special Risks

- 3000 — Cigarette smoker (one pack per day, lung cancer and heart disease), Amateur pilot
- 2000 — Parachutist
- 1840 — All cancers
- 1200 — Cigarette smoker (one pack per day, lung cancer risk only)
- 800 — Fire fighter, Hang gliding
- 600 — Mountaineer
- 500 — Digestive organ cancer, Respiratory cancer
- 320 — Breast cancer (woman only)
- 220 — Motor vehicle accident, Police officer
- 100 — Home accidents, Suicide,
- 50 — Homicide, Falls, Boating

Lower-Level Risks (Annual)

50	Falls while boating
0	
43	Women taking birth control pill (age 25–34)
38	Pedestrian
30	Fires, College football
20	Drowning
14	Accidental poisoning
10	Drinking one 12.5 oz. diet drink per day with saccharin, Bicyclist, Tuberculosis
5	Electrocution, Viral hepatitis
2	Tornadoes
1.0	Airline crash (one trip)
0.6	Floods
0.5	Lightning
0.2	Insect sting or bite
0.1	Hit by falling aircraft

21

termine whether metals might leach from the ash" (14:742).
This questioning alerted the public to the lack of sound, impartial evidence about risk.

Many times in risk messages, exact figures will be used to
represent a risk (4.9×10^{-6} is a typical example). It is important that you realize that the use of this type of figure
does not imply that scientists have certainty about the risk.
Risk numbers, as a risk analyst points out, are the product
of numerous estimated values. Because they represent estimates, risk figures are constantly being revised and disputed, as Fig. 11 indicates. Risk figures are uncertain, and
an honest risk presentation will acknowledge uncertainties in
information (see Fig. 12).

4.3 Evaluating the Source Organization

One goal of a risk communicator is to give the audience
the impression that the risk information is coming from a
credible source. As a recipient, you can combine your evaluation of the risk message with your prior knowledge of the
source organization to evaluate credibility.

Answering the following questions will help you to evaluate the credibility of the source of a risk message.

(1) Is the source advocating a position that appears inconsistent with the facts?

(2) Does the source organization have a history of deceit
   or misrepresentation in dealing with the public?

(3) Is the source contradicting any previous messages they
   have issued about the risk?

(4) Has the source used influence techniques, such as
   self-serving framing of information, in the risk communication?

(5) Is this source issuing a risk message that contradicts
   credible messages from other sources?

(6) Does the source organization lack the necessary professional competence to evaluate the risk?

22

Figure 11 Uncertainty of Risk Estimates. Risk estimates are
often disputed. The graph below illustrates an expert's new
analysis of a risk, which reveals that flying is not necessarily safer than driving.

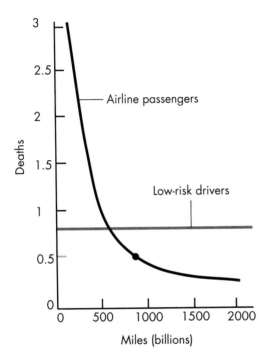

The risk for high-risk drivers (18 years old, intoxicated, no
seat belt) is more than 300 times the highest value on this
graph.

Source: Leonard Evans, Risk Analysis 10, no. 2, p. 243.

Figure 12 Uncertainty in a Risk Message. This chart compares
the risks involved in different types of energy production.
It reveals several areas of uncertainty in risk information.

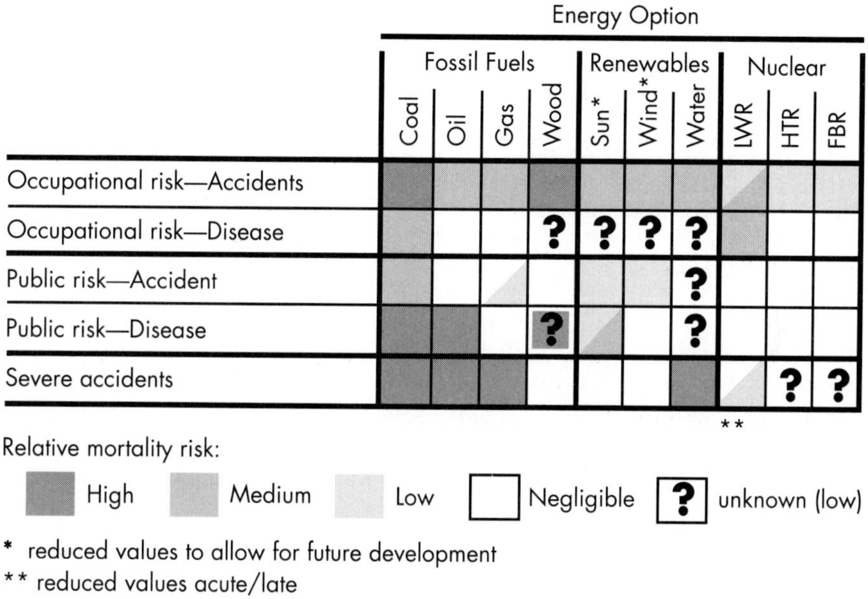

Energy Option

	Fossil Fuels				Renewables			Nuclear		
	Coal	Oil	Gas	Wood	Sun*	Wind*	Water	LWR	HTR	FBR
Occupational risk—Accidents										
Occupational risk—Disease				?	?	?	?			
Public risk—Accident							?			
Public risk—Disease				?			?			
Severe accidents									?	?

**

Relative mortality risk:

■ High   ▨ Medium   ▦ Low   ☐ Negligible   ？ unknown (low)

* reduced values to allow for future development
** reduced values acute/late

Source: Andrew Fritzsche, Risk Analysis 9, no. 4, p. 576.

24

An answer of "yes" to even one of these questions should cause you to be more careful about accepting this source's view of the risk (1:119). If there are credibility problems, this means that you will have to become an active participant in risk communication—asking questions and checking other sources of risk information—if you hope to make a sound decision about a risk.

4.4 Participating: Incentives

Environmental groups, trade unions, and you, as an individual, are at a disadvantage in evaluating risks to health or welfare. Private groups and individuals are usually unable to conduct the expensive research necessary to evaluate risks. It is important to remember, however, that in participating in the risk communication process, you are providing useful criticism of the information that has been developed by others (1:115). By questioning the source of the risk information, you may uncover some inadequacy in knowledge about the risk, as was the case with the L.A. garbage incinerator. In addition, you are adding the personal perspective to the study of the risk, by interpreting the risk in terms of your own values. Interpreting the risk in terms of values has the potential to change the way that some risks will be managed (6:128).

4.5 Participating Effectively

The manner in which you participate in the risk communication process will largely determine how your input is received. For example, when the director of a New Jersey government agency tried to explain how small a one in a million risk of cancer is (9:11), a citizens' group shouted at him, "We hope you're the one!" The citizens gained nothing and

25

probably ensured that their position would not be taken seriously. You can participate more effectively by treating the
positions taken by others about the risk carefully and
thoughtfully. You can also make it more likely that the risk
communicators will perceive your position as rational. This
is done by clearly and logically laying out your concerns,
premises, and assumptions about the risk (1:129).

4.6 Obtaining More Information

    After participating in a risk communication process, you
may have unanswered questions. There are several options for
obtaining more information. First, remember that you are not
alone in your concern about the risk. Other parties to a risk
issue may have incentives to sort out misrepresentations and
poor science which may be difficult for you alone to identify. For instance, in the early 1980s there was concern
about the risk of the pesticide EDB. Environmentalists, the
pesticide industry, and the EPA all had interests in the outcome of this risk decision (1:23). Because several parties
had their own incentives for investigating EDB, the facts
about the pesticide were more quickly gathered and sorted
out. Try to find out which organizations may have a stake in
the risk issue that concerns you and how they evaluate the
risk. It is also possible that more information about the
risk may be available from the source organization. In communicating the risk to a group, the source organization may
have elected not to cover in their presentation all that they
are willing to reveal about the risk. The organization may be
willing to provide supplementary documents upon request,
which reveal, for example, whether scientists agree about the
risk (1:171). Finally, if you want further information, you
may want to join an organization that monitors the risk and
reports developments about the risk to its members.

CONCLUSION

Opportunities are becoming available for consumer participation in risk communication. One highly successful example
of a dialogue approach to risk communication occurred in 1985
in Greensboro, North Carolina. The ECOFLO corporation wanted
to site a hazardous waste facility in Greensboro, but they
faced the obstacle of selling this idea to an established citizens' group that had opposed a similar facility only one
year before. Many citizens had experienced one-way persuasive
risk communication during the earlier siting attempt. Citizens were also well informed about hazardous waste issues.
These facts about the audience made it necessary for ECOFLO
to present the risk issues informatively rather than persuasively. ECOFLO also solicited extensive feedback from the citizens about the proposed waste facility. In response, ECOFLO
accepted the local residents' concerns about the adequacy of
their planned safety features and subsequently modified their
building plans. The new plans satisfied the community, as
well as ECOFLO's own technical experts (1:76 77) (see Fig.
13).

This case demonstrates that the interested public can participate successfully in risk communication when they are
well informed about risk communication. It also demonstrates
the reality of a new, interactive style of risk communication
and the value of participation—that action can result from
the involvement of the public.

27

Figure 13 The Case of ECOFLO. The case of ECOFLO demonstrates successful participation of the public in risk communication.

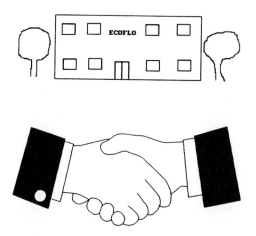

SOURCES CITED

1. National Research Council. 1989. <u>Improving Risk Commu-</u>
   <u>nication</u>. Washington: National Academy Press.

2. Handmer, John, and Edmund Penning-Rowsell. 1990. <u>Hazards</u>
   <u>and the Communication of Risk</u>. Hants, England: Gower
   Publishing.

3. Hadden, Susan G. 1989. <u>A Citizen's Right to Know: Risk</u>
   <u>Communication and Public Policy</u>. Boulder, Colo.: West-
   view Press.

4. Ahearne, John F. Telling the public about risks. 1990.
   <u>The Bulletin of the Atomic Scientists</u> (September):
   37–39.

5. Covello, V., Winterfeldt, Von, and Slovic, P. 1986. Risk
   communication: A review of the literature. <u>Risk Ab-</u>
   <u>stracts</u> 3:171–82.

6. Otway, Harry. 1987. Experts, risk communication, and democracy. <u>Risk Analysis</u> 7 (no. 2): 125–28.

7. Fiorino, Daniel J. 1989. Technical and democratic views
   in risk analysis. <u>Risk Analysis</u> 9 (no. 3): 293–99.

8. White, Douglas. 1991. Risk communication. A presentation
   for the January meeting of the Society for Technical
   Communication.

9. Chess, Caron, and Billie Jo Hance. 1989. Opening doors:
   Making risk communication agency reality. <u>Environment</u>
   (June): 11–39.

10. Kunreuther, Howard. 1990. Public attitudes toward siting
    a high-level nuclear waste repository in Utah. <u>Risk</u>
    <u>Analysis</u> 10 (no. 4): 469–73.

11. Morgan, M. Granger, and Lester Lave. 1990. Ethical consid-
    erations in risk communications practice and research.
    <u>Risk Analysis</u> 10 (no. 3): 355–58.

12. Roth, Emilie, et al. 1990. What do we know about making
    risk comparisons? <u>Risk Analysis</u> 10 (no. 3): 375–79.

13. Bord, Richard J. 1990. Risk communication, knowledge, and
    attitudes: Explaining reactions to a technology per-
    ceived as risky. <u>Risk Analysis</u> 10 (no. 4): 499–505.

14. Blumberg, Louis, and Robert Gottlieb. 1990. Citizens take
    on the experts. <u>The Nation</u> (May 28): 742–44.

SELECT BIBLIOGRAPHY

(Important sources not cited in the text.)

Laird, Frank N. 1989. The decline of deference: The political context of risk communication. Risk Analysis 9 (no. 4): 543–47.

Miller, James M., et al., comps. 1990. Instructions and Warnings: An Annotated Bibliography. Ann Arbor, Mich.: Fuller Technical Publications.

O'Riordan, Timothy, et al. 1990. Themes and tasks of risk communication: Report of an international conference. Risk Analysis 10 (no. 4): 515–16.

Russell, Milton (co-author of Improving Risk Communication). 1990. Telephone interview by Glenda Tolliver (March 26).

Stallen, Pieter Jan, and Rob Coppock. 1987. Risk communication and risky communication. Risk Analysis 7 (no. 4): 413–14.

United States Environmental Protection Agency Science Advisory Board. Strategic Options Subcommittee Relative Risk Reduction Project. 1990. Reducing Risk. Washington: Government Printing Office.

Weterings, Rob, and Jose Eijndhoven. 1989. Informing the public about uncertain risks. Risk Analysis 9 (no. 4): 445–47.

Figure 19.8   A Class B Professional Report

or nl

ORNL/TM-7904

OAK
RIDGE
NATIONAL
LABORATORY

The Electric Car—Is It Still the Vehicle of the Future?

R. L. Graves
C. D. West
E. C. Fox

Rauch & Lang Electric Car—1928

ORNL/TM-7904
Dist. Category UC-96

Contract No. W-7405-eng-26

Engineering Technology Division

THE ELECTRIC CAR—IS IT STILL THE
VEHICLE OF THE FUTURE?

R. L. Graves   C. D. West
E. C. Fox

Date Published—August 1981

Prepared by the
Oak Ridge National Laboratory
Oak Ridge, Tennessee 37830
operated by
Union Carbide Corporation
for the
Department of Energy

CONTENTS

THE ELECTRIC CAR—IS IT STILL THE VEHICLE OF THE FUTURE?

R. L. Graves   C. C. West

E. C. Fox

ABSTRACT

An analysis of electric and internal combustion engine (ICE) cars of equivalent performance shows that, even with advanced batteries, the electric vehicle would be much more costly to run (23¢/mile vs 16¢/mile) than the ICE car. The electric vehicle, of course, would not use gasoline, thus reducing the nation's dependence on imported oil; however, the cost of oil saved in this way would be about $190/bbl, and the same result could be achieved at about one-quarter the cost by manufacturing synfuels from domestic coal or oil shale. A similar analysis of some proposed hybrid electric vehicles indicates that they are also more costly to operate than an equivalent conventional vehicle, although by a smaller margin (25¢/mile vs 21¢/mile). The cost of oil saved by the use of hybrid vehicles is also lower ($95/bbl), although it is still much more than the projected cost of synthetic fuels. The key to improving the economics of the electric vehicle is to increase battery life or lower battery costs.

1. INTRODUCTION

With rising gas prices, many consumers would be interested in a vehicle that uses less gasoline and costs less to operate (witness the recent increase in small imported automobile sales). Further evidence of this is shown in the increasing public interest and news media coverage of the Department of Energy (DOE) electric car program. The electric car is often publicized as an effective way to decrease the nation's dependence on foreign oil. Also, because the electric vehicle uses no gasoline, many believe it could be cheaper to own and operate than a conventional automobile.

It is far from obvious that either of the above claims is true. Because of the uncertainty and because the issue is important to both the consumer and the nation, a study was undertaken to address, simply and clearly, these questions: are electric or hybrid vehicles cheaper to run than conventional cars, and are they an effective way to reduce the nation's dependence on oil?

In the early 1900s, about 40% of self-propelled vehicles were electric powered. Electric cars were particularly popular with women because the electrics were easier to start and operate than the internal combustion engine (ICE) vehicles of the time; the electrics were, however, relatively slow and had a very limited driving range. By the 1920s, the fate of the early electric automobile was sealed by the widespread introduction of the self-starter for IC engines.

Electric cars did not receive serious attention again until air pollution became an issue in the mid-1960s, but the ICE vehicle, even with the addition of expensive and power-robbing emission control devices, proved to hold the same advantages over the electric one as 60 years ago—longer driving range and superior performance. Interest in the electric car was heightened again after the 1973 Arab oil embargo which was a contributing factor in the formation of the Electric and Hybrid Vehicle Research, Development, and Demonstration Act of 1976 (Public Law 94-413). This act paved the way for expanded research and demonstration programs for electric and hybrid vehicles.

Gasoline prices have risen to the point where fueling an electric vehicle is usually less expensive than fueling an average family car, except in regions (such as New York) served by utilities that rely heavily on oil-fired generating capacity. Owning a vehicle, however, involves more than just fuel costs, and includes the initial capital cost, interest, insurance, and maintenance. All too often, economic comparisons of electric and combustion vehicles not only neglect these costs but also compare *typical* vehicles of both categories instead of *equivalent* vehicles. For example, the typical ICE car accelerates to 80 km/h (50 mph) in about 9 s, while the typical electric car accelerates to only about 40 km/h (25 mph) in the same time. Furthermore, an enormous disparity exists in the range for electric and combustion vehicles. These differences can be overcome by the use of "hybrids," which combine IC engines and electric propulsion systems in the same vehicle.

Recognizing the deficiencies in such comparisons, this study was performed (1) to compare electric and hybrid vehicles with their equiv-

2

alent ICE vehicle, (2) to account for more of the owners' cost factors, and (3) to evaluate the cost effectiveness of contributions made by hybrid and electric cars to reducing the nation's oil imports. Energy efficiencies were also evaluated for equivalent vehicles.

2.   COMPARISON OF EQUIVALENT ICE AND ELECTRIC CARS

The first tasks were the definition of a typical state-of-the-art electric vehicle and then the selection or development of characteristics of an equivalent ICE vehicle. (The opposite approach was rejected because it appears unlikely that electric vehicle performance can equal that of typical gasoline vehicles.)

The Electric and Hybrid Vehicle Act defines acceptable performance goals for electric vehicles (Table 1).[1] Typical values for today's electric car are compared with goals that were used to establish the performance required from the ICE vehicle.

A goal for the energy efficiency of the electric car is not set forth in the legislative act. This value depends on the car weight, battery/motor efficiency, and recharging rate and efficiency. The relation between weight, range, and efficiency is shown qualitatively in Fig. 1. Numerous tests indicate that the overall electric power use of a 950-kg (2100-lb) curb-weight electric car is at best 3 miles/kWh and more typically only 1.2 to 2.0 miles/kWh.[2,3]

Estimation of the fuel economy of an ICE vehicle of similar acceleration potential can be accomplished by correlating the fuel use and ac-

TABLE 1.   Summary of electric vehicle performance goals.

Parameter	Goal	State of art
Acceleration to 50 km/h, s	<15	8–15
Forward speed for 5 min, km/h	80	65–80
Range,[a] km	50	45–70
Recharge time from 80% discharge, h	<10	<10

[a] Using driving cycle in Society of Automotive Engineers Standard J227a.

Source: Ref. 1.

3

celeration of today's typical automobiles (Fig. 2); a direct comparison cannot be made, because no present-day automobile performs as poorly as the goals set for the electric vehicle. An extrapolation to very poor acceleration (0 to 50 km/h in about 12 s, which is still much better than the electric vehicle performance goal) shows that the corresponding fuel economy for a 950-kg (2100-lb) car would exceed 70 mpg over an Environmental Protection Agency (EPA) type driving mode. In fact, the equivalent ICE car weighs even less than this because it does not have the weight of the batteries. Nevertheless, a very conservative value of 60 mpg was used in this analysis. (Note that for a vehicle of this type, fuel consumption is already so low that further increases in efficiency make little difference to the running costs.)

The initial price of a mass-produced electric vehicle is difficult to estimate. Limited production two-passenger electric cars are selling for about $5500, but the price probably would be less if they were pro-

FIG. 1.   Illustration of trade-off between range and fuel efficiency.

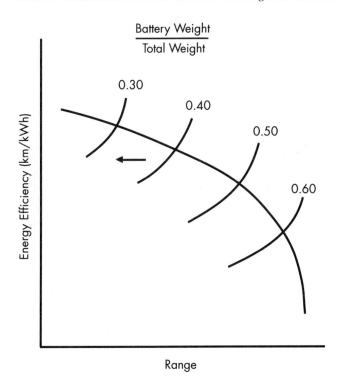

4

FIG. 2.   Correlation between fuel efficiency and acceleration capability.

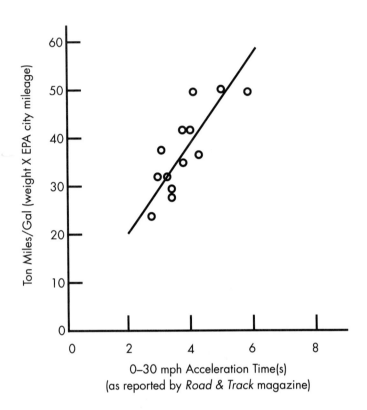

duced at a higher rate. There is no reason to believe that an electric car would be less expensive than a comparable ICE vehicle; both vehicles require the same basic components except for the power plant and its controls. In fact, the electric car might be more expensive because the additional weight of the batteries requires more structural strength and a larger power plant to yield the same performance as the lighter gasoline car (this is reflected in the weight comparisons shown in Table 2). Based on these arguments and some cost-estimating techniques offered by Hamilton[4] and others,[5] the electric car would appear to cost slightly more than an equivalent gasoline car, *excluding* the cost of the batteries. The price of gasoline-fueled cars in the 820-kg (1800-lb) weight range can be found in current markets.

5

TABLE 2.   Key characteristics of the equivalent electric and ICE
vehicles.

	Electric	ICE
Range, km	80	300
Top speed, km/h	80	85
Vehicle maximum payload, kg	270	270
Acceleration time (to 50 km/h), s	12.5	12.5
Weight, kg	1380[a]	816
Battery energy density, Wh/kg	50	
Battery life at 16,000 km/year use, years	1.6[b]	
Battery cost, 1980 $/kWh	60	
Vehicle price, 1980 $	6200[c]	4950
Fuel efficiency, miles/kWh, mpg	3	60

[a]Includes batteries, 395 kg.
[b]400 cycles.
[c]Includes batteries at $1200.

To be equivalent to an electric vehicle, the engine of the ICE car
would be down-sized from the norm, and vehicle options would be
minimal. Air-conditioning is excluded because it is not at all practical
for an electric car. The battery characteristics were taken from recent reports in the literature. A value for an energy-density characteristic of
an advanced lead-acid cell was used. The life of the battery is also typical of the near-term technology limit of about 300 to 400 deep cycles.[4,6]
The use of the vehicles was calculated under the assumption of an average 65 km/d, 250 d/year, corresponding to 16,000 km/year (10,000
miles/year) (the range of the electric vehicle is too limited, and the re-
charge time too long, to permit greater daily usage than this).

A summary of the key characteristics of the equivalent electric and
gasoline-fueled vehicles is presented in Table 2. These characteristics
were used throughout the economic and energy-use analyses.

The economic comparison of the two vehicle options was performed from a consumer viewpoint using a discounted cash flow analysis. The key parameters and assumptions in the economic analysis are
provided in Table 3. A number of nonequivalent (in fact, markedly superior) ICE vehicles were also included in the analysis, and their costs
and other data are shown in Table 4.

6

TABLE 3.   Parameters used in economic comparison.

Life of vehicles, years	8
Loan interest rate, %	12
Loan life, years	4
Tax rate, %	35
Debt fraction	0.8
Salvage value, %	20
Fixed-charge rate, %	16.7
General inflation rate, %	7
Oil inflation rate, %	9
General levelizing factor	1.33
Oil cost levelizing factor	1.44
Fuel cost	
Electricity, ¢/kWh	5.5[a]
Gasoline, $/gal	1.20[b]
Maintenance costs, ¢/mile	
ICE	3.5
Electric	0.0[c]

[a] Average electricity cost paid by residential consumers in 1980.
[b] Average gasoline cost paid by consumers in 1980.
[c] There will be some costs, of course, but they may be lower than for ICE vehicles: in this analysis, the most favorable assumption possible (zero costs) has been made for the electric vehicle.

TABLE 4.   Cost data for vehicles compared.

Car	Initial cost ($)	Fuel economy
Electric	6200	3 miles/kWh
ICE equivalent	4950	60 mpg
Ford Escort	5158	28 mpg[a]
Dodge Aries K	5880	24 mpg[a]
Buick Regal	7555	18 mpg[a]

[a] EPA-published city mileage, 1981 vehicles.

7

FIG. 3. Driving costs of electric and internal combustion vehicles.

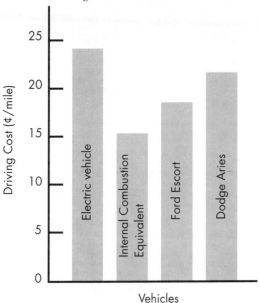

Figures 3 and 4 show the results of the economic comparison of these vehicles using the previously described ground rules and inputs, which include figures for the rate of escalation of oil prices. Figure 3 compares the cost per mile of various vehicle types; they were calculated on the basis of $1.20 gasoline in 1980; the cost does not include insurance and registration fees. The total annual costs of owning and operating the vehicles are plotted against first year gasoline prices (Fig. 4). At today's gasoline price, even a fairly large conventional car such as a Dodge Aries is economically equivalent to an electric car, while a gasoline vehicle equivalent in performance is clearly a very much better economic option than the electric car.

A closer examination of cost contributors (Table 5) indicates that the battery cost and life dominate the cost of owning and operating an electric vehicle. The sensitivity of the economic comparison to battery life is shown in Fig. 5. Obviously, with respect to consumer economics, development efforts should focus on this issue instead of attempting to increase energy density or power density, which affect primarily the range and acceleration of electric cars. To reduce the cost of the electric car to a competitive level would require that the battery be

8

FIG. 4.   Annual driving costs as a function of first-year gasoline costs.

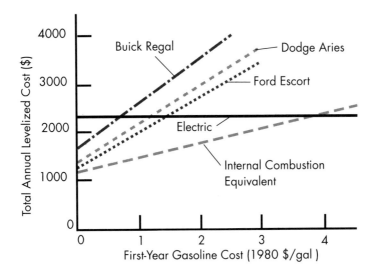

able to endure about 1300 cycles—over three times the state-of-the-art
lifetime—and cost no more than present battery systems.

Shorter driving cycles that discharge the battery less will lengthen
its life and, therefore, improve the economics of the electric vehicle.
To investigate the cost impact of this, the assumption was made that
the batteries would last about four years under light duty. The lesser
discharge also implies, of course, that the annual mileage of the car
must be lower. An important factor that should be included in the cost
analysis for short-trip driving is the inherently poor fuel economy of
the ICE vehicle during warm-up. A documented relation[7] between fuel
economy and trip length (Fig. 6) was used for this purpose. Although
shorter trips improve the electric car's position, the results (Fig. 7) indicate that it is still unattractive when compared with a conventional automobile of similar performance. Succinctly stated, if you don't drive
much, you can't save much.

The recent emphasis on deriving gasoline from coal presents an interesting comparison because coal can be used as the primary energy
source for either internal combustion or electric cars. An evaluation of
the overall fuel efficiency of the two options reveals that the equivalent
ICE and electric cars are, indeed, equivalent in this respect (Table 6).

9

TABLE 5.   Cost components of the two vehicle options.

Annual costs	Electric	ICE equivalent
Capital, %	45	52
Maintenance, %	0	29
Battery replacement, %	44	
Fuel, %	11	19
Total, %	100	100
Cost per mile, ¢	22.8	15.8
Total annual cost, $	2276	1580

FIG. 5.   Influence of battery life on electric vehicle cost based on 250 cycles per year, 10,000 miles per year.

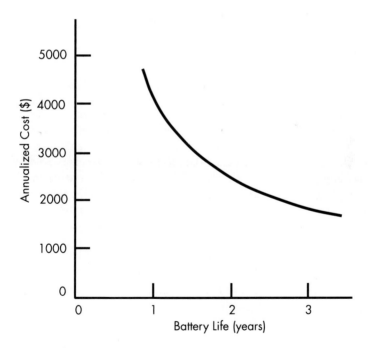

FIG. 6. Variation of automobile fuel economy with trip length due to engine warm-up.

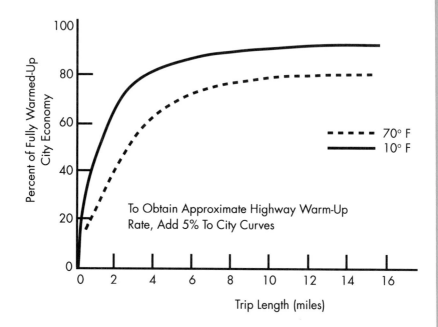

Finally, we recognize that the widespread use of electric vehicles would contribute to the solution of our most pressing energy problem: dependence on imported oil. The question is whether it is a cost-effective way of doing so. The equivalent ICE vehicle consumes 1/60 gal of gasoline per mile and costs 7¢/mile less to run than the electric vehicle (see Table 5). The cost per gallon of gasoline saved is therefore 7¢ × 60 = $4.20. The equivalent cost per barrel of crude oil saved is $190. This is very much more costly than producing the same amount of gasoline from coal or oil shale.

3. COMPARISON OF EQUIVALENT ICE AND HYBRID-ELECTRIC CARS

The extremely poor range and performance capabilities of the all-electric vehicles are among the several factors that could limit its use

FIG. 7.   Effect of daily mileage on annual costs of equivalent electric and internal combustion vehicles used for 250 d/year.

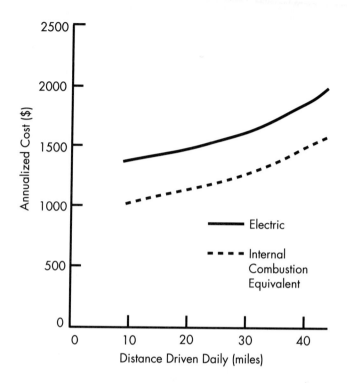

severely,[8] even if the life cycle costs were not so much higher than an equivalent ICE vehicle. The hybrid-electric vehicle may offer a way of overcoming many of these limitations and can, indeed, lead to a vehicle that equals the performance of a typical conventional automobile.

Extensive preliminary design studies of hybrid electric vehicles were carried out for DOE by four companies whose reports[9-12] provide much information on the expected costs and performance of their proposed hybrids. The design from General Electric Company (GE) was chosen by DOE as the basis for an experimental program involving construction and testing.

We have analyzed the GE design on the same basis as the all-electric vehicle; that is, by comparison with an ICE car of equivalent performance. Before making this comparison, however, it is instructive to consider the Near-Term Hybrid Vehicle Program objectives set by

12

TABLE 6. Comparison of overall fuel efficiency for ICE and electric cars.

Efficiency factors	ICE	Electric car
Mileage	60 mpg	3 miles/kWh
Fuel conversion	$\dfrac{24{,}000{,}000 \text{ Btu/ton of coal} \times 0.60}{120{,}000 \text{ Btu/gal}} = 120 \text{ gal/ton}$	$\dfrac{24{,}000{,}000 \text{ Btu/ton}}{10{,}000 \text{ Btu/kWh}} = 2{,}400 \text{ kWh/ton}$
Overall efficiency	7,200 miles/ton coal	7,200 miles/ton coal

13

DOE and the way in which these objectives have determined the outcome of the design exercise.

Six objectives of the DOE Near-Term Hybrid Vehicle Program are listed in the GE Phase-1 report.[9] Two of these are particularly crucial in determining the final design:

- Identify missions for hybrid vehicles that promise to yield high petroleum impact.
- Characterize the single vehicle concept which satisfies the mission or set of missions that provides the greatest potential reduction in petroleum consumption.

Neither these, nor any other of the objectives, included an evaluation of the cost effectiveness of saving petroleum in this way. The basic result of this omission is that all four hybrid designs from this program have very large batteries so that they may undertake short journeys with little or no use of the ICE and save all the gasoline that would otherwise be used on such trips. The resulting design is therefore basically a large short-range electric vehicle with a conventional engine added to extend the range and provide extra power for acceleration. Not surprisingly, the result is a very heavy and expensive vehicle.

The resulting economics may be approximately outlined very simply: the GE hybrid design costs about $2,300* more than the equivalent ICE vehicle, but it directly uses about 43% less petroleum fuel on the average (Table 3.4.1 of Ref. 9). Over a 100,000-mile life, the hybrid would use 2,200 gal of gasoline, and the conventional automobile would use 3,850 gal. The saving is therefore 1,650 gal, or the equivalent of about 36 bbl of crude oil over a lifetime of ~8-1/2 years. The $2300 capital investment therefore saves 36/8.5 × 365 = 0.012 bbl/d, equivalent to $2,300/0.012 = $200,000/bbl/d; this is very much more capital intensive than synthetic fuel production plants, which are estimated to involve an investment of less than $50,000/bbl/d. Note also that this calculation takes no account of the fact that 16% of the nation's electricity is generated from oil which, if the proportion remains unchanged over the life of the vehicle, would considerably reduce the lifetime oil saving.

The basis of the DOE reports[9-12] is a comparison of a hybrid car with a hypothetical 1985 equivalent ICE vehicle. Because the stated objective is to maximize the quantity of oil saved, the class of cars

------

* 1980 dollars.

TABLE 7.   GE hybrid vehicle parameters.

Parameter	Equivalent ICE vehicle	Hybrid-electric vehicle
Range, km	630[a]	400–550
Top speed, km/h		130[b]
Acceleration to 96 km/h, s	16	14
Engine type, cm³	Chevrolet V-6, 3800[c]	Volkswagen, 1600
Curb weight, kg	1180	1786[d]
Battery energy density, Wh/kg		40
Battery life, cycles		800[e]
Battery cost, 1980 $/kWh		60
Vehicle sticker price, 1980 $	7100	9400

[a] Assuming a 15-gal tank.
[b] Sustainable 1-min bursts of speed exceeding 150 km/h are possible.
[c] Engine used in 1979 model of equivalent ICE vehicle.
[d] Including 320 kg of batteries.
[e] Equal to 3.2 years if the battery is cycled once per day for 250 d/year. This battery cycle life, quoted from Ref. 9, is much larger than for the all-electric vehicle, possible because the use of the heat engine results in more favorable loading conditions on the battery.

selected by all four companies for the comparison is one that contributes significantly to the national gasoline consumption: five- or six-passenger capacity automobiles that are able to undertake both long and short trips. This was the inevitable choice because, to save a lot of gasoline, one must begin with something that uses a lot of gasoline. Table 7 lists the major characteristics of the ICE vehicle selected by GE as a reference and their equivalent hybrid.

The cost figures used in Table 7 deserve further explanation. The reference ICE vehicle chosen by GE was a hypothetical 1985 version of the Chevrolet Malibu with V-6 engine, air conditioning, automatic transmission, and power steering. The 1978 sticker price of such a vehicle was $5700, and the GE report assumed that the 1985 version of the same vehicle would carry the same price in constant dollars. This price would be $6670, expressed in 1980 dollars (actually, the sticker price of a Malibu with these options has already exceeded this by a large margin, being listed at $7430 at the end of 1980).

15

FIG. 8.   Correlations between vehicle weight and sticker price in two price ranges.

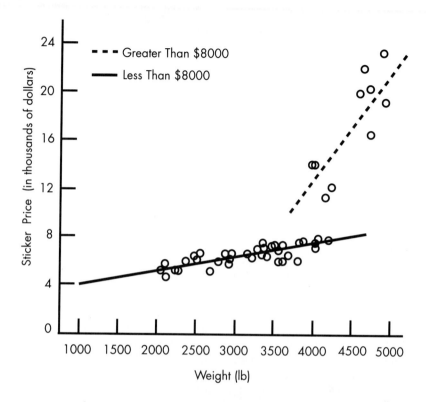

An alternative approach to calculating the initial cost is to apply the correlation shown in Fig. 8 between sticker price and car weight[13] to GE's estimate of 2600-lb weight for the 1985 version of their equivalent ICE vehicle, which yields a sticker price of $6050 for the basic car. The average cost of adding air conditioning, automatic transmission, and power steering to vehicles in this price range is $1050 (Ref. 14), which, added to the basic $6050, yields a total sticker price of $7100. This figure, which is about $350 below the current price of a Malibu and $450 higher than GE's projection, is the one used in Table 7.

The estimated price of the hybrid is greater than that of the reference ICE vehicle by $1967 (1978 dollars) (Table 3.5.1 in Ref. 9). Allowing for inflation between 1978 and 1980, this total becomes $2300 in 1980 dollars. Added to the $7100 estimated price of the ref-

16

TABLE 8. Energy use.

	Annual gasoline use (gal)	Annual electricity use (kWh)	Total annual energy use (MJ)
Hybrid vehicle	261	2,170	59,600[a]
Equivalent ICE vehicle	456		60,160

[a]Based on a 31% ratio between delivered electricity and primary energy, which takes into account conversion, transmission, and distribution losses. The 35% figure used in the GE report appears to be considerably overestimated.

erence vehicle, this gives a sticker price for the hybrid of $9400 ($7100 + $2300) (Table 7).

Table 8 gives some of the important power and energy performance characteristics of the two vehicles, based on Table 8 of Ref. 9. The annual mileage assumed by GE is 11,852. The total annual energy use is essentially the same for the two alternatives.

For the hybrid vehicle, as for the all-electric vehicle, the primary source of energy could be coal. Table 9 was constructed assuming that the energy content of the coal is 25 GJ/ton, and the conversion efficiency from coal to gasoline is 60%. This table indicates that if synfuels and electricity from coal are used, the hybrid vehicle makes somewhat more efficient use of the primary energy than does the equivalent ICE vehicle. This conclusion, however, is highly dependent on the specific vehicles compared and the ratio of ICE/electric operation in the hybrid.

Table 10 illustrates the financial implications of these energy-use figures, based on the average 1980 prices of 5.5¢/kWh for residential electricity and $1.20/gal for gasoline. Table 10 shows that the saving in fuel costs, at present prices, is much less than the cost of replacement batteries and is very much less than the extra initial cost of the hybrid vehicle. This comparison is, of course, grossly oversimplified because it ignores (1) the expected increases in oil prices relative to electricity prices and (2) the fact that some expenses (such as battery replacement costs) are incurred much later than others (such as first-year fuel costs). Nevertheless, the comparison is useful be-

17

TABLE 9. Primary energy (coal) use.

	Tons of coal per year			
	For electricity production	For gasoline production	Total	Miles per ton of coal
Hybrid vehicle	1.0	2.3	3.3	3,600
Equivalent ICE vehicle		4.0	4.0	3,000

TABLE 10. Annual costs at first year prices (1980 $).

		Annual cost		
	Initial cost	Maintenance[a]	Battery replacement[b]	Fuel
Hybrid vehicle	9400	310	240	430
Equivalent ICE	7100	410		550
Extra cost (saving)	2300	(100)	240	(120)

[a] Assuming 3.5¢/mile for the ICE vehicle (based on GE's estimate of 3¢/mile escalated to 1980 dollars) and 75% of this for the hybrid (based on GE's estimate of reduced maintenance costs due to part-time use of the heat engine).

[b] Assuming a 3.2-year average battery life (equivalent to 1 cycle/d for 250 d/year).

cause it indicates the basic reasons behind the results of a more thorough economic analysis that was conducted using the same ground rules and methods as for the electric vehicle. The results of this more realistic analysis, which did take these effects into account, are shown in Table 11.

From Tables 8 and 11, we see that the hybrid vehicle saves 195 gal/year of gasoline, equivalent to 4.24 bbl/year of crude oil, compared with the pure ICE vehicle. The extra running costs are 3.4¢/mile. The cost of oil saved is therefore 11,852 × $0.034/4.24 = $95/bbl. This greatly exceeds the estimated cost ($30–$50/bbl) of producing synthetic oil from domestic coal or oil shale resources.

18

TABLE 11. Comparison of equivalent ICE and hybrid cars using discounted cash flow analysis (All costs in 1980 $).

	ICE equivalent	Hybrid
Initial cost, $	7,100	9,400
Annual maintenance,[a] $	545	412
Annual battery cost,[a] $		319
Annual fuel cost,[a] $		
Gasoline	787	451
Electricity	0	159
Annual capital cost, $	1,186	1,570
Total annual cost, $	2,518	2,911
Mileage cost, ¢/mile	21.2	24.6

[a] Annual operating and maintenance costs are levelized over the life of the vehicle.

4. CONCLUSIONS

Primarily because of the initial cost of the battery and its necessary periodic replacement, the electric car is a significantly poorer economic choice from a consumer viewpoint than a high-fuel-economy conventional automobile. This is true for almost any driving scenario. The same conclusion has been reached for one type of hybrid-electric car currently being developed under DOE sponsorship.

From a national energy-use viewpoint, there is as much incentive for marketing low-powered combustion engined vehicles as there is for developing an electric or hybrid car industry, assuming that both would derive their respective fuels from coal.

This report is not intended to call for an end to research and development related to electric and hybrid vehicles, but rather to point out the deficiencies in the current technology and to emphasize what must be accomplished before these concepts will have a major impact on our fuel imports. Current battery technology is simply not good enough to compete with conventional cars available today, and even moderately advanced batteries are likely to result in an electric car that is uneconomical relative to an equivalently performing gasoline car. A novel approach to electric energy storage may be needed to raise battery lifetimes or lower battery costs and

19

"make the electric or hybrid car go." In the meantime, efforts to "demonstrate" the electric car with its present technical inadequacies may prove to be counterproductive in seeking its public acceptance in the future.

REFERENCES

1. *The Charge of the Future, An Introduction to Electric and Hybrid Vehicles*, DOE/CS-0107 (September 1979).

2. *Should We Have a New Engine—An Automobile Power Systems Evaluation*, Vol. II, Technical Reports, Jet Propulsion Laboratory (August 1975).

3. *EPRI-SCE Testing and Evaluation of Electric Vehicles: Lucas Van and Jet 007, 750, and 1400*, EPRI EM1723 (February 1981).

4. William Hamilton, *Electric Automobiles—Energy, Environmental, and Economic Prospects for the Future*, McGraw-Hill, 1980.

5. L. H. Gaines and K. Naeimek, *Development of Evaluation Techniques for Electrochemical Energy Storage Systems, Final Report*, DOE Contract EM-78-C-02-5157, Exxon Research and Engineering Co., March 1980.

6. *Electric and Hybrid Vehicle Program Quarterly Report, July–September 1980*, DOE/CS-0026/12 (November 1980).

7. T. Iura et al., *Research Plan for Achieving Reduced Automotive Energy Consumption*, Aerospace report No. ATR-76 (7467), National Science Foundation, Washington, D.C., 1975.

8. J. R. Wagner, *Vehicle Attributes Constraining Present Electric Car Applicability in the Fleet Market*, BNL 51099, December 1979.

9. General Electric Company, *Phase 1 of the Near-Term Hybrid Passenger Vehicle Development Program: Final Report*, DOE/JPL/955190-01, October 1980.

20

10. South Coast Technology, Inc., *Phase 1 of the Near-Term Hybrid Passenger Vehicle Development Program: Final Report*, DOE/JPL/955189-01, October 1980.

11. Minicars, Inc., *Phase 1 of the Near-Term Hybrid Passenger Vehicle Development Program: Final Report*, DOE/JPL/955188-01, October 1980.

12. Centro Ricerche Fiat, *Phase 1 of the Near-Term Hybrid Passenger Vehicle Development Program: Final Report*, DOE/JPL/955187-01, October 1980.

13. *Consumer Guide*, January 1981.

14. "Facts and Figures: Detroit's New Cars," *Changing Times*, pp. 25–32, January 1981.

ORNL/TM-7904
Dist. Category UC-96

Internal Distribution

1.	T. D. Anderson	24.	R. W. McCulloch
2.	J. T. Cockburn (Consultant)	25.	R. N. McGill
		26.	G. W. Oliphant
3.	J. C. Conklin	27–31.	T. W. Robinson, Jr.
4.	D. M. Eissenberg	32.	R. L. Rudman (Consultant)
5–9.	E. C. Fox	33.	G. Samuels
10.	W. Fulkerson	34.	I. Spiewak
11–15.	R. L. Graves	35.	J. B. Talbot
16.	D. Greene	36.	J. F. Thomas
17.	T. J. Hanratty (Consultant)	37–41.	H. E. Trammell
		42.	J. L. Wantland
18.	H. W. Hoffman	43–47.	C. D. West
19.	J. W. Jefferson	48.	ORNL Patent Office
20.	J. E. Jones	49.	Central Research Library
21.	H. F. Keesee	50.	Document Reference Section
22.	O. H. Klepper	51–52.	Laboratory Records Department
23.	L. F. Lischer (Consultant)	53.	Laboratory Records, RC

21

External Distribution

54. Paul Brown, Office of Vehicles and Engines Research and Development, Department of Energy, 1000 Independence Ave., Washington, DC 20585
55. A. M. Bueche, Vice President for Corporate Research, General Electric Company, 3135 Eastern Turnpike, Fairfield, CT 06431
56. C. Dean, Tennessee Valley Authority, 400 Commerce Ave., E12A7, Knoxville, TN 37902
57. G. Dochat, Mechanical Technology Inc., Latham, NY 12110
58. J. Franceschina, Chrysler Corp., P.O. Box 1118, Detroit, MI 48288
59. R. Freeman, Tennessee Valley Authority, 400 Commerce Ave., E12A11, Knoxville, TN 37902
60. S. David Freeman, Tennessee Valley Authority, 400 Commerce Ave., E12A9, Knoxville, TN 37902
61. R. J. Getz, CRBRP Project Office, P.O. Box U, Oak Ridge, TN 37830
62. L. King, 300 National Press Building, Washington, DC 20045
63. W. Martini, Martini Engineering, 2303 Harris, Richland, WA 99352
64. A. McEachern, Ingersoll-Rand Research Inc., Box 301, Princeton, NJ 08540
65. V. P. Roan, University of Florida, Gainesville, FL 32611
66. W. F. Rolf, CRBRP Project Office, P.O. Box U, Oak Ridge, TN 37830
67. J. Sandberg, Jet Propulsion Laboratory, Pasadena, CA 91103
68. P. Vickers, Fluid Dynamics Research Dept., General Motors Research Laboratories, Warren, MI 48090
69. Office of Assistant Manager for Energy Research and Development, Department of Energy, Oak Ridge Operations Office, Oak Ridge, TN 37830
70–216. Given distribution as shown in DOE/TIC-4500 under category UC-96

22

Source: U.S. Dept. of Commerce.

EXERCISES

1. As you complete your own long report for this class, arrange with another student (or students) in the class to trade next-to-last drafts prior to the final printing. Read and study the other report(s), and write a one-page report about what you learned from your own report based on examining someone else's.

2. Compare the student reports in this chapter with the professionally written reports in this and other chapters. How are the reports similar, and how are they different? Look at the full range of characteristics: style, format, use of headings, and so on. Write a one-page report on what you learn from this comparison.

3. Popularized science writing is one of the most important (and abundant) types of writing being done in the United States today. Find three popularized science articles from different sources (perhaps three different magazines, or a magazine, a brochure, or pamphlet), and compare their audiences, purposes, and messages. Write a one-page report on what you learn from this comparison.

4. Carefully study Glenda Tolliver's report on risk communication in this chapter (Fig. 19.9). How does it use the elements of effective professional writing discussed in this book? (Look particularly at Chapters 3, 4, 7–10, and 12–16.) Write a two- or three-page summary of what you discover.

5. Choose a non-textbook publication that has to do with a topic in your major field of study—perhaps a thesis, dissertation, pamphlet, journal article, or manual. Write a short report on how it does or does not exemplify the principles discussed in this book.

6. Risk communication, the subject of a student report in this chapter, is an important and sometimes controversial topic. Find three pieces of risk communication, and analyze them according to the material in Glenda Tolliver's paper (Fig. 19.9). Write the results of your analysis in a two- or three-page report; be sure to include copies of the pieces of risk communication in an appendix.

7. The sample Class B professional report, "The Electric Car" (Fig. 19.10) is by now quite dated, and the situation concerning the feasibility of electric cars has changed considerably. Work with a group of classmates to produce a report that parallels "The Electric Car" but updates its research. Emulate the style and page layout of the 1981 report as much as you can.

▶ 8. One interesting application of professional writing aimed at laypeople is in the advertising done to sell high-tech products to the public. Examples

---

A ▶ indicates a case study exercise.

include computer hardware and software advertising, automotive advertising (such as for antilock brakes, four-wheel steering, air bags, and so on), telecommunication equipment, and stereos and televisions. Find a piece of advertising aimed at the public that seems to attempt to explain how some such piece of high-tech equipment works, and write a short (two- or three-page) report about how selling is done via the medium of explaining how something works. Your audience is an employer who has questioned whether the public ever needs to be approached on the basis of "here's how this works" when the point of the communication is "buy this product."

20. *Oral Reports and Poster Presentations*

Once you've produced a well-written report that fulfills your readers' purposes and responds to their needs, how do you go about turning it into an effective *oral* presentation? For most people, the ability to make a five- to ten-minute oral presentation lies within easy reach. It's true that the art of public speaking has its own college courses, textbooks, and principles. As the kind of speaker who does mostly short, relatively informal presentations, however, you need only to remember a few key principles and to make sufficiently careful preparations, and you can make the kinds of oral presentations that are common in professional life a strong part of your communication skills.

Closely related to these short oral presentations is the *poster session* or *poster presentation*. In more and more professional fields today, one of the common ways to present information to others is by using a poster display, typically accompanied by a presenter who stands beside it to explain the poster's content to anyone who asks. In scientific or engineering meetings, these displays may be mounted on boards that are four feet by six feet or four feet by eight feet; in other areas such as sales or marketing, the use of commercially produced booths has become common. The second half of this chapter explains basic guidelines for designing and using the simpler kinds of poster displays.

In both the short oral presentation and the poster presentation, the important *intangibles* to remember are (1) the interaction between your purpose and your audience's purpose, (2) the essential elements of your message, (3) the nature of your audience, and (4) the nature of your role as speaker or presenter. As usual, the most important element is to focus squarely on adapting what you want to say to the specific audience that will hear it.

1. *Preparation and Organization of Oral Reports*

How should you prepare for your oral report? There are three key elements: keep your audience first, simplify your message's content, and reinforce the structure.

1.1 Keep Your Audience First

How many times have you seen speakers get so absorbed in something they are reading, or so caught up in the message they are delivering, that their eyes seem to glaze over and they lose all contact with the audience? In classrooms, professional meetings, boardrooms, and banquet halls, the speaker who gets totally absorbed in what he or she wants to say and totally ignores how the audience is responding is a recognizable species. This kind of speaking can be called "writing off the audience." (It's exactly the same phenomenon as the catalogical, writer-based kind of writing described in Chapter 11.) Although such speakers get through with their messages, their messages never get through to their audiences! No one likes to be ignored, and when an audience senses the speaker is writing them off, they will respond by writing off the speaker—making noise, dozing off, and in some cases, leaving. Such a speaker has violated the first rule of making oral presentations: *Keep the audience first.*

In order to keep your audience first, tailor your presentation just for them. Think about whether they are there voluntarily or on orders, whether they may feel threatened by anything you may say, whether they will be tired of sitting and listening, and what their questions might be. Think through these points, and deal with them both in the content of your talk and in the way you present it. The generalizations about audience in earlier chapters are equally true of audiences for oral reports, and the analysis techniques offered in Chapter 1 work equally well for audiences of oral presentations.

Don't get locked into a particular amount of detail or method of presentation before you analyze your audience. For a five- to ten-minute presentation, you should have a very few key points in mind that you want to make, and the way you make them should depend on your audience. You cannot make the best decision about how to approach your audience until you have analyzed that audience.

One specific mistake speakers often make deserves particular mention here. Especially in a short presentation, there is no reason to read a speech word for word or to deliver it verbatim from memory. A few lucky people can do that successfully, but most of us cannot avoid seeming like robots when we follow a prepared speech word for word. All the best ways to deliver a short presentation involve knowing the subject matter thoroughly and preparing for delivery by practicing—but not memorizing or reading the speech. Know what to say because you know the subject thoroughly. Then you can use either a brief outline on note cards or a more detailed outline on paper to help you remember what point to go to next. If you know your subject well enough to make a speech about it, you should be able to talk your speech through, working from no more than an outline. Reading a speech word for word is the surest way to write your audience off—and to invite them to write *you* off.

1.2   Simplify the Content

How many key points can you make, effectively, in seven minutes? Probably no more than five. And if the subject matter you're speaking on is in an area the audience is not familiar with, thus requiring more introduction and explanation, probably no more than three.

It does not matter much how important *you* think all your topic's parts are; limited time means a limited number of points you can make. Most people agree that it makes much more sense to present five points well, in a way that the audience understands, appreciates, and remembers, than it does to rush through 10 or 15 or 20 (and have your audience grasp few or none of them). Simplifying the content of your presentation is an important way to keep your audience first. It shows that you understand the difference between an oral presentation and a written presentation and are willing to modify the way you express yourself based on your audience's needs.

One good way to simplify your report's content is to make an outline of it, with upper-case roman numerals (or decimal numbering) for each major point and with supporting elements arranged appropriately under those major points. Then choose the five or six major points that are most essential and eliminate the rest. For supporting material under those major points, select as much as you think you can cover easily in the time allotted, eliminating levels from the lowest up, to adjust to your time requirements.

If you don't have time to do all of the supporting material under I, first take out everything at the lowest level (maybe a, b, c, d); if you need to cut more, take out the next lowest level (1, 2, 3, 4); and keep trimming under the various uppercase roman numerals until you've adjusted the presentation to the time you have. While you're cutting you may feel some pain, but the result will be worth it. Successfully cutting your speech to fit your time limit tells your audience that you are more interested in their understanding what you *do* cover than in your need to tell them *everything* you know about your topic.

1.3   Reinforce the Structure

Because your presentation is oral rather than written, your audience can neither scan ahead to see where you're going nor look back to see where you've been. Because of that, you should reinforce the structure of your presentation. The easiest way is to put your presentation in the form of a number of key points, and to use visual aids (discussed in Section 2.2) to emphasize these points. Don't hesitate to use the *cpo* introduction (see Chapter 8): forecast the key points, work through them, and then summarize the points in your conclusion. *The more technical, specialized, or possibly confusing your subject is, the more important structural reinforcement becomes.* First, build a strong structure into your presentation; second, use visual aids to reinforce that structure in the minds of your audience.

2.   Presentation of Oral Reports

If you do a good enough job preparing for your report and then you practice it enough, presenting it can be both enjoyable and rewarding. If you are nervous or unsure about your upcoming presentation, prepare it some more, and practice it some more. The best solution for nerves is overpreparation.

2.1   Making It Easy On Yourself

You can make your presentation better by making yourself more comfortable; preparing more thoroughly and practicing it is one way to do this. Another way is to make sure that you know the environment you'll be speaking in—when, where, and to whom. Find out exactly when you are to speak, and get a look at the room in advance. If you can, while the room is empty, stand in the front of it and try out the sound. And the more you know about how many and what kind of people will be in your audience, the better your adaptation will be. Practice and preparation are two good ways to make delivering your oral presentation more comfortable for you and therefore better.

When you practice your presentation, be careful to practice it out loud, at the same speed and volume and with the same expressiveness you intend to use for the actual presentation. Practicing saying the words over in your mind, or mumbling the words in a monotone, is too different from public speaking to be very useful as practice. And if you can get a chance to practice in the room you will speak in, your practice will be much more productive.

2.2 Using Props and Visuals

One of the ways the talks you may give in professional life will usually be different from speaking situations you may have experienced in school is that you will use props and visuals much more frequently. Props include models, pieces of equipment, or anything else you use as an example. Visuals include overhead-projector transparencies, slides, flipcharts, poster boards, chalkboard drawings, and any of a number of media you might use to help present information. Used widely, props and visuals can be a speaker's best friends.

You can get your points across clearly and more vividly to your audience by using models and actual examples as illustrations in your speech. You might have scale models or mockups of the subjects you are discussing, you might have actual examples of the electrical device in question, or you might have the implements used in the process you are explaining. If your audience can actually see what you're talking about, your speech will be much more effective.

Although props are nearly always a good idea, passing a prop around in the audience while you speak is generally not a good idea. This technique inevitably causes more distraction than the knowledge the audience gains from it is worth. A good alternative to passing the prop around is to invite interested people in the audience to come up and examine the prop after the presentation is finished.

A hidden bonus of using props is that nervous speakers often find handling the props calms their nerves. For example, an architect whose voice has quavered through the first two minutes of speaking picks up the architectural scale model of the house he is discussing and suddenly becomes self-assured and a much better speaker. Again, if you know your topic thoroughly, your speech will be much better; if you can bring a part of that topic you know so well up to the front of the room with you, your speech can be better still.

Visuals. Visuals may be even more common than props among speeches in business and industry. Types of visuals can range from the simplest poster boards and flipcharts to the most complex multimedia presentations, but for this chapter we will concern ourselves only with the simpler forms: poster boards, flipcharts, overhead projectors, and slide projectors. Use of the more complex forms of visuals, especially when attempted by people who are not professional speakers, usually demonstrates an important truth: the more complicated the form or medium you use for visual aids, the more you can impress your audience if everything goes right, but the more chances there are for everything not to go right. *The more complicated forms of visual aids bring possibilities for greater success; they also bring possibilities for greater failure.*

Many speakers are comfortable using *poster board visuals* (see Fig. 20.1). Using poster boards—either on a stand especially designed for them or, in a classroom, on the chalk trays at the blackboard—can help a speaker reinforce the structure of a short presentation very effectively. For example, to describe a process, you can have the speech's outline on one poster board, with subsequent steps and a summary each on separate boards. Using chalk trays you can display the entire process right across the front of the room. Poster board is cheap, available in a variety of colors, and relatively foolproof.

Figure 20.1   Poster Boards
To explain a four-step process, use five boards—one for an overview and one for each step.

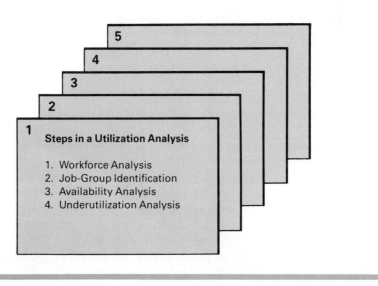

Steps in a Utilization Analysis

1. Workforce Analysis
2. Job-Group Identification
3. Availability Analysis
4. Underutilization Analysis

Figure 20.2   A Flipchart
This may be the most versatile form of visual aid.

1. Definition Phase

●Problem Analysis
●Project Planning

Poster boards have several inherent disadvantages. The poster board rigid enough to stand on its own is often fairly small, and thus the size of your visuals—and your audience—is relatively limited. The size of the poster board also limits the complexity of the visuals; everything has to be very simplified, and words come across better than all but the simplest drawings. (See Section 4 of this chapter for more on poster boards.)

Flipcharts are a time-tested and popular medium for speakers (see Fig. 20.2). The flipchart should be big enough for all of your audience to see its message, and you need a stand especially designed for its use. Using a flipchart allows you to use a number of different visuals—as many pages as the flip-chart has; the visuals can be any color and can be words, simple cartoons, or fairly complicated drawings. The flipchart has a nice simplicity, even a spontaneity to it, that many other kinds of visuals lack. Many people like to use contrasting colored markers to highlight the flipchart pages as they speak. Although a little less portable than poster board, the flipchart's increased size also makes it usable in more situations.

The flipchart's main disadvantages are that it is relatively large and requires a proper stand. Some people claim to be bothered by the sound of the pages flipping, but they are in the minority. Other people think using a flipchart is trite or old-fashioned. But most people believe flipcharts have been used so much and so long precisely because they are so effective.

Overhead projectors bring us to higher-technology visuals, those with more possibilities and more problems. In addition to the machine itself, the overhead projector requires transparencies, a screen, and a power source. These in turn require extension cords, bulbs, and the apparatus for constructing the transparencies. Many things can go wrong, but also a great deal can go right. Using the overhead projector gives you light, color, a very large image, the possibility for as many visuals as you want, and the ability to be very detailed—and you retain the ability to mark on the visuals as you go.

Overhead projectors have other disadvantages besides their complexity. Inevitably you lose some eye contact with your audience. Many overheads are noisy in operation. Transparencies can be difficult to handle, especially if you use them without frames. And some people think that for a five- to ten-minute talk the overhead projector is, in general, too much trouble.

Slide projectors give you beautiful light and colors, big images, fast-changing images, and a chance to impress an audience with the kind and amount of preparation you have done. However, slides take careful preparation, slide projectors can malfunction, and you cannot use one without a screen. You lose eye contact with your audience, and again, this may be just too much trouble for a short talk.

No matter what kinds of visuals or props you use, certain principles apply, as seen in the following list.

Principles for Using Visuals or Props in Oral Presentations

- Use visuals or props when the subject cries out for it, when a point is particularly complex, or when a point is particularly important.
- Never fail to *double*-check the availability and working order of your visual or prop. Ensure that the projector is there and works, that the screen is there, that the cord is long enough, and, if necessary, that there is someone to run the machine while you speak. Ensure that the flipchart and poster boards are on hand and that the kind of stand you want is available. Never take for granted that someone else will do or has done these things for you. As the speaker, they are *your* responsibility.
- Whatever kind of visual you use, look at your audience, not at the visual. If you use an overhead projector, look at the audience, not at the machine or its screen. This usually means that you need someone else to change the transparencies.

- If you use a slide projector, position yourself in front of the audience, looking at them. If you want to operate the machine yourself, use a long extension cord and a remote control switch. If you don't want to, arrange for someone else to operate it for you.
- Finally, regardless of the kind of visuals you use, specifically adapt them to the situation you are in. Consider the room, the nature of your audience, their purpose, and the nature of your subject. And consider whether the amount of energy you invest in preparing your visuals is justified by what they could add to your speech.

2.3   Tips on Making Oral Presentations

A number of elements of making oral presentations can be learned through practice. Like skill at writing, skill at public speaking comes naturally only to a few; the rest of us must work at it. The most important thing to remember in any kind of public-speaking situation is that *long after your audience has forgotten what it is you said, they will remember what kind of person you seemed to be.* In many ways, the way you present yourself may be as important in the long run as any particular thing you say. Over and again, you will hear people say things like, "Yes, I heard Joe's talk last week. He seemed like a pretty sharp fellow." Or "Yes, I heard Joe's talk last week. He didn't really seem very interested in what he was doing." So think carefully about the way you present your material. The tips for speakers in the next list are like tips about writing: you can learn them, but only *experience* can really teach you their value.

Tips on Making Oral Presentations

- Bring enthusiasm to your talk; your audience's attitude will mirror your own.
- Be well dressed and well groomed; look the way you want your audience to envision you when you're at your best.
- Save your mouth for talking; don't chew gum or smoke while you're speaking. Some people find it merely distracting, and others find it offensive.
- Decide in advance what to do with your hands. It's fine to rest them on a lectern, but don't put a death grip on it.
- Plan how you will stand. Don't fall into shifting from one foot to another or standing only on one foot.
- Concentrate on speaking slowly. Most speakers go too fast, trying to do too much.
- If you use notes, prepare them neatly. Speaking from tatters and scraps of paper isn't effective.
- Beware of passing things around while you're speaking. Handouts and props are distracting to you and your audience if they're moving around the room while you're speaking.

- Establish and maintain eye contact with your audience. People like to feel that you are talking to *them,* and the way to do that is to look at their eyes. Look briefly at your visual aids occasionally, but spend most of your time looking at your audience.
- Monitor your word choice. Use words appropriate to your audience in both formality and technical level.
- Ensure that your visuals are ready. If you're using a chalkboard, draw the figures on it in advance. If you're using poster board or a flipchart, set up the stand in advance. If you're using an overhead or slide projector, focus the machine ahead of time.
- Avoid asking opening questions of a "cold" audience. It works only on talk shows, where a comic has been out "warming up" the audience for fifteen minutes before the cameras are turned on.
- Never start by saying, "This will be easy" or "This will be hard." The first statement invites the audience not to listen, and the second invites them not to understand.
- Be aware of the importance of your listener's questions. Anticipate the obvious ones and have answers ready. And answer all questions carefully (discussed in the next section of this chapter).

2.4   Answering Questions

One of the most important parts of any presentation in a professional setting is answering questions. Your audience has listened to you, and now you must listen to them. To answer questions well, you must first *listen* well. Establish eye contact with your questioner, and don't begin to answer until the person has finished asking the question. Give the question as thoughtful an answer as you can. Give special treatment to all questions, but especially to difficult questions, those you can't answer, or those that are challenging or threatening.

Dealing with Questions That You Can't Answer. Sometimes you will be asked a question you just can't answer. Usually this means you don't have the right information. You can deal with the situation by saying, "I'm sorry—I just don't know the answer to that right now, but I'll find the answer and give you a call on it as soon as I do." Or you can refer the question to someone else who you think may have the answer.

Another way to deal with a question you can't answer is to deflect it back to the person who asked it: "That's a question I hadn't anticipated at all, and I don't know the answer right now." Then say either "What do you think the answer is?" or "It's an unusual question—why do you ask it?" Many times a questioner who asks an unusual or very particular question will either have his or her own idea of the answer or a specific reason for asking. Given a chance to express their idea of the answer or their reason for asking it, some of these questioners will be satisfied.

Dealing with Threatening or Challenging Questions. Occasionally, you will be asked a question that tells you the questioner is challenging or threatening you. When you detect a hostile attitude in a questioner, you can deal with it more effectively if you follow this advice. Listen carefully to the questioner, keeping eye contact and a closed mouth. When the questioner finishes, the first thing you should do is to make sure you understand the question: "Let me be sure I understand your question. As I understand you, you want to know . . ." Then do your best to restate and rephrase the question. This technique gives you a number of advantages. By carefully *listening* to the entire question, you show the questioner that you take him or her seriously. By carefully *restating* the question, you ensure that the question you *think* you heard is really the one the questioner asked. And you also give yourself extra time to formulate a satisfactory answer. Your answer should show that you take seriously

1. the *question.*
2. the *questioner.*
3. the questioner's *attitude.*

If the person is frustrated, you need to say something like, "I can understand how you feel frustrated that . . ." Of course, at some point you have to answer the question, but this approach will establish some goodwill in your audience, so that both the questioner and the rest of the audience will more readily accept your answer.

3.   *Characteristics of Poster Presentations**

Poster presentations offer several advantages over the more traditional ten-minute speech. A poster displayed for eight hours in a common viewing area may well reach more people than a speech— whose life span is only ten minutes—ever could. And the well-thought-out poster automatically adjusts the level of its content to varying levels of interest among viewers: those who have only passing interest can glance at it and pass on by; those who have casual interest can stop, quickly scan the visuals and text, pick up an information sheet, and keep walking; and those who are really interested in what you're presenting can stop and talk with you. This last point—that people who are really interested can stop and talk with the presenter—is a major advantage of poster presentations.. Finally, for viewers, an exhibit hall full of poster presentations allows them to

---

* Much of the information about poster presentations in this chapter is adapted from Diane L. Matthews, "The Scientific Poster: Guidelines for Effective Visual Communication," *Technical Communication* 37 (August 1990): 225–232.

sample 50 or 100 different products or ideas in a way no day-long series of short talks could possibly match.

The heart of the poster presentation is the poster board itself, along with the visual and text elements fastened to it. While many professional meetings have their own guidelines for poster presentations, and in some settings elaborate displays are the norm, not the exception, generally the basic poster presentation will be on a background that is four feet by six feet or four feet by eight feet (cork board, poster board, or some other similar material), in one of four possible configurations:

1. *The pin-up format.* Each element (text or visual) is separately matted, and the matted elements are pinned to the poster board when the presentation is needed.
2. *The stand-alone tabletop format.* Three panels that are two feet by four feet are fastened together at the back allow the poster presentation to stand on a tabletop or be folded together to be moved.
3. *The floor-standing format.* Generally this is a professionally produced booth, elaborate to produce and requiring custom transportation.
4. *The roll-up format.* This is one continuous sheet made to be rolled up when it needs to be moved and then unrolled and pinned to a board for display.

For students, the pin-up or stand-alone tabletop formats are easiest to produce. Whichever format is used, the display is usually accompanied by a presenter (at least part of the time it's up) and handouts providing more information for interested viewers.

4. Design of Posters

Your poster presentation must be designed to tell its story *at a glance,* so that people passing by can get the gist of your work while, literally, passing by. Because of this need for the presentation to be visually absorbed quickly, the most important part of any poster presentation's design is the relationship among its text elements and its visual elements. Figure 20.3 shows several possibilities for arranging elements on the poster board.

For poster presentations, the critical elements of graphic design are (1) the flow of the information on the board and (2) the design, contrast, and placement of the text and visual elements. The *flow*—the way a viewer's eyes scan the presentation—usually will start at the upper left-hand corner and then proceed either to the right (as in Figure 20.4) or down (as in Figure 20.5). There must be some logical kind of flow, and whatever pattern you begin, you must continue. Once in a while, a poster presentation appears in which there is no

Figure 20.3   Sample Layouts for Poster Presentations

Figure 20.4   Left-to-Right Visual Flow

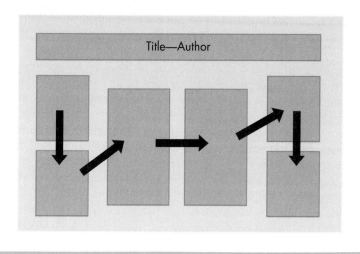

Figure 20.5   Top-to-Bottom Visual Flow

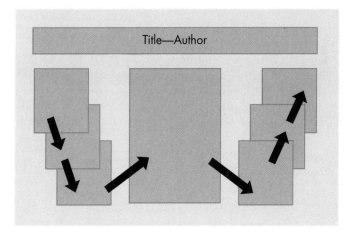

particular relationship between one element and another. If you watch viewers looking at such a presentation, what you'll see on their faces is a brief expression of puzzlement—and then they move on.

The *design, contrast, and placement of the text and visual elements* play a major role in viewers' responses to poster presentations. The following sections discuss the design of individual verbal and visual elements; the concern here is how all of the elements fit together in design terms. Beyond the flow of the elements, the way they all work together visually is critical—their sizes, shapes, uses of color, framing, the white space between them—everything that determines how they are seen by the viewers. Given the limitations established by the necessity for the poster to flow, coupled with the 24 or 32 square feet the poster yields to work with, the possible arrangements of the elements are limited. You need to keep experimenting with the overall layout until its appearance satisfies you; then test it on three other people.

Each element on the poster board needs to be framed with a mat and outlined inside the mat with trim tape. Generally the color of all the poster's mats will be the same; if you use contrasting colors, use them very conservatively. In most cases, it's easy to use too much color and safer to risk using too little. The mat sizes need to be regular and proportional to the sizes of the various elements being framed, but the mat sizes also need to allow for ample white space to separate the various elements.

4.1   Text Elements

Successful poster presentations are always mixes of verbal and visual elements. Even the verbal elements need to be presented highly visually. In both professional typesetting and most word processing programs, the size of type is designated in *points*. The higher the point size of a letter, the larger it is, as seen in the following list.

9 Point Type

12 Point Type

14 Point Type

18 Point Type

24 Point Type

In poster presentations, use of larger type can help headings stand out.

For the title, use

72 Point Type

For headings, use

48 Point Type

The title should be done in *all caps,* and not more than seven or eight words long. The headings in a poster presentation should not exceed four words in length and there should be no more than six or seven major headings in the presentation.

Margins within the mat need to be at least one inch all around, and the right edge of the text needs to be "ragged" (not justified) for greater readability.

The text itself needs to feature short sentences and, whenever possible, lists. Listed items should be in "summary" (or "phrase") style, in which items begin with the infinitive, *-ed,* or *-ing* forms of verbs; especially avoid the passive voice. At all costs, avoid solid pages of text. The content needs to be expressed in topic-based chunks, not paragraphs and pages. Figures 20.6 and 20.7 show before-and-after views of a short technical pasasage being adapted to a poster presentation format.

It's true that the version designed for poster presentation loses a bit of the meaning of the original. That more detailed and more

Figure 20.6   A "Before" Passage, from a Research Report

Controller Tuning Results

This section presents the results of simulating the first-order process using the nonlinear optimal control algorithm with varying control horizons and prediction horizons. The nonlinear optimal control algorithm model uses the same gain and time concept as the process, and the simulations contain no unmeasured disturbances. Because ideal model parameters are used and no unmeasured disturbances are present, the feedback cultivation algorithm is not implemented in these simulations.

Too low a value for the control horizon results in a very sluggish response of the controller to the setpoint changes. Too high a value for the control horizon results in a faster but more oscillatory response of the controller. Changes in the prediction horizon affect the degree of oscillation of the controller's response.

Figure 20.7   The Passage from Figure 20.6, Adapted for Poster Presentation
This would be one verbal element in a six- or seven-element poster presentation.

Controller Tuning—Simulating First-Order Process

Conditions: Nonlinear optimal control algorithm
                 Varying control horizons
                 Varying prediction horizons
                 Using ideal model parameters
                 Using no unmeasured disturbances
                 Not implementing feedback cultivation

Results: * Low value for control horizon →
                     Sluggish controller response
                 * High value for control horizon →
                     Oscillatory controller response
                 * Prediction horizon changes →
                     Controller oscillation changes

subtle area of meaning is appropriately reflected in the informational handouts available at the exhibit and in the comments the presenter makes in conversation with interested passers-by.

For all of this text, you need to remember people will be looking at the presentation from a distance of two to four feet. Thus, the letters in the text generally need to be one-third of an inch high (36-point type).

4.2 Visual Elements

The visual elements of poster presentations can include any of the kinds of visuals described in Chapter 10: graphs, tables, line drawings, and so on. The important thing to remember, once again, is the distance from which and speed with which they will be viewed. Keep your visuals simple (line graphs, for example, should not have more than three or four lines); make sure the lines in the visuals are big enough and dark enough to be seen easily; use contrasting colors or shading to make elements clearer; and don't overload visuals with more than one message—don't try to do too much with one visual.

The easier-to-read kinds of visuals are generally more successful in poster presentations: pie graphs, bar graphs, and line drawings. Line graphs and tables take too long to read to be effective unless they are very simple. Remember that the whole poster presentation has to make basic sense *at a glance;* don't count on things like, "Well, they can read it if they really try." The level of detail that elicits a comment like that is more appropriate for informational handouts than for a poster presentation.

5. *The Presenter's Roles*

In most poster exhibits there will be a certain number of specified hours during which the person who did the work reflected in that presentation (or a representative of that team) will be available at the exhibit to discuss the work with interested people. When this is your role, remember to

- Distribute informational handouts freely. Make sure they have your name and the date on them, in case there is any question later about who did what work when. If you have business cards, bring a supply of them to distribute as well.

- Be prepared to talk about the work the exhibit describes. If you think you might need to review the full research report before you go to the exhibit, by all means do so; you're there as the expert, so you need to be loaded with appropriate information.

- Keep a list of the names and addresses of people who stop to ask you questions. The list will be handy when you need to recommend reviewers for follow-up work, prove to others how many people are interested in the project, or have professional contacts in other places.

EXERCISES

1. Give a short, informal introduction speech to your classmates and teacher. Topics to cover might include your name, the name and location of your home town and one interesting fact about it, your major, your area of specialization within that major, your reasons for choosing that area of specialization, your job plans, and one personal fact (your hobby, for example).

2. Adapt one of the papers you have already written for this class as the basis of a five- to seven-minute oral presentation to your classmates. If your teacher was the only audience for the original paper, remember you may have to change your approach for this presentation because the audience is different. Plan on using at least one visual aid or prop.

3. Adapt one of the papers you have already written for this class as the basis of a poster presentation to your classmates. If your teacher was the only audience for the original paper, you may have to change your approach for this presentation because the audience is different. Plan on using either a pin-up or stand-alone poster format.

4. Choose a professional or scholarly article from a periodical in your field to adapt for poster presentation. Your audience should be other students and faculty in your major. Plan on using either a pin-up or stand-alone format.

5. Adapt the content of the major report you are writing for this class for presentation in a five- to eight-minute informal talk, with your classmates as the audience. Use at least one visual aid. Either summarize your whole report, or focus on one aspect of it in detail.

6. Adapt the content of the major report you are writing for this class for poster presentation, with your classmates as the audience. Either summarize your whole report or focus on one aspect of it in detail.

7. Make modified versions of the audience analysis and audience adaptation lists in Chapter 3, tailoring them specifically for speaking (rather than writing) situations.

───────

A ▶ indicates a case study exercise.

➤ 8. Attend a speech given on campus by a speaker who is a professional, and write a critical analysis of the presentation. Because the readers of your analysis (your teacher and your classmates) will not have attended the speech, decribe in your report what the speaker said and did in addition to analyzing it.

➤ 9. Interview three professionals in your field about the requirement in that field for public speaking skills. Give a short speech to your class about what you learn.

APPENDIX
A Short Course in Writing

Whehen you look at the demands professional life will make on your intelligence and ability, you will probably see there are two kinds of tasks you have to perform successfully:

- Technical tasks
- Communication tasks

Up to this point in your college education, you've probably worked almost exclusively on learning how to handle the technical tasks professionals in your field are required to perform. The requirement that you successfully complete a senior project or long report in your major may be your first exposure to the *necessity* for combining solid technical knowledge with performing communication tasks successfully. It will not be your last.

A hundred years ago, publication was seen as the culmination of a years-long scholarly endeavor—something that happened slowly, carefully, and rarely. Today, in very nearly every field, publication (or at least writing) is a condition of continued employment. A hundred years ago, professionals in business, industry, and government knew that communication skills were essential for success; the courses we know today as freshman English were created in part as a response to a perceived need for college graduates to have better communication skills. Today's international marketplace and information economy require any professional who wants to succeed to be able to count on solid communication skills as an important professional asset. We've all heard that any professional who is not computer literate is rapidly becoming a technological second-class citizen. As you face the problems of writing that taking upper-division courses and entering professional life present to you, you should be beginning to realize that any professional who cannot count *writing* as an *asset* is well on the way to becoming another kind of second-class citizen in professional life.

What can you do to become a better writer? Ultimately, no amount of reading about writing, and no one book (however good) can do much to improve your writing. More than anything else, making a significant and lasting improvement in your writing requires *determination* on your part—something only you can provide. Beyond that, it requires practice, then feedback, then more practice, more feedback, and so on. And if you can sequence the types of practice you do and the levels of feedback you get, the process works even better.

1. Axioms

1. You can only get better at writing by writing.

- The reason is that writing is a complex, cognitive skill, and not a body of knowledge.

- The *formula* is Practice/Feedback/Practice/Feedback/Practice/Repetition.

2. You have to commit to multiple drafts.

- The reason is the combination of the pressure you're under when you're doing this kind of writing, the degree of complexity of your subject matter, and the length of the final documents you need to produce. Some things you simply cannot do at an acceptable level in one attempt—that includes effective professional writing.

- The formula is to start writing early, and to write often. In your first attempts at writing something down, don't even try to get your writing anywhere nearly perfect, just try to get the content (the subject matter, however much of it you know) down on paper. You can make it good later. *Nothing you write as a professional will ever be good enough in just one draft (or—usually—even in two).* Accept that fact and use it to your advantage.

3. Attack the first draft!

- The *reason* is that you're only human. Left to yourself, you'll put the writing off as long as you can. If you put this kind of work off until you feel like doing it, until you've done everything else, you've already violated axioms 1 and 2, and you're starting out in a crisis mode, leaving yourself no room for error, no chance for quality, no good options at all. So take that strongly proactive stance—attack the first draft!

- The *formula* is to inventory your situation, assemble your materials, then attack! Don't dawdle over which word is the exactly right one, over the spelling, over the punctuation; push that technical subject matter out of your head and onto the page in front of you. Once you commit to multiple drafts, you don't have to worry about getting the first one perfect. (Now, you can't *attack* when you're worn out at the end of the day—you've got to set aside some time when your mind and body are fresh. And you've got to use that time well.) Set a goal of a certain number of words or pages (or a certain section) to get done in that time period, and attack until you achieve that goal.

4. Use a word processor.

- The reason is that by getting rid of as much of the drudgery of writing as possible, you free your energy for the important things—like making your writing as clear as you can.

- The formula is IBM or Mac, Word or WordPerfect.

2. *The Formula for Writing*

Writing is a *process*. You can't get what you want out of your writing simply by focussing on the finished product; you've got to pay attention to the steps along the way. The most productive way to look at the kind of writing you need to be doing is to see it as having these steps.

Planning. Every successful writer plans. The plans take many different forms, from conceptual designs to traditional outlines, but every successful writer plans. In your situation, planning should have two phases: a task analysis, and a structural map. Those phases are explained on the following pages.

Drafting. Attack the first draft. Don't wait until you have all the material gathered. Write early and often. Don't worry about perfection (yet). Get the material down on paper. How else can you know if you have all the material you need? How else can you know whether you understand the material? How can you know what you think until you see what you say?

Revising. It's during revising that you make your writing good. Revising should be done in five waves or cycles:

- Revise for overall rightness—do you have the right material?
- Revise for overall structure—does the letter or report as a whole, and does every section, have a beginning, middle, and end? Is there a clear structure, and will your reader see it, recognize it, and find it right?
- Revise for paragraph structure.
- Revise for sentence structure.
- Revise for your own, idiosyncratic errors.

(This five-wave revision process is explained in more detail later on in this appendix.)

Editing. During editing you are preparing the final draft of your project for submission. Professionals discuss editing as having at least ten forms. These are explained later in this appendix.

In fact, the process of writing is not nearly so neat as it is described here. The various activities involved hardly ever sort themselves out quite this way. But you *can* improve your writing if you go about it in a more orderly way, and the order you need to impose on it is the one described on the following page.

2.1 Planning

In this model of the writing process, planning has two parts: doing a task analysis, and making some kind of structural map.

Do a Task Analysis. Answer these questions:

1. What are your goals, as a writer, on this project?

2. What are your reader's needs from this project? What things does your reader need to see in this piece of writing? What can your reader get out of it?

3. What are the characteristics of the subject matter, of the message you are trying to communicate? Is it highly controversial, very abstract, highly mathematical, very technical, hard to define, etc.?

4. What are the characteristics of the *forum*, the setting your writing will appear in? How do you plan to adapt your work to them?

Make Some Kind of Structural Map. Early in any writing project you need to come up with some idea of what structure you think you'll be using for your writing. Often that structure is dictated to you by tradition in your field, by the situation, by your reader, or by the nature of the material you're writing about. You've probably all seen this structure:

Introduction
Literature review
Purpose
Procedure
Results
Discussion
Bibliography

Such a pattern is essentially a *chronological* pattern, loosely imitating the sequence of a piece of research. In fact, there's a fairly small number of typical patterns. Here's a list:

Chronological
Simple to Complex
General to Particular (or vice versa)
Functional Organization
Analytical
Spatial
Order of Importance
Pros and Cons
Classification
Task-oriented Organization

2.2   Drafting

Set aside one hour each day for dealing with the communication task. What time of day is best for you—when is your generative, creative energy at its peak? What kinds of things do you need to do to make sure you're at that peak? Can you do your most important work when you're tired, when your stomach is full, when you're still asleep? For students as well as professionals, many times the key is to make writing the first thing they do in the morning rather than the last thing they do at night.

2.3   Revising in Waves

Because writing with a word processor makes doing multiple drafts so easy, more and more people are coming to view revising as just one wave in a writing process comprised of a number of waves.

The first wave in writing becomes gathering information and making some kind of structural plan. That structural plan could come from some pattern you see as inherent in the material you're writing about, or some pattern required externally for that kind of document, or some combination of the two.

The second wave is to compose a first draft (often called a zero draft) at the keyboard as fast as possible, not worrying about grammar and punctuation, skipping over parts that might bog you down, just trying to get the basic information down onto the pages, to see if you have enough information, and to have a beginning, middle, and end.

The third wave is revision on the level of overall rightness and overall structure. First, make sure what you've done meets the criteria set forth in your task analysis. Then print out a draft and mark it up with an eye toward improving its overall structure—you may need to move whole sections, delete parts, and add new parts. Then make those changes in the on-line version of your report. You want to ensure that the structure you're using is one that your reader will recognize easily, and that it fits the subject matter.

The fourth wave is to revise on the level of paragraph structure. Print out a copy and look at each paragraph individually—does each paragraph have a strong and easy-to-see structure? Does each paragraph have enough detail of the right kind? Mark the appropriate changes on your hard copy, and then make those changes in your on-line version. Here's how to figure out paragraph structure. Pick a paragraph. Number each sentence. Now make a grid with five columns, and label the columns from A to E. Put the number of the most general sentence in the paragraph under A, and arrange the rest accordingly. Make sure you do shift levels, that you do so in an orderly manner, and that you use a predominately right-branching structure.

The fifth wave of revision involves sentence-level revision. Print out a copy of your report and work on each sentence (described in more detail in the following pages) to make each one tighter and clearer. Mark your hard copy, and then make those changes in your report. Sentence-level revision may well be the single most important kind of revising, the most time consuming and the most rewarding. You want to control both the length of your sentences and the ordering of their elements:

- *Check your sentence length.* Be especially suspicious of any sentence longer than 15 words (about a line and a half). Those sentences, in particular, must meet the *ordering of elements* criterion.

- *Check the ordering of elements in your sentences.* You need to show a very strong preference for what we call *normal* order: subject—verb—(optional) object or complement.

Frequently the sentences that are real problems violate both of those criteria. Remember, you shouldn't be worrying about these things at all when you are *drafting* (if you do, those creative juices will dry right up); we're talking about *revising* now.

The last wave of revision is the one when you look for punctuation, grammar, spelling, and usage errors—you become an editor (or English teacher). Once again, print out a copy, mark it, and make those changes on your on-line file. Here, you worry about those things your English teacher used to circle in red, especially your own idiosyncratic, grammatical errors. (The sections on editing and on grammar and usage problems, later in this appendix, present this wave in more detail.)

More on Sentence-Level Revision. When you are doing sentence-level revision, you need to work toward three goals: clarity, economy, and straightforwardness.

- *Clarity* means the sentence has a meaning that presents itself to the reader (the reader doesn't have to dig it out), and the sentence has the exact meaning you want it to have (it cannot be easily misunderstood).

- *Economy* means your sentence gets its meaning across in an economical manner—one that does not use more words than necessary.

- *Straightforwardness* means that the reader can process the sentence from beginning to end without having to loop back and re-examine some elements—usually this means the vast majority of your sentences use "subject—verb—(optional) complement or object" word order.

Effective sentence-level revision is not so much a matter of knowledge as it is a matter of taking the time and effort to put a meticulous, smooth finish on your work. Although getting the *content* and larger *structural* elements of your document right is essential if your writing is to be able to stand up to the challenges of difficult subject matter and demanding readers, doing careful and thorough *sentence-level revision* is what makes your writing shine.

Here is a more detailed explanation of the criteria of clarity, economy, and straightforwardness:

Clarity. Does the sentence have one and only one meaning, and does that meaning present itself to the reader with a minimum of effort on the reader's part? Often sentence length is critical here. Is the sentence length excessive? Is the sentence structure too complex? Does the sentence use too much nominalization (for example, too many words that end in *-tion*, or use too many four-noun modifier clusters)?

Economy. Does the sentence accomplish its meaning with as few words as reasonably possible? (Note this does not mean the rock-bottom minimum, just the absence of deadwood.) Specifically check the verbs—are too many passive, are too many weak?

Straightforwardness. Can the reader process the sentence straight through from beginning to end without having to loop back, to re-read earlier elements? In particular, check sentences that are longer than 15 or so words to make sure that if they *don't* have "normal" order (subject—verb—(optional) object or complement) they nonetheless are straightforward and clear.

Examples of Clarity

ORIGINAL: The corrosion behavior of selected stainless steel welds (and conditions) will be measured in the laboratory and compared with the field results.

REWRITE: Laboratory tests will measure the corrosion behavior and conditions of selected stainless steel welds; lab results will be compared with field results.

ORIGINAL: The amount of alloying element added is important because depending on whether the element added is an austenite or a ferrite former, the tendency to form delta ferrite at the solution treatment temperature will increase or decrease.

REWRITE: The tendency to form delta ferrite at the solution treatment temperature will increase or decrease depending on whether the element added is an austenite or a ferrite former; thus the amount of alloying element added is important.

ORIGINAL: The problem with wild garlic is not that it competes with wheat but in its reduction of wheat grain quality at harvest.

REWRITE: The problem with wild garlic is not that it competes with wheat, but that it reduces the quality of wheat grain at harvest.

Examples of Economy

ORIGINAL: A synthetic input generator will randomly select inputs which have the same distribution as found in the input space analysis. The synthetic input generator is usually a software package that generates random numbers.

REWRITE: A synthetic input generator, usually a software package that generates random numbers, will randomly select inputs with the same distribution as found in the input space analysis.

ORIGINAL: It is during this time that scapigerous wild garlic plants produce aerial bulblets at approximately the same height as the grain head in wheat.

REWRITE: During this time scapigerous wild garlic plants produce aerial bulblets at approximately the same height as the grain head in wheat.

Examples of Straightforwardness

ORIGINAL: In general, the austenitic types have high ductility, low yield strengths, and the highest corrosion resistance of all the stainless if properly treated by annealing and quenching.

REWRITE: If properly treated by annealing and quenching, the austenitic types generally have high ductility, low yield strengths, and the highest corrosion resistance of all the stainless.

ORIGINAL: During solidification, segregation of residual elements such as sulfur and phosphorous to grain boundaries contributes to cracking.

REWRITE: During solidification, residual elements such as sulfur and phosphorous segregate to grain boundaries. This segregation contributes to cracking.

Here are some sentences to practice on.

1. Explorations of the importance, physiology, economies, equipment, media, method, requirements during growth, what is happening now, and what the future holds for aseptic methods of propagation are included.

2. The total drag for such craft when consideration is limited to craft with fully submerged hydrofoils and hydrofoil wave drag is represented in drag polar form can, with a precision that matches that of the laboratory, be expressed analytically.

3. Subsequent configuration finalization in concurrence with postulated hardware interrelationships and full utilization of integral criteria qualifications implies anticipation of future growth dependencies.

4. Factors such as an extension of the market or changes in the income of people, caused by population growth or other external factors such as changing tastes or product preferences, can cause a shift in demand curves.

5. Feces-urine radionuclide concentrations decrease significantly during the first 10 days and then level out over time after removal from a contaminated pond.

6. These levels would probably not present a hazard to man since it is unlikely that a transient duck would retain high levels of radioactivity and that he would consume a duck with such high levels.

7. This thesis project included an extensive analysis of recycled surplus school projects drawing parallels between successful and unsuccessful projects based on end use.

8. In this example, the ability of a feedforward network with a hard-limiter (step function) transfer function to perform pattern classification is examined for single-, double-, and triple-layer networks.

2.4   Editing

Editing is the very last stage of writing. You should not get involved in it at all until you are absolutely certain the content of your project is exactly right. Here is a list of ten different kinds of editorial practices. (They are explained in more detail in Chapter 11). Different projects will require different combinations of these; some will require as few as one or two, while other projects need them all.

- *Coordination Edit.* Make sure all the pieces that need to be available in a certain place at a certain time are indeed available. (Results of questionnaires, permission forms, interlibrary loan materials, sections from other people.)

- *Policy Edit.* Make sure no institutional policies are being violated by what this document says or does. (Research on human subjects, copyrights, etc.)

- *Integrity Edit.* Make sure that all elements referred to in the text—charts, tables, visuals, and so on—are actually there.

- *Consistency Edit.* Make sure what you say in one chapter does not contradict what you say in another.

- *Screening Edit.* Check for spelling errors, subject-verb agreement problems, incomplete sentences, and obviously incomprehensible statements (where words have obviously been left out, or word order is garbled).

- *Copy Clarification Edit.* Make sure whoever is preparing the final draft can read and interpret everything you submit to that person—handwritten notes, Greek letters, subscripts and superscripts, equations (are the breaks clearly marked?), etc.

- *Format and Mechanical Style Edit.* Make sure everything is done the way the recommended style manual requires.

- *Language Edit.* A full treatment of grammar, punctuation, usage, parallelism, and other similar features of your writing.

- *Substantive Edit.* Make sure the material is grouped and subdivided in a rational manner, that apparent contradictions or inconsistencies are resolved, and that the presentation is complete and coherent.

- *Usability Edit.* If your project tells someone how to do something, make sure your directions have actually been tested by that kind of person in that kind of setting.

Doing effective editing means taking complete responsibility for every feature of the document, from its typing to its meaning. Don't get involved in that kind of editing until you have gone through all the other parts of the writing process described here, and don't underestimate the time it will take!

3. *A Brief Review of Sentence Grammar*

Sooner or later we all have to learn grammar. You can't hope to write at a professional level in any field if you still make basic errors in such areas as sentence structure, punctuation, and usage. Although it may not seem logical, it is nonetheless true that professional audiences frequently make decisions about the technical merit of reports based on the grammatical correctness of the reports.

3.1 Parts of Sentences

The grammar presented here is designed to help you learn to use the basic elements of sentence construction more correctly in order to produce more effective sentences. This grammar, a combination of traditional and generative approaches, divides sentences into sentence bases, openers, closers, and interrupters.

Sentence Bases. The basic building block of any sentence is a sentence base—a noun/verb unit (also called a *clause*).

A sentence base consists of a noun, a verb, and (optionally) an object or other kind of complement.

$$SB = SV(O)$$

Sentence bases can be independent or dependent. Independent sentence bases can stand alone as sentences; dependent sentence bases cannot.

Independent sentence bases (complete sentences)

The information is in a different data base.
Enter a carriage return and the monitor returns.
The user must have a unique PPN number.

Dependent sentence bases (sentence fragments)

that the program identified
where you were before using the Select command
which can be obtained from the CSD

The difference between the two kinds of sentence bases is easy for some people to judge by intuition. An English instructor might say, "The first three strings of words can stand alone as sentences; the next three can't. It's common sense." In case your common sense doesn't work that way yet, here's how to tell the difference: dependent sentence bases begin with words that signal their dependency. Those words may be *relative pronouns or subordinate conjunctions.*

If the sentence base you are looking at begins with one of the words in the following list of relative pronouns and subordinate conjunctions, it is a dependent or subordinate one, and it cannot stand alone as a sentence.

Relative Pronouns

which	that
what	where
when	how
who	whoever
whom	whomever
whose	

Subordinate Conjunctions

if	because
after	although
as	because
before	if
once	since
that	though
till	unless
until	when
whenever	where
wherever	while
as if	as soon as
as though	even though
in order that	in that
no matter how	so that

If you have a series of words that you want to test to determine whether or not it constitutes a grammatical sentence, there are two tests you must apply:

Two Tests for Distinguishing Sentences from Sentence Fragments

1. Ensure that the series of words has a noun and a verb.
2. Ensure that the series *does not* have a subordinate conjunction or a relative pronoun at its beginning.

Only if the string of words passes *both* tests can it be called (and punctuated as) a sentence.

Several things can be done to sentence bases. We can add a group of words

- to the front of the sentence base—an *opener.*
- to the end of the sentence base—a *closer.*
- to the middle of the sentence base—an *interrupter.*

If the addition is an essential part of the sentence (if its meaning is necessary), and if it is not an opener, joining it to the rest of the sentence requires no punctuation. Otherwise (if the addition's meaning is not necessary, or if it is an opener), it must be spliced on with a comma. (Combining two independent sentence bases has its own rules, discussed in Section 3.3.)

Openers. An *opener* is a word or group of words added to the beginning of a sentence base. Openers are usually connected to the sentence base with a comma. The longer the opener, the stronger the need for a comma. With more than one opener, each one gets a comma.

Examples

A sentence with an opener: With more than one opener, each opener gets a comma.

A sentence with more than one opener: With more than one opener, and with long openers, be sure to use commas.

A sentence with a short opener and no comma: Today you need to learn grammar.

Closers. A *closer* is a word or group of words added to the end of a sentence base. Closers can be spliced on with a comma, a dash, or parentheses. Although there is no theoretical limit to the number of closers you can use, as a practical generalization you probably shouldn't use more than one at a time.

Examples:

A sentence with a closer: A closer is something added at the end, often to pack more information into the sentence.

A sentence with too many closers: A closer is something added at the end, often an afterthought, sometimes an important qualification, probably a place for revision, and especially a nuisance when there are too many of them.

When the closer is a group of words (such as a dependent sentence base), you must decide whether to join it to the rest of the sentence with a comma. This depends on whether the group of words to be added is *restrictive* (essential) or *nonrestrictive* (nonessential). Restrictive means that the group of words is essential to the meaning of the sentence. Nonrestrictive means that the group of words is nonessential. Sometimes this is a judgment call on your part, and sometimes the distinction is clear cut. In one notable instance, custom rules: In most scientific and technical writing, if you introduce the string with *which,* it's nonrestrictive. If you introduce the string with *that,* it's restrictive.

Examples:

Nonrestrictive: You next must access the HELP program, which tells you the allowable input.

Restrictive: You next must access the program that tells you the allowable input.

Interrupters. An *interrupter* is a word or group of words placed within the sentence base to clarify and specify the ideas in the base. Interrupters are spliced in with paired commas, parentheses, or dashes. Interrupters also can be classed as restrictive or nonrestrictive, and once again the restrictive ones are not set off from the rest of the sentence. *Which* again typically signals that the interrupter is nonrestrictive, and *that* signals that it is restrictive.

Examples:

Nonrestrictive: This file, which can be accessed at any level, tells you the current status of the system.

Restrictive: A file that can be accessed at any level is called a common file.

When the interrupter is nonrestrictive, commas, dashes, or parentheses may be used.

Examples:

Some elements of sentences—such as interrupters—can be punctuated in several ways.

Some elements of sentences, such as interrupters, can be punctuated in several ways.

Some elements of sentences (such as interrupters) can be punctuated in several ways.

Sentence-Base Rules. As Chapter 2 explained, the time to consider style is when you are revising. Don't worry too much about the elements presented here while you are writing your first draft. When you turn your attention to revising, you will find that with only the building blocks of sentence base, opener, closer, and interrupter, you can already begin to make better sentences. The following list gives four rules for building better sentences, using the changes in sentence bases suggested here.

Four Rules for Building Better Sentences

1. Prefer straightforward sentence bases and sentences (SVO order).
2. Shorten your sentence bases and sentences.
3. Use openers; prefer them over closers and interrupters.
4. Use grammatical parallelism whenever you can.

Prefer Straightforward Sentence Bases and Sentences. Written English is primarily a language that relies on the order of subject, verb, and (optional) object or complement, abbreviated "SV(O)" here. Although having every sentence written in the same order would be very tedious for readers, many times problem sentences have at their heart an accidental and unwise alteration of the SV(O) order. Such alterations can come from sentence inversion—putting an object (or other complement) first—or from making too big an interruption between subject and verb. Another unwise alteration of SV(O) straightforwardness is placing a part of the sentence last when that part is necessary for the reader to understand the first part of the sentence.

> *Examples:*
>
> ORIGINAL: The GAO reports that after examining 111 federal consulting contracts valued at $20 million these services weren't needed.
>
> REVISED: After examining 111 federal consulting contracts valued at $20 million, the GAO reports that the services in those contracts weren't needed.
>
> ORIGINAL: There will be more regulations as oil prices continue to rise.
>
> REVISED: As oil prices continue to rise, there will be more regulations.
>
> ORIGINAL: Into Zoffer's category of administrative skills fit creativity, clarification, organization, and invention.
>
> REVISED: Creativity, clarification, organization, and invention fit into Zoffer's category of administrative skills.

Shorten Your Sentence Bases and Sentences. Analysis of confusing and jumbled sentences written on topics in business and industry shows again and again that length is frequently a major factor in the problems in such sentences. Although such sentences may have other problems as well, it is their length that usually makes them unintelligible. It would be an incredibly hasty generalization to suggest a maximum number of words per sentence, but a good rule of thumb to follow is: *If your rereading of your first draft reveals a sentence that is confusing, you should first suspect its length, especially if it's over fifteen words long.* Try breaking the sentence into two shorter sentences; that may well solve the problem.

Examples:

ORIGINAL: The judicial system contains the symbols of all the great principles that give dignity and honor to the individual, independence to the businessman, and that gives the title of Righteous Protector to the state while performing the dual duty of keeping it in its proper place.

REVISED: The judicial system contains the symbols of all the great principles that give dignity and honor to the individual and independence to the businessman. The judicial system gives the title of Righteous Protector to the state at the same time that it keeps the state in its proper place.

ORIGINAL: Viewing this statement of a man who has been in the insurance business for over thirty-five years and being pro–term insurance indicates that there should be more comparative information given to the consumer and letting him make the decision between whole- and term-life insurance for himself.

REVISED: This man has been in the insurance business for over thirty-five years and is in favor of term insurance. This indicates that more comparative information should be given to the consumer, so that the consumer can make the decision between whole- and term-life insurance for himself.

ORIGINAL: Pyrimidine bases contain a single heterocyclic ring, and purine bases contain two rings that are fused together, heterocyclic being a molecule with a circular arrangement of atomic elements.

REVISED: A heterocyclic molecule has a circular arrangement of atomic elements. Pyrimidine bases contain a single heterocyclic ring. Purine bases contain two rings that are fused together.

Use Openers; Prefer Them Over Interrupters and Closers. First-draft writing often arranges the elements in each sentence in the order in which the writer thought of them. That sequence frequently is not the best order for the reader to read those elements in. As a writer, you frequently will find an interrupter in one of your sentences that would be better as an opener or a closer, or a closer that should be an opener. The question to ask about interrupters is whether putting something in between subject and verb sidetracks the reader's thoughts; if you're in doubt at all, move the interrupter to the beginning or end of the sentence. If the interrupter or closer contains information that the reader must know in order to understand the rest of the sentence (such as a logical, temporal, or causal precondition), it would probably be better as an opener. The extent to which you reorder such elements for your reader's convenience is one important measure of your concern for audience adaptation. Once again, the time to make such decisions is when you are revising.

Examples:

ORIGINAL: These supplies can be purchased separately, because they are not on public display.

REVISED: Because they are not on public display, these supplies can be purchased separately.

ORIGINAL: The combine's reliability will suffer if daily preventive maintenance is inconvenient to do.

REVISED: If daily preventive maintenance is inconvenient to do, the combine's reliability will suffer.

ORIGINAL: Please let me know if I can provide you with any additional information if the position is still open.

REVISED: If the position is still open, please let me know if I can provide you with any additional information.

Use Grammatical Parallelism Whenever You Can. Using grammatical parallelism means putting sentence elements that are roughly equal in importance and logically parallel (such as items in a series) into similar (parallel) grammatical structures. Using parallelism allows you to say more with fewer words and greater clarity.

Examples:

ORIGINAL: My background in communication includes several job interviews and writing letters to get information.

REVISED: My background in communication includes having several job interviews and writing letters to get information.

ORIGINAL: The rewards to the hobbyist are money saved, satisfaction of accomplishment, and a unique piece of furniture.

REVISED: The rewards to the hobbyist are saving money, satisfying a desire for accomplishment, and owning a unique piece of furniture.

3.2 Kinds of Sentence Bases

There are four different kinds of sentence bases:

Active bases   *It . . . that* bases

Passive bases   *Is* bases

During revising, work on using more *active* bases and fewer of the other kinds.

Active Bases. An active sentence is one in which the subject does something and the verb tells what the subject did. The action is straightforward through the sentence, from beginning to end:

S to V to (O)

As the typical sentence in written English for business and industry, and as the most easily understood structure, active sentences are the ones you should use most.

Examples:

A search strategy makes your research process more efficient.

Thousands of larvae hatch simultaneously.

The chief programmer codes ten to twenty times faster than the other programmers do.

Passive Bases. The passive sentence is the reverse of the active. In the passive sentence, the verb explains something that is, or was, done to the subject. The *doer* (called the *agent*) may or may not be named in the sentence. The flow of the passive sentence is usually backwards:

Subject (S) having something done to it (V) by (agent)

The file   was coded   by the client.

Passive sentences require more words than active ones do, take readers longer to read, and increase the chances of misunderstanding. They increase the possibility for grammatical mistakes, and they encourage vagueness. An occasional passive verb is acceptable, but avoid long strings of passives. If you tend to write such long strings of passives, build one step into your revising process in which you check only for them. Then eliminate them.

Recognizing Passives. Some students have trouble recognizing passives. First, passives have nothing to do with present and past *tense*. A passive can be in any tense. The following explanation tells how to recognize a passive sentence.

How to Recognize a Passive Sentence

All passives use a form of the verb *to be* (*is, are, was,* etc.) plus a past participle. (A *past participle* is anything that can fill the empty slot in this sentence: "I had _____ it.") Some passives also have agents expressed in the sentence.

Passive = subject + form of *to be* + past participle (+ agent)

Revising Passives. Once you have recognized a sentence as a passive and decided to change it, how can you revise it into an active sentence? First you have to find the agent (insert one if it isn't there to begin with but is understood or implied). The agent is the person or entity by whom (or by which) the action of the sentence is being done. With the "by + agent" in place at the end of the passive sentence, you can then change the sentence from passive to active by rotating it around the verb:

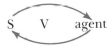

By positioning the (former) agent first and the (former) subject last, and then reshaping the verb to match, you can change a passive sentence to an active one.

Examples:

ORIGINAL: Houses were destroyed by the storm.

REVISED: The storm destroyed houses.

ORIGINAL: Your proposal has been turned down.

REVISED: Our committee has turned down your proposal.

ORIGINAL: The experimental rationale was summarized in the Foreword and detailed in Chapter One.

REVISED: The Foreword summarizes the experimental rationale, and Chapter One explains it.

Necessary Passives. One special situation does justify using the passive. When there is no actor or agent, or when the actor or agent is either unimportant or unknown, you may need to use the passive. Here is an example:

The wind flows were divided into three categories.

To introduce an actor or agent ("we" or "the research group," for example) into this sentence would place the emphasis in the wrong place. In addition, in the organization for which this sentence was written, it is customary for quarterly progress reports (from which this sentence was taken) not to mention human actors or agents in such sentences.

"It . . . That" Bases. People sometimes write "throat clearing constructions"—that is, sentences that wander around aimlessly before they decide what to say. These sentences often begin with either an "*it . . . that*" structure or some similar unnecessary construction. You can nearly always revise such a sentence for economy by deleting that entire construction.

> *Examples:*
>
> ORIGINAL: It goes without saying that productivity measures the economic progress for a whole nation.
>
> REVISED: Productivity measures the economic progress for a whole nation.
>
> ORIGINAL: The reason for this is that the strip of unplowed land will help prevent water erosion.
>
> REVISED: The strip of unplowed land will prevent water erosion.
>
> ORIGINAL: What many people do not realize is that there are many benefits derived from using fire in the forests.
>
> REVISED: Many benefits are derived from using fire in the forests.

The following is a list of frequent "*it . . . that*" openers and the alternatives to them.

Avoid	*Use*
It can be seen that	Note that
It can be shown that	We have shown that
It is apparent that	Apparently
It is assumed that	We assume that
It is clear that	Clearly
It is natural to expect that	One could expect
It is possible that	Possibly
It is proposed that	We propose that
It is shown herein that	We show that
It is worth noting that	Note that
It might be thought that	One might think that
It should be noted that	Note that
It will be remembered that	Remember that; recall that

"Is" Bases. Written English draws most of its strength and vigor from the verbs it uses. Inexperienced writers often produce first drafts that contain many weak verbs, especially forms of *to be, to have,* and *seem.* We call sentences that employ these weak verbs *"is"*-base sentences. They are very tedious to read, especially when they occur in chains (see Figure A.1).

When you see such a chain of weak verbs in your writing, revise at least some of those sentences—the easy ones—into action bases. Revising an *is* base into an action base can happen in several different ways. Sometimes simply rotating the sentence around the verb will make the new verb obvious. Sometimes if you look past the *is* verb you can find another word in the sentence that is a verb that has been changed into another part of speech. Convert that word back into a verb, and you usually have a strong verb. This same pattern for revision will also alert you to passives, which can then also be revised in the ways discussed here under *is* bases.

Figure A.1 A Paragraph Containing a Chain of Weak Verbs

In foods, the rate of evaporation is dependent upon the moisture content of the foods. Because no food is composed of a free surface of water, there is great variation in the amounts of water available for evaporation in foods. Some sort of measuring tool for the amounts of evaporatable water present is needed. This tool is known as free water content. Free water is all moisture in a product which is not "bound" to the proteins or other molecular structures in the food, and can be evaporated. Bound water is defined as all water that remains unfrozen at −30 C. This water is closely tied up with the substrate and cannot be evaporated out. So free water is all evaporative water within a product or cell, except the bound water in the food.

Examples:

ORIGINAL: The profit is dependent on the government's stimulation of the economy.

REVISED: The profit depends on the government's stimulation of the economy.

ORIGINAL: The purpose of this paper is to measure the potential usefulness of replacement-cost data.

REVISED: This paper measures the potential usefulness of replacement-cost data.

ORIGINAL: When a Harley-Davidson Sportster engine is in need of major repair, it most commonly is the piston that must be fixed.

REVISED: When a Harley-Davidson Sportster engine needs major repair, commonly the piston must be fixed.

By revising "*it . . . that,*" "*is,*" and passive bases into active bases, you will make your writing clearer, easier to read, and more vigorous stylistically. You do not have to make *every* sentence in your writing active, but a strong majority of them should be.

3.3 Combining Sentence Bases

You can combine independent sentence bases in two different ways: by putting them together side by side, or by putting one inside the other (using a relative pronoun or subordinate conjunction, as demonstrated in Section 3.1). Combining independent sentence bases on an equal level can be done in two ways (see Figure A.2):

1. Use a comma plus a linking word such as *and, but,* or *or* (coordinate conjunctions).
2. Use a semicolon with either no linking word or a word such as *therefore, thus,* or *however* (conjunctive adverbs).

Examples:

ORIGINAL: Designing large components is challenging. Much time is required to develop such skills.

REVISED: Designing large components is challenging, and much time is required to develop such skills.

ORIGINAL: Turbocharging can provide more passing power. Diesels are traditionally slow to accelerate.

REVISED: Diesels are traditionally slow to accelerate; however, turbocharging can provide more passing power.

Figure A.2 Combining Two Sentence Bases

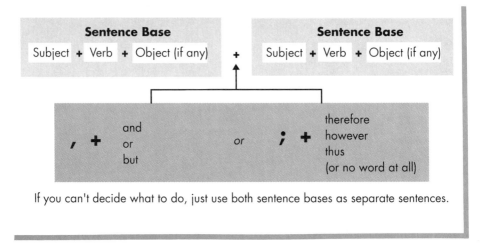

If you can't decide what to do, just use both sentence bases as separate sentences.

ORIGINAL: Some expenses can be related to revenue. These expenses are called product costs.

REVISED: Some expenses can be related to revenue, and these expenses are called product costs.

You can also combine independent sentence bases by placing one inside the other, which is called *embedding*. Doing this puts the embedded sentence in a particular subordinate relationship to the main sentence. Embedding uses words such as *that, which, when,* or *where* to introduce the newly embedded base. This is the same process (using relative pronouns or subordinate conjunctions) discussed in Section 3.1.

svo embeds in SVO:
S ([*that* or *which*, etc.] vo) VO

There are too many other embedding patterns to list here, but they all follow essentially the same method.

Examples:

ORIGINAL: *Embed (a) in (b):*
   (a) The device requires a source of potential heat to produce power.
   (b) The words *heat energy* imply something.

REVISED: The words *heat energy* imply that the device requires a source of potential heat to produce power.

ORIGINAL: *Embed (a) in (b):*

    (a) When young the mass has no organization.

    (b) The callus is a mass of cells.

REVISED: The callus is a mass of cells that when young has no organization.

ORIGINAL: *Embed (a) in (b):*

    (a) Tasks were once thought difficult.

    (b) Hydraulic systems make tasks easy to perform.

REVISED: Hydraulic systems make tasks that were once thought difficult easy to perform.

3.4 Separating Sentence Bases

Sometimes during revision you will find a sentence (or at least a string of words beginning with a capital letter and ending with a period) that just doesn't seem to make sense. To make such a string make sense, find the subject and put it first, put the verb next to it, and the object (or complement), if any, next. With the sentence base united, sort out the rest of the sentence as openers, closers, and interrupters, and place them accordingly.

Examples:

ORIGINAL: He will be able to see potential problems that might result from too great a reliance on short-term financing, an excessive investment in inventory or accounts receivable, or an over-generous dividend policy may be spotted readily through analysis of a sources-and-uses statement.

REVISED: Through analysis of a sources-and-uses statement, he will be able to see potential problems that might result from too great a reliance on short-term financing, an excessive investment in inventory or accounts receivable, or an over-generous dividend policy.

ORIGINAL: Cytoplasmic male sterility is a genetic character in which male sterility is carried in the female cytoplasm, as opposed to the normal situation of genetic characters being carried in the male and female germ plasm.

REVISED: In the normal situation the genetic characters are carried in the male and female germ plasm. In cytoplasmic male sterility the male sterility character is carried in the female cytoplasm.

ORIGINAL: Computer system documentation is a relatively new idea and lacked real direction until in 1974, four years after it was organized, the Documentation Subcommittee of the American Society for Testing Materials formed the concept of CSD.

REVISED: Four years after it was organized, the Documentation Subcommittee of the American Society for Testing Materials formed the concept of computer-system documentation, a relatively new idea that had lacked real direction until that year (1974).

4. Common Grammar and Usage Problems

A number of kinds of minor grammar problems and usage conventions can become major problems if left uncorrected or handled incorrectly. The following sections discuss how to deal with ten of those kinds of problems.

4.1 Sentence Fragments

A sentence fragment has no independent sentence base in it; it's a string of words that begins with a capital letter and ends with a period, question mark, or exclamation point, but that lacks a subject, or a verb, or both. If a sentence fragment has a noun/verb unit, the unit has usually been introduced by a relative pronoun or a subordinate conjunction, which disqualifies it from being a sentence. When upper-division students write sentence fragments, the fragment usually conforms to one of two patterns:

1. The *-ing* pattern (Example: Being a pilot.)
2. The noun-phrase pattern (Example: That the thruster rotates.)

The *-ing* pattern is a sentence fragment because *-ing* forms of words cannot function as verbs without help from another word, such as *is* or *are*. Carefully scrutinize any "sentence" that begins with an *-ing* form of a verb and consider splicing that group of words (the fragment) onto the previous sentence.

Example:

Being a pilot, I have travelled extensively around the country.

Recognize *noun-phrase* fragments as being introduced by relative pronouns (*that, which, where,* etc.) or subordinate conjunctions (*after, because, when,* etc.)—words that make whatever follows them depend on some other sentence; the noun-phrase fragment just doesn't make sense standing on its own:

Examples:

That the thruster rotates.

What the manual says.

Where the moisture persists.

As with the *-ing* pattern of fragments, noun-phrase fragments must be connected to another sentence in order to become grammatically complete.

Example:

The Harrier jet gets its unique abilities from one fact about its engine arrangement—that the thruster rotates.

4.2 Fused Sentences and Comma Splices

When you join independent sentence bases on a side-by-side level, connect them with a conjunction, such as *and, but,* or *or,* and insert a comma before the linking word. Failure to do so results in the error called a *fused sentence.*

Example:

The heavy fighting that characterized this conflict has now ceased the war-weary population now can begin to assess the damages.

(The fusing of the two sentence bases occurs between *ceased* and *the.*)

To remedy the fused sentence, merely insert a comma and the word *and.* (If you use the comma but no conjunction, you still have an error called a *comma splice.*) Be sensitive to the length of your sentences when you revise, and check long sentences for these problems.

4.3 Semicolons

There are two situations in which you should use semicolons:

1. When you join two independent sentence bases on a side-by-side basis and do not use a linking word, or when you join them with a word like *thus, therefore,* or *however,* use a semicolon between the bases (and *before* the linking word).

Example:

The image is transmitted in digital code; therefore it must go through a computer to appear as a visual image for analysis. The image can also be enhanced by the computer; the result can be something clearer than any normal camera could produce.

2. When you have a list of long items in a series, and some of those items have commas in them, use semicolons to separate the items from each other, for clarity:

   Example:

   The following students will work together in groups: Smith, Wilson, and Nunley; Thompson, Davidson, and Derrick; and Freid, Wells, and Hernandez.

4.4 Colons

Use a colon to introduce a quote, a list, or an example. When you use a colon, be sure that the words in front of it comprise a complete sentence base. The following paired examples illustrate this point:

Examples:

Ungrammatical: Two possible problem sources are: overlubrication and lubricant contamination.

Grammatical: There are two possible problem sources: overlubrication and lubricant contamination.
or
Two possible problem sources are overlubrication and lubricant contamination.

4.5 Hyphens

The use of hyphens is not standardized in English; authorities disagree about when to use them, when not to use them, and when using them is optional. In a recent article in *Technical Communication*, Lindsay Murdock (an editor for the National Oceanic and Atmospheric Administration) suggested the following guidelines.

Guidelines for Using Hyphens. Hyphens are properly used in some kinds of "unit modifiers"—pairs of words that work together to modify another word.

1. If the unit modifier consists of two nouns, do not use a hyphen.
   Examples: *avalanche prevention program, heat exchange capacity.*

2. When the unit modifier consists of an adjective and a noun, or a noun and an adjective, or an adverb and an adjective, use a hyphen.
   Examples: *happy-face shirt, hand-blown glass, round-bottom flask.*

   Exception: If the first element of the unit modifier is a comparative or a superlative, omit the hyphen, as in *lower order problems.*

3. Use a hyphen when you combine two color terms to make a unit modifier.
   Example: *rosy-red heat.*

4. Use a hyphen in unit modifiers that contain numbers.
   Examples: *four-way intersection, 10-m board.*

5. When a connecting word between the words in the unit modifier is implied, use a hyphen.
   Example: *east-west air traffic.*

6. When the unit modifier is a one-of-a-kind construction and contains a verb, or when it contains three or more words, use a hyphen.
   Example: *his run-for-glory attitude.*

7. When the unit modifier contains a present or past participle, or when it contains words ending in *-ed* or *-ing*, use a hyphen.
   Examples: *free-running river, forced-convection flow.*

8. Proper names (words that are capitalized) don't use hyphens, even though the same words when not capitalized would use hyphens.
   Examples: *a First Stage Alert* versus *a first-stage-alert, the New York office.*

9. If the unit modifier contains an adverb ending in *-ly*, do not use a hyphen. Use a hyphen in unit modifiers containing *well, still,* or *ever.*
   Examples: *a happily married man, a well-worn path, wholly owned subsidiary.*

   Exception: Do not use a hyphen with *well, still,* or *ever* when the word is modified by another adverb.
   Example: *fairly well worn tires* (not *fairly well-worn tires*).

As you may have guessed, the rules for hyphenation are complex (in addition to varying from field to field). A good guide to consult is your dictionary. An even better guide is the appropriate style manual for your field or firm (see Chapter 16).

4.6 Subject-Verb Number Agreement

In English, subjects and verbs must agree in number; if the subject is singular, the verb must also be singular. The most common problem that advanced students have with subject-verb number agreement is when another noun occurs between the subject and the verb, and the second noun is of a different number.

Examples:

Ungrammatical: The angle of the vanes are crucial to the turbojet's operation.

Grammatical: The angle of the vanes is crucial to the turbojet's
operation.

Even though the plural noun *vanes* is closer to the verb, the other
noun, *angle,* is still the subject of the sentence, and it still controls the
number of the verb: singular.

4.7 Abbreviations and Acronyms

Professional writing in business and industry abounds with abbreviations and acronyms. Some are standard throughout entire fields, and
seldom require explanation within their particular fields. Others,
such as the ever-changing ones in government and the military,
should nearly always be explained at least once (typically at their first
use, sometimes in a glossary) in each document that uses them.

Examples:

The annual meeting of the Southeastern Conference of Teachers of English in the Two-Year College (SECTETYC) will be
in Atlanta this year.

The Executive committee put Paul Blakely in charge of the Strategic Planning Committee (STRAPLCOM).

The operation of the Long Island Lighting Company's
(LILCO's) power plant is frequently studied.

Whether or not you explain the abbreviations you use is another
measure of how carefully you adapt your writing to its audience.
If there is any doubt in your mind about whether your reader will
immediately know what your abbreviation means, explain the abbreviation.

4.8 Confusing Words

Some words are very easily misused because they look or sound very
much like other words. Consider the following list of words and make
sure you choose the right one for the meaning you want.

Often-Confused Words

advice / advise	capital / capitol
adapt / adept / adopt	complement / compliment
allusion / illusion	consists in / consists of
affect / effect	continuous / continual
principal / principle	device / devise

cite / site	discreet / discrete
to / too	farther / further
its / it's	imply / infer
alternate / alternative	personal / personnel
among / between	practical / practicable
less / fewer	precede / proceed
among / number	respectively / respectfully
assure / insure / ensure	stationary / stationery

4.9 Complex Words and Phrases Versus Simple Ones

Avoid many word traps by using simple, concrete, specific words. Avoid cliches, complex phrases, and pompous expressions. In the following list, avoid the words in the left column.

Avoid	*Use*
agree with the idea	agree
at the present time	now
at this point in time	now
by means of	by
demonstrates that there is	shows
during the time that	while
for the purpose of	for
for this reason	therefore
if the developments are such that	if
in all cases	always
in a similar fashion	similarly
in order to	to
in the course of	during
in the event that	if
in the neighborhood of	about

4.10 Lists

The kind of writing that people do on the job tends to employ lots of lists. One particular kind of list that especially concerns writers and editors is called a "where" list. The following example shows how to do a "where" list.

Example:

In one computer program, the user needs to type in

ATT   RO   VAL

where

ATT = any attribute,
RO = any relational operator, and
VAL = a specific value.

Several important things should be noted about the way this list is done. Because the sentence that introduced it is incomplete, no colon is used. When you introduce a list with a complete sentence, use a colon.

Example:

In one computer program, the user needs to type in the following:

ATT   RO   VAL

The subject of the "where" list here is a piece of computer syntax; it could equally easily be a chemical formula or a mathematical equation. In any case, the subject of the "where" list is centered on the page. The "where" itself is back at the left margin, and the explanations of each term are indented five spaces from the left margin. When the explanations are complete sentences, they are followed by periods. Otherwise they are followed by commas or semicolons. Authorities disagree on whether or not to use *and* before the last line.

The "where" discussed in the previous paragraph is one example of a larger group of lists called "displayed" lists. Displayed lists include any kind of list that is set off from the rest of the text. This could also include bullet lists (in which each item is introduced by a "bullet" (•), or alphabetical lists (items introduced by a, b, c, etc.), or numbered lists (items introduced by 1, 2, 3, etc.). Displaying your lists is especially useful if the list contains information that your reader may want to be able to find easily by skimming your material.

EXERCISES

Correct the grammar and style errors in the following sentences.

1. There are two types of combines. The pull-type combine which is pulled by a tractor. The power to operate the combine is taken from the tractor. The second type is the self-propelled type. The operator rides on the combine itself. It has its own engine to move it through the field and harvest the grain.

2. To maintain productivity, farming requires modern, efficient machinery. Machinery which means big investment for the farmer.

3. The self-propelled combine is the type most commonly seen in grain fields today. And the self-propelled combine is the type I will be describing.

4. Needed proteins can be separated from a donor's blood. When a person's body cannot produce needed proteins.

5. About 48% of all the consumers surveyed indicated that they like Coors. Only 26% indicated that they disliked Coors. A surprisingly large amount of those surveyed indicated that they had no opinion about Coors, 25%.

6. Advertisers are able, by use of various methods such as sociopsychological tactics, to force opinions and beliefs on others that is against their own better judgment.

7. The social and economic impact, together with the environmental effects of airport development and operations, should be evaluated in order to guide development to make the airport environs compatible with airport operations and physical development and use of airports compatible with existing and proposed patterns of use.

8. Unfortunately, there is not much that consumers can do to prevent credit-life insurance abuses. The reasons being that the ordinary borrower is so pressed for time and the transactions are so small that the fuss and trouble of trying to evade the credit-life scandal is not worth the trouble.

9. The Douglas Amendment directs that no application for approval of a non-exempt transaction shall be approved which will permit any bank holding company or subsidiary thereof to acquire, directly or indirectly, any voting shares of, interest in, or all or substantially all of the assets of any additional bank located outside the State in which such bank holding company maintains its principal office and place of business, or in which it conducts its principal operations; unless such acquisition is specifically authorized by the statute laws of the State in which such bank is located, "by language to that effect and not merely by implication."

10. The saprophytic *Escherichia coli* can turn into pathogenic *Escherichia coli* under certain conditions and produce toxins.

11. Rather than dealing heavily with lobbyists, through previous contacts and selective advertising, sales are directed toward private industry.

12. This corporation needs an experienced (with corporate finances) accountant to insure proper financial records.

13. The government operates on a relatively fixed budget whereas a private company's budget is flexible. Flexible in the sense that if more funds are needed, sales can be increased to increase revenues.

14. The restrictions removed were primarily those requiring research with small amounts of bacteria, that would not grow outside the laboratory, to be reported.

15. Two varieties of costing, the historical method and the predetermined method, exist.

16. By defining the term "bid," giving bid contract information sources, and analyzing an actual government bid solicitation package; businessmen will have a clearer view of the bid process.

17. The estimate of credit losses are usually based on past experience, with some consideration given to projected sales.

18. Pyrimidine bases contain a single heterocyclic ring, and purine bases contain two rings which are fused together. Heterocyclic being a molecule with a circular arrangement of atomic elements.

19. By purchasing software, almost any business activity can be handled by a computer.

20. It goes without saying that to a certain extent productivity indicates the economic progress for a whole nation or for a particular industry.

21. The other category called fixed assets are assets that are valued at their depreciated cost.

22. I am presently attending State University in the capacity of a senior. My graduation date is May, 1992. My major being Engineering Technology.

23. I spent two quarters studying the government securities market and three quarters of computer (Basic, Fortran, and Cobol) which will allow me to handle your data processing.

24. An indication of a food's textural characteristics is obtained from the measurement of the distance that a penetrometer's probe falls through or against a food material during a specific period of time.

25. Computer system documentation is a relatively new idea and lacked real direction until in 1974, four years after it was organized, the Documentation Subcommittee of the American Society for Testing Materials formed the concept of CSD.

INDEX